Ockham Algebras

Ockham Algebras

T. S. BLYTH

Mathematical Institute, University of St Andrews, Scotland

and

J. C. VARLET

Institut de Mathématique, Université de Liège, Belgique

Oxford New York Tokyo

OXFORD UNIVERSITY PRESS

1994

This book has been printed digitally and produced to a standard design in order to ensure its continuing availability

OXFORD
UNIVERSITY PRESS

Great Clarendon Street, Oxford OX2 6DP

Oxford University Press is a department of the University of Oxford.
It furthers the University's objective of excellence in research, scholarship,
and education by publishing worldwide in

Oxford New York

Auckland Bangkok Buenos Aires Cape Town Chennai
Dar es Salaam Delhi Hong Kong Istanbul Karachi Kolkata
Kuala Lumpur Madrid Melbourne Mexico City Mumbai Nairobi
São Paulo Shanghai Singapore Taipei Tokyo Toronto

with an associated company in Berlin

Oxford is a registered trade mark of Oxford University Press
in the UK and in certain other countries

Published in the United States
by Oxford University Press Inc., New York

Reprinted 2002

A catalogue record for this book is available from the British Library

Library of Congress Cataloging in Publication Data
(Data available)

ISBN 0-19-859938-2

Preface

An *Ockham algebra* is a bounded distributive lattice with a dual endomorphism, the nomenclature being chosen since the notion of de Morgan negation has been attributed to the logician William of Ockham (c1290–c1349). The class of such algebras is vast, containing in particular the well-known classes of boolean algebras, de Morgan algebras, Kleene algebras, and Stone algebras. Pioneering work by Berman in 1977 has shown the importance of Ockham algebras in general, and has since stimulated much research in this area, notably by Urquhart, Goldberg, Adams, Priestley, and Davey.

Here our objective is to provide a reasonably self-contained and readable account of some of this research. Our collaboration began in 1982 in the consideration of a common abstraction of de Morgan algebras and Stone algebras which we called *MS-algebras*. This class of Ockham algebras is characterised by the fact that the dual endomorphism f satisfies $f^0 \leqslant f^2$, which implies that $f = f^3$. The subvariety $\mathbf{M}$ of de Morgan algebras is characterised by $f^0 = f^2$. In general, it seems an impossible task to describe all the subvarieties of Ockham algebras. The subvarieties of paramount importance are those in which $f^q = f^{2p+q}$ for some p, q; these are denoted by $\mathbf{K}_{p,q}$ and are called the *Berman varieties*. Of these, the most significant seems to be $\mathbf{K}_{1,1}$ in which each algebra L is such that $f(L) \in \mathbf{M}$. Here we concentrate particularly on $\mathbf{K}_{1,1}$, its subvarieties, subdirectly irreducibles, and congruences.

No study of Ockham algebras can be considered complete without mention of the theory of duality, in which the work of Priestley is fundamental. We make full use of Priestley duality in considering the subvariety $\mathbf{K}_{1,1}$.

Chapters 0–11 deal entirely with Ockham algebras whereas Chapters 12–15 are devoted to a brief study of double algebras. More precisely, we consider algebras $(L; {}^\circ, {}^+)$ for which $(L; {}^\circ)$ is an MS-algebra and $(L; {}^+)$ is a dual MS-algebra with the unary operations $^\circ$ and $^+$ linked by the identities $a^{\circ +} = a^{\circ\circ}$ and $a^{+\circ} = a^{++}$. Particular subvarieties of double MS-algebras are those of double Stone algebras and trivalent Lukasiewicz algebras.

As we have written this text with beginning graduate students in mind, we have included many illustrative examples and diagrams, as well as several useful tabulations. We would add that we make no claim that the list of references is complete, nor that what is due to Cæsar has not been attributed to Brutus.

T.S.B., J.C.V.

Acknowledgments

The entire text was set by the first author on a Macintosh Quadra, using a test copy of a System 7 version of MacTEX. We express here our deep gratitude to Lian Zerafa and mARK bLOORE for making this magnificent program available for this purpose.

Our grateful thanks are also due to the Scientific Affairs Division of the North Atlantic Treaty Organisation for Collaborative Research Grant 850532 from 1985–87, and its renewal for 1987–89.

Contents

0 Ordered sets, lattices, and universal algebra

It is of course impossible to give a full account of ordered sets, lattices, and universal algebra in a few pages, so we refer the reader to the various books cited in the bibliography. Nevertheless, in order to make this monograph reasonably self-contained, we shall summarise in this introductory chapter the fundamental notions that we shall use throughout. More specific concepts that we shall require will be defined as necessary. As far as notation is concerned, we provide an index of the various symbols that are used throughout.

The concept of *order* plays in mathematics a very prominent rôle, probably as important as that of size, though its importance has only rather recently been recognised. It is probably the success of the work of George Boole in the first half of the last century that has acted as a catalyst in producing a new area of research, namely that of ordered sets and, more particularly, lattices.

An *ordered set* (or *partially ordered set* or *poset*) is a set S on which there is defined a binary relation R which is *reflexive* (aRa for all $a \in S$), *transitive* (for all $a, b \in S$ the relations aRb and bRc imply aRc), and *anti-symmetric* (for all $a, b \in S$ the relations aRb and bRa imply $a = b$). Mathematics is replete with examples of such *order relations* ; for example, the relation of magnitude on the set of real numbers, the relation $\subseteq$ of inclusion on the power set $\mathbb{P}(E)$ of any set E, the relation $|$ of divisibility on the set $\mathbb{N}_0$ of strictly positive integers, etc.. Usually, an order relation is denoted by $\leqslant$ and its converse by $\geqslant$. Two elements x, y of an ordered set are said to be *comparable* (in symbols, $x \not\parallel y$) if $x \leqslant y$ or $y \leqslant x$, and *incomparable* (in symbols, $x \parallel y$) if $x \not\leqslant y$ and $y \not\leqslant x$. The (order) *dual* S^{op} of an ordered set S is the same set equipped with the converse order. We write $x \prec y$ if $x \leqslant y$ and $\{z \mid x < z < y\} = \emptyset$. If $x \prec y$ then we say that x is *covered by* y, or that y *covers* x. The relation $\preceq$ is clearly an order.

There are several useful ways in which disjoint ordered sets P, Q can be combined to produce a third ordered set. In particular, the *disjoint union* $P \mathbin{\dot\cup} Q$ consists of the set $P \cup Q$ with the order defined by

$$x \leqslant y \iff (x, y \in P \text{ with } x \leqslant y \text{ in } P) \text{ or } (x, y \in Q \text{ with } x \leqslant y \text{ in } Q).$$

The *linear sum* $P \oplus Q$ consists of $P \cup Q$ with the order

$$x \leqslant y \iff (x, y \in P \text{ with } x \leqslant y \text{ in } P) \text{ or } (x, y \in Q \text{ with } x \leqslant y \text{ in } Q)$$
$$\text{or } (x \in P \text{ and } x \in Q).$$

Finally, the *vertical sum* $P\overline{\oplus}Q$ is defined only when P has a biggest element a and Q has a smallest element b, and is obtained from $P \oplus Q$ by identifying a and b.

If A, B are ordered sets then a mapping $f : A \to B$ is said to be *order-preserving* or *isotone* if it is such that

$$(\forall x, y \in A) \qquad x \leqslant y \Rightarrow f(x) \leqslant f(y),$$

and *order-reversing* or *antitone* if it is such that

$$(\forall x, y \in A) \qquad x \leqslant y \Rightarrow f(x) \geqslant f(y).$$

A and B are (*order-*)*isomorphic* if there is a surjection $f : A \to B$ such that

$$(\forall x, y \in A) \qquad x \leqslant y \iff f(x) \leqslant f(y),$$

and *dually* (*order-*)*isomorphic* if there is a surjection $f : A \to B$ such that

$$(\forall x, y \in A) \qquad x \leqslant y \iff f(x) \geqslant f(y).$$

An ordered set that is dually isomorphic to itself is said to be *self dual*. A mapping $f : A \to A$ such that $f^2 = \mathrm{id}_A$ is called an *involution* (of period two). An order-reversing involution is called a *polarity*.

Very often (and this is so for the three examples mentioned above) any pair of elements x, y of an ordered set have a *greatest lower bound* (or *meet*, or *infimum*) which is denoted by $x \wedge y$; and a *least upper bound* (or *join*, or *supremum*) which is denoted by $x \vee y$. Such ordered sets are called *lattices*. For example, $(\mathbb{R}; \leqslant)$ is a lattice in which

$$x \wedge y = \min\{x, y\}, \quad x \vee y = \max\{x, y\};$$

$(\mathbb{P}(E); \subseteq)$ is a lattice in which

$$X \wedge Y = X \cap Y, \quad X \vee Y = X \cup Y;$$

and $(\mathbb{N}_0; |)$ is a lattice in which

$$m \wedge n = \gcd\{m, n\}, \quad m \vee n = \mathrm{lcm}\{m, n\}.$$

Clearly, any finite subset of a lattice L has a meet and a join. If *every* subset of L has a meet (resp. join) then L is said to be *meet-complete* (resp. *join-complete*). By a *complete lattice* we mean a lattice which is both meet-complete and join-complete.

The concept of a lattice was introduced at the end of the last century by C. S. Peirce and E. Schröder, but the study of lattices became really systematic with G. Birkhoff's first paper **[30]** in 1933 and his book **[2]**, the first edition of which appeared in 1940 and was for several decades the bible of lattice theoretists. In recent years, lattice theory has grown considerably. Lattices appear in all branches of mathematics : for any algebra the subalgebras, the equivalence relations, the congruence relations form lattices; in a topological space the open sets, the closed sets, the clopen (i.e. closed and open) sets form lattices; the convex subsets of a vector space form a lattice; in classical logic the propositions form a lattice in a way that we shall make precise below.

But what are lattices from the point of view of universal algebra? As is well known, the aim of *universal algebra* is to highlight the properties that various algebraic systems (e.g. groups, rings, fields, modules, lattices, ...) have in common. If we leave aside some early papers of Whitehead, we might say that the first pioneer of universal algebra was also G. Birkhoff.

Fundamental to universal algebra is the notion of an *operation*. If n is a non-negative integer then an *n-ary operation* on a set A is a mapping $f : A^n \to A$. The integer n is called the *arity* of the operation. We shall be mainly concerned with the cases where $n = 0, 1, 2$ which give respectively a *nullary* operation (this simply picks out an element of A), a *unary* operation, and a *binary* operation. An *algebra of type* $(n_1, \dots, n_\alpha)$ is a pair (A, F) where A is a non-empty set and F is an α-tuple $(f_1, \dots, f_\alpha)$ such that, for each i with $1 \leqslant i \leqslant \alpha$, f_i is an n_i-ary operation on A. Thus, for example, a lattice is an algebra of type $(2, 2)$, the two binary operations being meet and join, and satisfying the following identities :

$$\begin{array}{ll} x \wedge y = y \wedge x; & x \vee y = y \vee x; \\ x \wedge x = x; & x \vee x = x; \\ x \wedge (y \wedge z) = (x \wedge y) \wedge z; & x \vee (y \vee z) = (x \vee y) \vee z; \\ x \wedge (x \vee y) = x; & x \vee (x \wedge y) = x. \end{array}$$

If a lattice is *bounded*, i.e. if it has a least element 0 and a greatest element 1, then it can be considered as an algebra of type $(2, 2, 0, 0)$.

If A and B are algebras of the same type $(n_1, \dots, n_\alpha)$ then a mapping $\varphi : A \to B$ is a *morphism* if, for each i such that $1 \leqslant i \leqslant \alpha$,

$$f_i((\varphi(a_1), \dots, \varphi(a_{n_i})) = \varphi(f_i(a_1, \dots, a_{n_i}))$$

whenever $(a_1, \dots, a_{n_i}) \in A^{n_i}$. If, in addition, the mapping φ is surjective then φ is said to be an *epimorphism* with B an *epimorphic image* of A; if it is injective then it is a *monomorphism* ; and if it is both then it is an *isomorphism*. A morphism $f : A \to A$ is called an *endomorphism* on A; and an isomorphism $f : A \to A$ is called an *automorphism* on A.

Note that if L and M are bounded lattices then any morphism $\varphi : L \to M$ has to satisfy $\varphi(0_L) = 0_M$ and $\varphi(1_L) = 1_M$.

As we have mentioned above, lattice theory began properly with the work of George Boole **[4]** in formal deductive logic with an attempt to codify the laws of thought. In fact, Boole considered very special (but also very important) lattices in which, originally, the meet and the join were the binary connectives called *conjunction* ('and') and *disjunction* ('or') respectively, with an additional unary operation called *negation* ('not'). These so-called *boolean lattices* have turned out to be very useful in many areas of science and mathematics : in electrical engineering, in computer science, in axiomatic set theory, in model theory, and so on. Precisely, a *boolean lattice* L has three characteristics :

(1) it is bounded;

(2) it is *distributive* in the sense that

$$(\forall x, y, z \in L) \qquad x \wedge (y \vee z) = (x \wedge y) \vee (x \wedge z).$$

It is quite remarkable that this equality is equivalent to its dual

$$(\forall x, y, z \in L) \qquad x \vee (y \wedge z) = (x \vee y) \wedge (x \vee z).$$

(3) it is *complemented* in the sense that for every $a \in L$ there exists $a' \in L$ (called the *complement* of a) such that $a \wedge a' = 0$ and $a \vee a' = 1$; in other words, the *centre* $Z(L)$ of L, i.e. the set of complemented elements, is L itself.

The property of distributivity is shared by many lattices. For instance, each of the three examples given above is distributive. Since all the lattices that we shall deal with will be distributive, we shall say nothing about the various forms of weak or restricted distributivity. On the contrary, the notion of complement is very strong and many weakened forms of it have been considered.

Note that in a boolean lattice L the operation $x \mapsto x'$ of complementation is a polarity and satisfies the so-called *de Morgan laws*

$$(\forall x, y \in L) \qquad (x \wedge y)' = x' \vee y', \quad (x \vee y)' = x' \wedge y'.$$

As observed by H. B. Curry **[8]**, 'the term is customary despite its historical inaccuracy. According to Bochenski, the formulas were known in the Middle Ages'.

If, in a bounded distributive lattice, we can define a polarity that satisfies the above de Morgan laws then we obtain what is called a *de Morgan algebra*. More precisely, this is an algebra $(L; \wedge, \vee, f, 0, 1)$ of type $(2,2,1,0,0)$ where $(L; \wedge, \vee, 0, 1)$ is a bounded distributive lattice and $f : L \to L$ is a unary operation that satisfies the identities

$$f(x \wedge y) = f(x) \vee f(y), \quad f(x \vee y) = f(x) \wedge f(y), \quad f^2(x) = x.$$

From these identities it follows that $f(0) = 1$ and $f(1) = 0$. De Morgan algebras were introduced by G. C. Moisil **[75]** and investigated by A. Monteiro **[76]** and his school. A *Kleene algebra* is a de Morgan algebra satisfying the inequality $x \wedge f(x) \leqslant y \vee f(y)$.

Another way of generalising the notion of complementation is to retain the identity $a \wedge a' = 0$ and to drop the other. In this manner we define a *semicomplementation*. A lattice L that is bounded below is said to be *semicomplemented* if every $a \in L$ has a semicomplement, i.e. a non-zero element that is disjoint from a. Here, of course, the second lattice operation $\vee$ plays no part, so that the notion of semicomplementation can be defined on a meet semilattice. Of considerable interest are those lattices (or meet semilattices) in which, for any element a, the subset of elements disjoint from a has a greatest element. This is called the *pseudocomplement* of a and is denoted by $a^\star$. Thus $a^\star = \max\{x \in L \mid a \wedge x = 0\}$. Pseudocomplemented lattices are necessarily bounded. Examples of these are : the lattice of open subsets of a topological space, the pseudocomplement of an open set being the interior of its complement; and the lattice of ideals of a distributive lattice that is bounded below.

Of course, if we require that $a \vee a^\star = 1$ for every $a \in L$, then $a^\star$ becomes the complement of a and, when L is distributive, L is then a boolean lattice. Guided by what occurs in many examples, Stone **[92]** suggested a restriction of the identity $a \vee a^\star$ to those elements a that are pseudocomplements, i.e. that it would be fruitful to consider the identity $a^\star \vee a^{\star\star} = 1$ for all $a \in L$. Distributive pseudocomplemented lattices that satisfy this identity are therefore called *Stone lattices*. When the unary operation $a \mapsto a^\star$ is considered as a fundamental operation of the algebraic system, we shall use the term *Stone algebra*. Note, therefore, that whereas a Stone lattice is of type $(2, 2, 0, 0)$, a Stone algebra is of type $(2, 2, 1, 0, 0)$. This distinction is essential not only with regard to morphisms, but also with regard to subalgebras and congruences.

A *subalgebra* B of an algebra A is a non-empty subset of A which is closed under all the operations of A. The operations on B are then those of A restricted to B. An algebra A and its subalgebras are of the same type. For example, a subalgebra of a Stone lattice is just a sublattice, whereas a subalgebra of a Stone algebra is a sublattice that is closed under $a \mapsto a^\star$.

A *congruence relation* on an algebra A of type $(n_1, \dots, n_\alpha)$ is an equivalence relation ϑ on A which satisfies the *substitution property* : for each $i \in \{1, \dots, \alpha\}$,

$$(a_j, b_j) \in \vartheta \ (j = 1, \dots, n_i) \ \Rightarrow \ \big(f_i(a_1, \dots, a_{n_i}), f_i(b_1, \dots, b_{n_i})\big) \in \vartheta.$$

For example, in a Stone lattice L an equivalence relation ϑ is a congruence if $(a,b) \in \vartheta$ and $(c,d) \in \vartheta$ imply

$$(a \wedge c, b \wedge d) \in \vartheta \quad \text{and} \quad (a \vee c, b \vee d) \in \vartheta.$$

Note that, by the commutativity of $\wedge$ and $\vee$, this can be simplified to

$$(a,b) \in \vartheta \Rightarrow (\forall c \in L)\ (a \wedge c, b \wedge c) \in \vartheta,\ (a \vee c, b \vee c) \in \vartheta.$$

In a Stone algebra, however, there is the supplementary requirement

$$(a,b) \in \vartheta \Rightarrow (a^{\star}, b^{\star}) \in \vartheta.$$

The set of congruences on an algebra A, ordered by set inclusion, is a lattice with smallest element $\omega = \{(x,x) \mid x \in A\}$ and biggest element $\iota = A \times A$. This lattice is called the *congruence lattice* of A and is denoted by Con A.

If, on a bounded distributive lattice L, there is defined a unary operation f which satisfies the de Morgan laws and is such that $f(1) = 0$ and $f(0) = 1$ (i.e. if we drop the assumption that $f^2 = \mathrm{id}_L$, so that f becomes a dual lattice endomorphism, but not necessarily a dual lattice automorphism) then we obtain what is called an *Ockham algebra*. This idea goes back essentially to 1977 in a short but very deep paper by J. Berman **[28]**. Two years later, A. Urquhart **[94]** developed a topological duality theory for this type of algebra, gave a logical motivation for his study, and introduced the name *Ockham lattices* with the justification : 'the term *Ockham lattice* was chosen because the so-called *de Morgan laws* are due (at least in the case of propositional logic) to William of Ockham'. The name *Ockham algebra* has since become classical and was used in the thorough doctoral thesis of M. Goldberg **[68]** and in a subsequent paper **[69]**.

Since 1981 many papers have been published on Ockham algebras. The objective of this book is therefore to develop the general properties of this class of algebras and to consider more particularly some important subclasses which are interesting not only in the framework of universal algebra but also for their significance in the algebra of logic. At this point, it is not superfluous to recall that de Morgan algebras arose in the researches on the algebraic treatment of constructive logic with strong negation. The operation f that is involved in an Ockham algebra can also be interpreted as a negation (and for this reason $f(a)$ is often written as $\sim a$), though this does not in general satisfy the law of double negation. If we impose on f the restriction that $f^n = \mathrm{id}_L$ for some $n \in \mathbb{N} \backslash \{1,2\}$ then we obtain a new logic whose interpretation, as far as we know, has still to be made explicit. In this connection, interesting work has been done by D. Schweigert and M. Szymanska **[88]** on those Ockham algebras that belong to the class $\mathbf{P}_{n,0}$ (n odd) described in Chapter 4. The class of algebras they deal with is shown to be the semantic for a propositional calculus called correlation logic.

The reader will see from the examples of Ockham algebras given in Chapter 1 that the study of Ockham algebras is far from being gratuitous.

We close this brief introduction by observing that all the classes of algebra that we have mentioned, and indeed all that we shall consider later, are *equational*, in the sense that they can be defined by a set of identities. It is a celebrated theorem of Birkhoff that the equational classes of algebras are precisely those that are closed under the formation of subalgebras, epimorphic images, and direct products, i.e. are *varieties*.

1 Examples of Ockham algebras; the Berman classes

Recall that a *distributive Ockham algebra* is an algebra $(L; \wedge, \vee, f, 0, 1)$ of type $(2,2,1,0,0)$ in which $(L; \wedge, \vee, 0, 1)$ is a bounded distributive lattice and $x \mapsto f(x)$ is a unary operation such that $f(0) = 1$, $f(1) = 0$ and

$$(\forall x, y \in L) \qquad f(x \wedge y) = f(x) \vee f(y), \quad f(x \vee y) = f(x) \wedge f(y).$$

Without explicit mention to the contrary, all the Ockham algebras that we shall deal with will be distributive as lattices so we shall agree to drop the adjective 'distributive' and talk of an *Ockham algebra*. We shall often also denote this by the simpler notation $(L; f)$.

The class of Ockham algebras is equational (in other words, a variety), and will be denoted by $\mathbf{O}$. As mentioned in Chapter 0, the concept of an Ockham algebra arose from successive attempts to generalise the notion of a boolean algebra. Important steps in this long history are the *de Morgan algebras* and the *Stone algebras*, these forming important subvarieties of the variety $\mathbf{O}$. In 1979, A. Urquhart **[94]** observed that 'an outstanding open problem is that of determining all equational subclasses of the class of Ockham lattices'. To this day, the problem remains unsolved. However, very important subclasses of $\mathbf{O}$ were introduced by J. Berman **[28]** and we shall call them the *Berman classes*. These are obtained by placing restrictions on the dual endomorphism f. Precisely, if we let $f^0 = \mathrm{id}$ and define f^n recursively by $f^n(x) = f[f^{n-1}(x)]$ for $n \geqslant 1$, then for $p, q \in \mathbb{N}$ with $p \geqslant 1$ and $q \geqslant 0$ we define the *Berman class* $\mathbf{K}_{p,q}$ to be the subclass of $\mathbf{O}$ obtained by adjoining the equation

$$f^{2p+q} = f^q.$$

The Berman classes are related as follows :

$$\mathbf{K}_{p,q} \subseteq \mathbf{K}_{p',q'} \iff p \mid p' \text{ and } q \leqslant q'.$$

We shall give a simple proof of this later. It follows that the smallest Berman class is the class $\mathbf{K}_{1,0}$, which is determined by the equation $f^2 = \mathrm{id}$. This is none other than the class $\mathbf{M}$ of de Morgan algebras.

The importance of the Berman classes is partly justified by the following property.

Theorem 1.1 *Every finite Ockham algebra belongs to a Berman class.*

Proof Let $(L;f)$ be a finite Ockham algebra, and consider the sets

$$\{f, f^3, f^5, \ldots\}, \quad \{f^0, f^2, f^4, f^6, \ldots\}$$

of, respectively, dual endomorphisms and endomorphisms on L. Since both of these must be finite, we have that $f^q = f^{2p+q}$ for some p, q. ◇

Note that Theorem 1.1 is no longer true if the algebra in question is infinite, as the following example shows.

Example 1.1 Let $L = \{0,1\} \cup \{a_i \mid i \in \mathbb{Z}\} \cup \{b_i \mid i \in \mathbb{Z}\}$, ordered linearly by $a_i < a_j$ for $i < j$; $b_i < b_j$ for $i < j$; $0 < a_i < b_j < 1$ for all i, j. Define $f : L \to L$ by

$$f(0) = 1, \quad f(1) = 0, \quad f(a_i) = b_{-i}, \quad f(b_i) = a_{-i-1}.$$

Then $(L;f)$ is an Ockham algebra. We can depict the effect of f as follows :

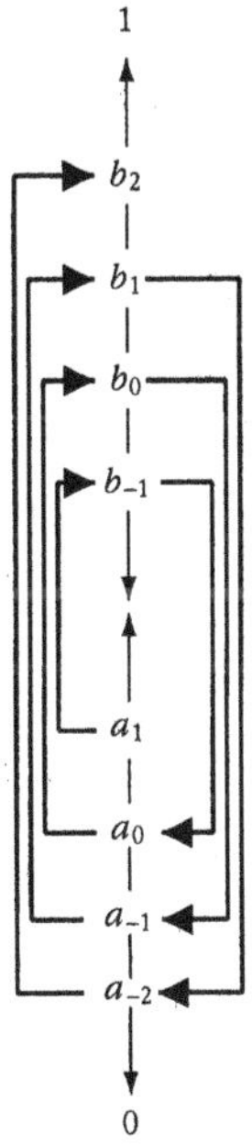

For every $i \in \mathbb{Z}$ and $n \in \mathbb{N}$, we have the chains

$$\cdots < f^{2n}(a_i) < \cdots < f^2(a_i) < a_i < f(a_i) < \cdots < f^{2n+1}(a_i) < \cdots$$

$$\cdots < f^{2n+1}(b_i) < \cdots < f(b_i) < b_i < f^2(b_i) < \cdots < f^{2n}(b_i) < \cdots$$

It follows that L does not belong to any Berman class. Note also that in this example f is a bijection.

The following three examples are by no means surprising.

Example 1.2 In a de Morgan algebra it is traditional to write the unary operation as $x \mapsto \overline{x}$. Every de Morgan algebra $(L; ^-)$ belongs to the Berman class $\mathbf{K}_{1,0}$. In fact, since $\overline{\overline{x}} = x$ for every $x \in L$, the operation $x \mapsto \overline{x}$ is a dual automorphism.

Example 1.3 In a boolean algebra it is traditional to write the unary operation as $x \mapsto x'$. Every boolean algebra belongs to the subclass of $\mathbf{K}_{1,0}$ obtained by adjoining the equation $x \wedge f(x) = 0$. In fact, from this it follows that $f(x) \vee f^2(x) = 1$, i.e. $f(x) \vee x = 1$, so that $f(x) = x'$ is the complement of x.

Example 1.4 In a Stone algebra it is traditional to write the unary operation as $x \mapsto x^\star$. Every Stone algebra $(L; ^\star)$ belongs to the subclass of $\mathbf{K}_{1,1}$ obtained by adjoining the equation $x \wedge f(x) = 0$. In fact, by the properties of the pseudocomplementation $x \mapsto x^\star$ we have $(x \wedge y)^\star = x^\star \vee y^\star$, $(x \vee y)^\star = x^\star \wedge y^\star$ with $0^\star = 1$ and $1^\star = 0$, so that $(L; ^\star) \in \mathbf{O}$. Since moreover $x^\star = x^{\star\star\star}$ it follows that $(L; ^\star) \in \mathbf{K}_{1,1}$. Finally, $x \wedge x^\star = 0$ by the definition of the pseudocomplement.

Less trivial are the following examples.

Example 1.5 Let $(S; ^\star)$ be a Stone algebra, let A be a distributive lattice that is bounded below, and let B be a distributive lattice that is bounded above with a dual isomorphism $\vartheta : A \to B$. On the linear sum $L = A \oplus S \oplus B$ define a unary operation f by

$$f(x) = \begin{cases} \vartheta(x) & \text{if } x \in A; \\ x^\star & \text{if } x \in S; \\ \vartheta^{-1}(x) & \text{if } x \in B. \end{cases}$$

Then $(L; f)$ is an Ockham algebra that belongs to $\mathbf{K}_{1,1}$.

Example 1.6 Consider the set F of mappings $p : \mathbb{R}_+ \to [0,1]$ under the usual order, namely that given by

$$p \leqslant q \iff (\forall x \in \mathbb{R}_+)\ p(x) \leqslant q(x).$$

It is clear that F is a bounded lattice; the smallest element is the constant map $\mathbf{0} : x \mapsto 0$, the greatest element is the constant map $\mathbf{1} : x \mapsto 1$, and for $p, q \in F$ their infimum and supremum are respectively the lower and upper envelopes $p \wedge q$ and $p \vee q$ given by the prescriptions

$$(p \wedge q)(x) = \min\{p(x), q(x)\}, \quad (p \vee q)(x) = \max\{p(x), q(x)\}.$$

Moreover, this lattice F is distributive.

Now let $a \in \mathbb{R}_+$ be fixed and for every $p \in F$ define $f(p)$ by setting

$$(\forall x \in \mathbb{R}_+) \quad \big(f(p)\big)(x) = 1 - p(x+a).$$

Clearly, we have $f(\mathbf{0}) = \mathbf{1}$ and $f(\mathbf{1}) = \mathbf{0}$. Also, since

$$1 - \min\{p(x+a), q(x+a)\} = \max\{1 - p(x+a), 1 - q(x+a)\}$$

we see that $f(p \wedge q) = f(p) \vee f(q)$, and likewise $f(p \vee q) = f(p) \wedge f(q)$. Thus $(F; f)$ is an Ockham algebra. When $a = 0$ we have $\big(f(p)\big)(x) = 1 - p(x)$ and $\big(f^2(p)\big)(x) = p(x)$. It follows that in this case $(F; f) \in \mathbf{K}_{1,0}$. Moreover, since

$$\min\{1, 1-p\} \leqslant \tfrac{1}{2} \leqslant \max\{q, 1-q\}$$

we have $p \wedge f(p) \leqslant q \vee f(q)$ and $(F; f)$ is a Kleene algebra.

Example 1.7 Consider the bounded distributive lattice consisting of the interval $I = [0, 1]$ of real numbers under the usual order. Every $x \in I$ with $x \neq 1$ has a unique decimal representation

$$x = 0 \cdot x_1 x_2 x_3 \dots$$

where each $x_i \in \{0, 1, \dots, 9\}$. For our convenience here, we shall write this as $x = (x_i)_{i \geqslant 1}$. Let a be a fixed positive integer and for each such x define

$$f(x) = (9 - x_{i+a})_{i \geqslant 1},$$

with $f(1) = 0$. Then $(I; f)$ is an Ockham algebra.

In every Ockham algebra $(L; f)$ the subset

$$S(L) = \{f(x) \mid x \in L\}$$

is clearly a subalgebra of L which we shall call the *skeleton* of L. The skeleton of L is a de Morgan algebra precisely when $f^3(x) = f(x)$ for every $x \in L$, i.e. precisely when L belongs to the Berman class $\mathbf{K}_{1,1}$. When this is the case, we shall say that L *has a de Morgan skeleton*. Note that in this case we also have

$$S(L) = \{f^2(x) \mid x \in L\}.$$

Every Ockham algebra $(L; f)$ contains a subalgebra that has a de Morgan skeleton. The greatest such subalgebra is

$$M(L) = \{x \in L \mid f^3(x) = f(x)\}.$$

For the Ockham algebra $(F; f)$ of Example 1.6, $M(F)$ is the set of those mappings p such that $p(x+a) = p(x+3a)$ for all $x \in \mathbb{R}_+$, i.e. those that are of period $2a$ after the point $x = a$. The skeleton of $M(F)$ consists of those mappings that are of period $2a$.

In the Ockham algebra $(I;f)$ of Example 1.7, $M(I)$ is the set of real numbers $x = (x_i)_{i\geqslant 1}$ in $[0,1]$ with $x_{i+p} = x_{i+3p}$, i.e. those that are $2p$-repeating after the p-th digit. The skeleton of $M(I)$ consists of those x that are $2p$-repeating.

We now describe a useful way of making bounded distributive lattices into Ockham algebras with de Morgan skeletons. First observe that if $(L;f)$ is an Ockham algebra then the mapping $f^2 : L \to L$ is a lattice morphism; and if L has a de Morgan skeleton then f^2 is idempotent. These observations yield the following result which, as we shall see, is very useful in constructing examples.

Theorem 1.2 *Let L be a bounded distributive lattice and let $\varphi : L \to L$ be an idempotent $\{0,1\}$-lattice morphism such that the sublattice* $\operatorname{Im}\varphi$ *admits a polarity p. Define a unary operation $f : L \to L$ by the prescription*

$$(\forall x \in L) \qquad f(x) = p[\varphi(x)].$$

Then $(L;f)$ is an Ockham algebra with a de Morgan skeleton. Moreover, every such Ockham algebra arises in this way.

Proof Since φ preserves 0 and 1, we have $f(0) = 1$ and $f(1) = 0$. Now

$$\begin{aligned} f(x \wedge y) = p[\varphi(x \wedge y)] &= p[\varphi(x) \wedge \varphi(y)] \\ &= p[\varphi(x)] \vee p[\varphi(y)] \\ &= f(x) \vee f(y), \end{aligned}$$

and similarly $f(x \vee y) = f(x) \wedge f(y)$, so $(L;f)$ is an Ockham algebra.

Since φ is idempotent we have that φ acts as the identity on $\operatorname{Im}\varphi = \operatorname{Im}p$. It follows from this that $\varphi p \varphi(x) = p\varphi(x)$ for every $x \in L$, so that

$$f^3(x) = p\varphi p\varphi p\varphi(x) = p\varphi p^2\varphi(x) = p\varphi^2(x) = p\varphi(x) = f(x).$$

Thus $(L;f)$ has a de Morgan skeleton.

Conversely, if $(L;f)$ is an Ockham algebra with a de Morgan skeleton then the mapping $\varphi : x \mapsto f^2(x)$ describes a $\{0,1\}$-lattice morphism on L, and from $f^3 = f$ we deduce that $\varphi^2 = \varphi$ and that f is a polarity on $\operatorname{Im}\varphi = \{f^2(x) \mid x \in L\}$. ◇

As an application of Theorem 1.2, we shall obtain an affirmative answer to the quite natural question of whether, given a bounded distributive lattice L, it is always possible to make L into an Ockham algebra with a de Morgan skeleton. For this purpose, we recall some definitions.

A subset Q of an ordered set P is said to be a *down-set* if it is decreasing, in the sense that $i \in Q$ and $j \leqslant i$ imply that $j \in Q$. The *down-set generated by a subset* X of P is defined by

$$X^{\downarrow} = \{y \in P \mid (\exists x \in X)\ y \leqslant x\}.$$

In particular, when $X = \{x\}$ we write $X^{\downarrow}$ as $x^{\downarrow}$. An *ideal* of a lattice L is a sublattice I of L which is also a down-set; and an ideal of the form $x^{\downarrow}$ is called a *principal ideal*. Dually, a subset Q of an ordered set P is said to be an *up-set* if it is increasing, in the sense that $i \in Q$ and $j \geqslant i$ imply that $j \in Q$. The *up-set generated by a subset* X of P is defined by

$$X^{\uparrow} = \{y \in P \mid (\exists x \in X)\ y \geqslant x\}.$$

In particular, when $X = \{x\}$ we write $X^{\uparrow}$ as $x^{\uparrow}$. A *filter* of a lattice L is a sublattice F of L which is also an up-set; and a filter of the form $x^{\uparrow}$ is called a *principal filter*. Ideals and filters are *convex*, in the sense that if $a, b \in I$ (resp. F) and $a \leqslant c \leqslant b$ then $c \in I$ (resp. F). An ideal or a filter is said to be *proper* if it is not the whole lattice. A proper ideal I of L is said to be *prime* if $a, b \in L$ and $a \wedge b \in I$ imply that $a \in I$ or $b \in I$. The notion of a prime filter is defined dually. The set-theoretic complement of a prime ideal is a prime filter. In a distributive lattice L there are sufficiently many prime ideals, so that any two elements can be 'separated' by a prime ideal, in the sense that if $a, b \in L$ with $a \neq b$ then there is a prime ideal I of L that contains one of these elements and not the other.

Theorem 1.3 *Every bounded distributive lattice can be made into an Ockham algebra with a de Morgan skeleton.*

Proof Let L be a bounded distributive lattice. Then we can find in L a finite chain of prime ideals

$$I_0 \subset I_1 \subset \cdots \subset I_n$$

and a chain C of elements

$$0 = a_0 < a_1 < \cdots < a_n < 1$$

such that $a_i \in I_i \setminus I_{i-1}$ for $1 \leqslant i \leqslant n$. Consider the mapping $\varphi : L \to L$ defined by

$$\varphi(x) = \begin{cases} 0 & \text{if } x \in I_0; \\ a_i & \text{if } x \in I_i \setminus I_{i-1}; \\ 1 & \text{if } x \notin I_n. \end{cases}$$

Clearly, φ is an idempotent $\{0,1\}$-lattice morphism with $\operatorname{Im}\varphi = C$. Now on the chain C a polarity p is uniquely defined. By Theorem 1.2, therefore, with

$f(x) = p[\varphi(x)]$ for every $x \in L$, we see that $(L; f)$ is an Ockham algebra with a de Morgan skeleton. ◇

The preceding proof shows in particular that many non-isomorphic Ockham algebras can be defined on the same distributive lattice; for we can attribute different values to n, and even for a fixed value of n there can exist different chains of prime ideals.

Example 1.8 Let $\mathbf{n}$ denote the n-element chain

$$0 = a_0 < a_1 < \cdots < a_{n-1} = 1.$$

Every antitone mapping $\varphi : \mathbf{n} \to \mathbf{n}$ such that $f(0) = 1$ and $f(1) = 0$ determines an Ockham algebra. Thus the number of non-isomorphic Ockham algebras definable on $\mathbf{n}$ is equal to the number of antitone mappings from $\mathbf{n-2}$ to $\mathbf{n}$, which is known to be

$$\alpha_n = \binom{2n-3}{n-2}.$$

For small values of n, this number is given as follows :

$$\begin{array}{rcllllll} n & = & 2 & 3 & 4 & 5 & 6 & 7 \\ \alpha_n & = & 1 & 3 & 10 & 35 & 126 & 462 \end{array}$$

Example 1.9 Consider the unit cube

$$C = \{(x, y, z) \in \mathrm{IR}^3 \mid x, y, z \in [0, 1]\}.$$

With respect to the cartesian order, C is a bounded distributive lattice. The mapping $\varphi : C \to C$ given by

$$\varphi(x, y, z) = (z, y, z)$$

is a $\{0, 1\}$-lattice morphism and is idempotent. Moreover, $\operatorname{Im}\varphi$ admits the polarity p given by

$$p(x, y, z) = (1 - z, 1 - y, 1 - z).$$

By Theorem 1.2 we can therefore make C into an Ockham algebra with a de Morgan skeleton by defining

$$f(x, y, z) = (1 - z, 1 - y, 1 - z).$$

The skeleton of C is the set of elements (x, y, z) for which $x = z$, i.e. a diagonal plane.

Now there is another polarity p' that can be defined on $\operatorname{Im}\varphi$, namely that given by

$$p'(x, y, z) = (1 - y, 1 - z, 1 - y).$$

In this case we obtain

$$f'(x, y, z) = (1 - y, 1 - z, 1 - y),$$

which makes C into a different Ockham algebra with the same de Morgan skeleton.

Example 1.10 Let B be a boolean lattice and let $L = \{0\} \oplus B \oplus \{1\}$. Let $a, b \in B$ with $b < a$ and define $\varphi : L \to L$ by $\varphi(0) = 1$, $\varphi(1) = 0$ and

$$(\forall x \in B) \qquad \varphi(x) = (b \vee x) \wedge a = b \vee (x \wedge a).$$

Clearly, φ is an idempotent $\{0, 1\}$-lattice morphism. We can define a polarity p on $\operatorname{Im}\varphi = \{0\} \cup [b, a] \cup \{1\}$ by $p(0) = 1$, $p(1) = 0$ and

$$(\forall x \in [b, a]) \qquad p(x) = \varphi(x')$$

where x' is the complement of x in B. It is easy to verify that $p(x)$ is the relative complement of x in $[b, a]$. It follows from Theorem 1.2, with $f(x) = p[\varphi(x)]$ for every $x \in L$, that $(L; f)$ is an Ockham algebra with a de Morgan skeleton. Here we have $f(0) = 1$, $f(1) = 0$ and, for $x \in B$,

$$f(x) = p[\varphi(x)] = \varphi([\varphi(x)]') = \varphi[(b' \wedge x') \vee a'] = b \vee (a \wedge x') = \varphi(x').$$

Example 1.11 Let L consist of the (possibly infinite) lazy tongs lattice with a new smallest element 0 and a new greatest element 1 adjoined. Consider the mapping $\varphi : L \to L$ given by

$$\varphi(x) = \begin{cases} x & \text{if } x \in \{0, 1, a_0, b_0\} \cup \{a_{2n+1}\}_{n \in \mathbb{Z}}; \\ a_{2n-1} & \text{if } x = a_{2n}\ (n \neq 0); \\ a_{2n+1} & \text{if } x = b_{2n}\ (n \neq 0). \end{cases}$$

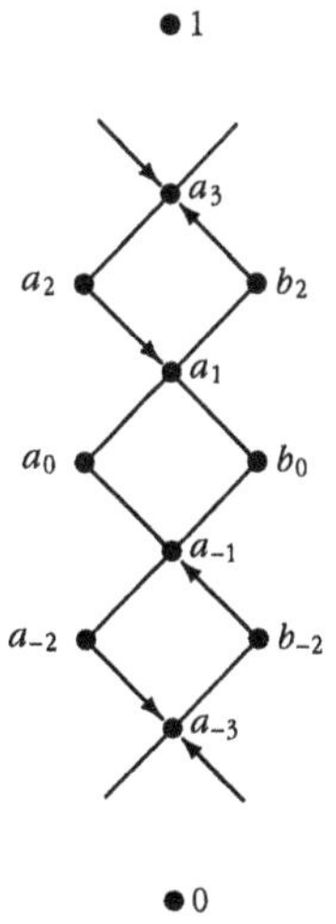

In the Hasse diagram opposite, the arrowheads indicate the effect of φ. It is readily seen that φ is an idempotent $\{0, 1\}$-lattice morphism. There are two polarities on $\operatorname{Im}\varphi$, namely p, p' given by

$$\begin{aligned} &p(0) = p'(0) = 1; \\ &p(1) = p'(1) = 0; \\ &p(a_{2n+1}) = p'(a_{2n+1}) = a_{-2n-1}; \\ &p(a_0) = a_0, \quad p(b_0) = b_0; \\ &p'(a_0) = b_0, \quad p'(b_0) = a_0. \end{aligned}$$

These polarities give rise to different Ockham algebras with the same de Morgan skeleton.

Example 1.12 Let E be a set and let a, b be distinct elements of E. Consider the mapping $\varphi : \mathbb{P}(E) \to \mathbb{P}(E)$ given by

$$\varphi(X) = \begin{cases} X \cup \{a\} & \text{if } b \in X; \\ X \cap \{a\}' & \text{if } b \notin X. \end{cases}$$

Roughly speaking, φ adds a if X already contains b, and removes a if X does not contain b. Clearly, $\varphi(\emptyset) = \emptyset$ and $\varphi(E) = E$. It is readily seen that φ is both a $\cup$-morphism and an $\cap$-morphism. Moreover, φ is idempotent. Now

$$\operatorname{Im}\varphi = [\emptyset, \{a, b\}'] \cup [\{a, b\}, E]$$

and so admits the polarity p provided by complementation. We can therefore make $\mathbb{P}(E)$ into an Ockham algebra with a de Morgan skeleton by defining

$$f(X) = [\varphi(X)]' = \begin{cases} X' \cap \{a\}' & \text{if } b \in X; \\ X' \cup \{a\} & \text{if } b \notin X. \end{cases}$$

Example 1.13 Let B be a boolean lattice. Given $n \geqslant 3$, define

$$B^n_\star = \{(x_1, \dots, x_n) \in B^n \mid x_1 \leqslant x_n\}.$$

Then $B^n_\star$ is a sublattice of B^n, with $(0, \dots, 0)$ as smallest element and $(1, \dots, 1)$ as greatest element. Define $\varphi : B^n_\star \to B^n_\star$ by

$$\varphi(x_1, x_2, \dots, x_n) = (x_1, x_n, \dots, x_n).$$

Clearly, φ is an idempotent $\{0, 1\}$-lattice morphism. A polarity p on $\operatorname{Im}\varphi$ is

$$p(x_1, x_n, \dots, x_n) = (x'_n, x'_1, \dots, x'_1).$$

We can therefore make $B^n_\star$ into an Ockham algebra with a de Morgan skeleton by defining

$$f(x_1, x_2, \dots, x_n) = (x'_n, x'_1, \dots, x'_1).$$

We have seen above that if $(L; f) \in \mathbf{K}_{1,1}$ then the mapping $x \mapsto f^2(x)$ is an idempotent lattice morphism. We shall now investigate an important special case of this, namely when the mapping $x \mapsto f^2(x)$ is a closure. We recall that a *closure* on an ordered set E is an isotone mapping $f : E \to E$ such that $f = f^2 \geqslant \mathrm{id}_E$. Thus $x \mapsto f^2(x)$ is a closure precisely when

$(\forall x, y \in L) \quad x \leqslant y \Rightarrow f^2(x) \leqslant f^2(y);$
$(\forall x \in L) \quad x \leqslant f^2(x);$
$(\forall x \in L) \quad f^2(x) = f^4(x).$

Note that the first of these properties is satisfied by all $L \in \mathbf{O}$, and that the third is satisfied by all $L \in \mathbf{K}_{1,1}$.

Definition By a *de Morgan–Stone algebra*, or an *MS-algebra*, we mean an algebra $(L; \wedge, \vee, ^\circ, 0, 1)$ of type $(2,2,1,0,0)$ such that $(L; \wedge, \vee, 0, 1)$ is a bounded distributive lattice and $x \mapsto x^\circ$ is a unary operation on L such that

(MS1) $1^\circ = 0$;
(MS2) $(\forall x, y \in L) \quad (x \wedge y)^\circ = x^\circ \vee y^\circ$;
(MS3) $(\forall x \in L) \quad x \leqslant x^{\circ\circ}$.

Clearly, de Morgan algebras and Stone algebras are MS-algebras; hence the terminology. In fact, when we introduced the notion of an MS-algebra in 1983 **[33]** our objective was to stress the numerous similarities between these two classes of algebras. In a de Morgan algebra, $x \mapsto \overline{x}$ is a dual automorphism and $x \mapsto \overline{\overline{x}}$ is the identity. In a Stone algebra, $x \mapsto x^\star$ is a dual endomorphism and $x \mapsto x^{\star\star}$ is a closure. So the notion of an MS-algebra arises quite naturally by retaining the properties that are common to these two classes of algebras.

The class **MS** of MS-algebras is equational; it is the subclass of $\mathbf{K}_{1,1}$ obtained by adjoining the equation $x \wedge f^2(x) = x$. In this connection, we note that M. Ramalho and M. Sequeira **[82]** have considered more generally the subvarieties of **O** defined by $x \wedge f^{2n}(x) = x$.

The relation of MS-algebras to Ockham algebras is as follows.

Theorem 1.4 *Every MS-algebra is an Ockham algebra with a de Morgan skeleton. An Ockham algebra $(L; f)$ is an MS-algebra if and only if $x \leqslant f^2(x)$ for every $x \in L$.*

Proof Let $(L; ^\circ)$ be an MS-algebra. Then, by (MS1) and (MS3), we have $0^\circ = 1^{\circ\circ} \geqslant 1$ and so $0^\circ = 1$. By (MS2), the mapping $x \mapsto x^\circ$ is antitone. So

$$(x \vee y)^\circ \leqslant x^\circ \wedge y^\circ \leqslant (x^\circ \wedge y^\circ)^{\circ\circ} = (x^{\circ\circ} \vee y^{\circ\circ})^\circ.$$

Since clearly $x \vee y \leqslant x^{\circ\circ} \vee y^{\circ\circ}$, which implies that $(x \vee y)^\circ \geqslant (x^{\circ\circ} \vee y^{\circ\circ})^\circ$, we deduce that $(x \vee y)^\circ = x^\circ \wedge y^\circ$. It follows that $(L; ^\circ)$ is an Ockham algebra. Now by (MS3) we have $x \leqslant x^{\circ\circ}$ and so $x^\circ \geqslant x^{\circ\circ\circ}$. But, again by (MS3), $x^\circ \leqslant x^{\circ\circ\circ}$. Consequently, we have $x^\circ = x^{\circ\circ\circ}$ and so $(L; ^\circ)$ has a de Morgan skeleton. The second statement is immediate from the definitions. ◇

In an Ockham algebra $(L; f)$ the biggest MS-subalgebra is

$$\mathrm{MS}(L) = \{x \in L \mid x \leqslant f^2(x)\}.$$

In the case where $(L; f)$ belongs to $\mathbf{K}_{1,1}$ and is obtained as in Theorem 1.2,

$$\mathrm{MS}(L) = \{x \in L \mid x \leqslant \varphi(x)\}.$$

Thus, in Example 1.9 we have (for both of the algebras described)

$$\mathrm{MS}(C) = \{(x, y, z) \in C \mid x \leqslant z\}.$$

In Example 1.10, we have $x \leqslant \varphi(x)$ if and only if $x \in [b, a]$, and so

$$\mathrm{MS}(L) = \mathrm{Im}\,\varphi.$$

In Example 1.11, we have

$$\mathrm{MS}(L) = \{0, 1, a_0\} \cup \{a_{2n+1}, b_{2n} \mid n \in \mathbb{Z}\}.$$

In Example 1.12, we have

$$\mathrm{MS}(\mathbb{P}(E)) = [\emptyset, \{a\}'] \cup [\{b\}, E].$$

In Example 1.13, we have

$$\mathrm{MS}(B^n_\star) = \{(x_1, \dots, x_n) \mid (\forall i)\; x_i \leqslant x_n\}.$$

Theorem 1.5 *Let L be a bounded distributive lattice and let $\varphi : L \to L$ be a 0-preserving closure morphism such that the sublattice $\mathrm{Im}\,\varphi$ admits a polarity p. Define a unary operation $^\circ : L \to L$ by the prescription*

$$(\forall x \in L) \qquad x^\circ = p[\varphi(x)].$$

Then $(L; {}^\circ)$ is an MS-algebra. Moreover, every MS-algebra arises in this way.

Proof Every closure map must also be 1-preserving and so by Theorem 1.2 we see that $(L; {}^\circ)$ belongs to $\mathbf{K}_{1,1}$. Since $x^{\circ\circ} = \varphi(x) \geqslant x$, it follows by Theorem 1.4 that $(L; {}^\circ)$ is an MS-algebra.

Conversely, if $(L; {}^\circ)$ is an MS-algebra then the mapping $\varphi : x \mapsto x^{\circ\circ}$ describes a 0-preserving closure morphism on L, and $p : x \mapsto x^\circ$ is a polarity on $\mathrm{Im}\,\varphi = \{x^{\circ\circ} \mid x \in L\}$. ◇

Definition A bounded distributive lattice L together with a unary operation $^+ : L \to L$ will be called a *dual MS-algebra* if $(L^{\mathrm{op}}, {}^+)$ is an MS-algebra, where L^{op} denotes the order dual of the lattice L.

It is clear that we can construct dual MS-algebras by using the dual of Theorem 1.5, which involves a 1-preserving dual closure morphism.

Example 1.14 Let $\mathbb{N}_0$ be ordered by the relation of divisibility and let $L = \{0\} \oplus \mathbb{N}_0 \oplus \{\infty\}$. Let $n = \prod_{i \in I} p_i^{\alpha_i}$ be the decomposition into prime factors of a fixed positive integer n. Define $\varphi_n : L \to L$ by

$$\varphi_n(x) = \begin{cases} \infty & \text{if } x = \infty; \\ \gcd\{n, x\} & \text{if } x \in \mathbb{N}_0; \\ 0 & \text{if } x = 0. \end{cases}$$

Then φ_n is a $\{0,\infty\}$-preserving dual closure morphism with

$$\operatorname{Im}\varphi_n = [0,n] \cup \{\infty\}.$$

A polarity on $\operatorname{Im}\varphi_n$ is given by $0 \mapsto \infty$, $\infty \mapsto 0$ and, for $x \in [1,n]$,

$$x = \prod_{i \in J} p_i^{\beta_i} \mapsto \tfrac{n}{x} = \prod_{i \in I \cup J} p_i^{\alpha_i - \beta_i}.$$

It follows that $(L;^{+})$ is a dual MS-algebra in which $0^{+} = \infty$, $\infty^{+} = 0$ and, for $x = \prod_{i \in K} p_i^{\gamma_i} \in \mathbb{N}_0$,

$$x^{+} = \frac{n}{\gcd\{n,x\}} = \prod_{i \in I \cup K} p_i^{\alpha_i - \min\{\alpha_i,\gamma_i\}}.$$

2 Congruence relations

Let $(L; f)$ be an Ockham algebra. Then an *Ockham algebra congruence* (or, briefly, a *congruence*) on L is an equivalence relation that has the substitution property for both the lattice operations and for the unary operation f. It follows that every congruence is in particular a *lattice congruence* and it is essential to distinguish these two types. In order to do so, we shall use the subscript 'lat' to denote a lattice congruence.

If $a, b \in L$ and $a \leqslant b$ then the *principal congruence* $\vartheta(a, b)$ generated by a, b is defined by

$$\vartheta(a, b) = \bigwedge\{\varphi \in \operatorname{Con} L \mid (a, b) \in \varphi\}.$$

In other words, it is the smallest congruence that identifies a and b. Similarly, the *principal lattice congruence* generated by a, b is

$$\vartheta_{\text{lat}}(a, b) = \bigwedge\{\varphi \in \operatorname{Con}_{\text{lat}} L \mid (a, b) \in \varphi\}.$$

Note that we then have

$$\vartheta_{\text{lat}}(a, b) \leqslant \vartheta(a, b).$$

We recall that, in a distributive lattice,

$$(x, y) \in \vartheta_{\text{lat}}(a, b) \iff x \wedge a = y \wedge a \text{ and } x \vee b = y \vee b,$$

and that the intersection of two principal lattice congruences is again a principal lattice congruence; in fact, if $a \leqslant b$ and $c \leqslant d$ then

$$\vartheta_{\text{lat}}(a, b) \wedge \vartheta_{\text{lat}}(c, d) = \vartheta_{\text{lat}}((a \vee c) \wedge b \wedge d, b \wedge d).$$

A fundamental result concerning congruences that we shall require is that if L is an algebra and ϑ is a congruence on L then for any congruence φ of L such that $\varphi \geqslant \vartheta$ the relation φ/ϑ defined on L/ϑ by

$$([x]\vartheta, [y]\vartheta) \in \varphi/\vartheta \iff (x, y) \in \varphi$$

is a congruence on L/ϑ; and every congruence on L/ϑ can be uniquely represented as φ/ϑ for some congruence $\varphi \geqslant \vartheta$. Moreover, $\operatorname{Con} L/\vartheta$ is isomorphic to the filter $[\vartheta, \iota]$ of $\operatorname{Con} L$.

The following result, due to J. Berman **[28]**, is fundamental in the investigation of congruences.

Theorem 2.1 *Let (L, f) be an Ockham algebra. If $a \leqslant b$ in L then*

$$\vartheta(a, b) = \bigvee_{n \geqslant 0} \vartheta_{\text{lat}}(f^n(a), f^n(b)).$$

Proof Let $\varphi = \bigvee_{n\geqslant 0} \vartheta_{\text{lat}}\big(f^n(a), f^n(b)\big)$. We show as follows that $\varphi \in \text{Con}\, L$. Suppose that $(x,y) \in \varphi$, so that there are integers $i_1,\ldots,i_m$ and elements $z_0, z_1, \ldots, z_m$ such that

$$x = z_0 \overset{\varphi_{i_1}}{\equiv} z_1 \overset{\varphi_{i_2}}{\equiv} z_2 \equiv \cdots \equiv z_{m-1} \overset{\varphi_{i_m}}{\equiv} z_m = y,$$

where $\varphi_{i_k} = \vartheta_{\text{lat}}\big(f^{i_k}(a), f^{i_k}(b)\big)$. Now if i_k is even we have $f^{i_k}(a) \leqslant f^{i_k}(b)$ and so

$$z_{k-1} \wedge f^{i_k}(a) = z_k \wedge f^{i_k}(a), \quad z_{k-1} \vee f^{i_k}(b) = z_k \vee f^{i_k}(b).$$

Applying f to each of these equalities, we obtain

$$(\star) \qquad \big(f(z_{k-1}), f(z_k)\big) \in \vartheta_{\text{lat}}\big(f^{i_k+1}(a), f^{i_k+1}(b)\big) = \varphi_{i_k+1}.$$

If, on the other hand, i_k is odd then $f^{i_k}(b) \leqslant f^{i_k}(a)$, and by a similar argument $(\star)$ also holds in this case. The integers $i_1+1,\ldots,i_m+1$ and the elements $f(z_0),\ldots,f(z_m)$ together with $(\star)$ now give $\big(f(x), f(y)\big) \in \varphi$. Consequently, $\varphi \in \text{Con}\, L$.

Clearly, $\vartheta_{\text{lat}}(a,b) \leqslant \varphi$ and hence $(a,b) \in \varphi$. Since $\vartheta(a,b)$ is, by definition, the smallest congruence to identify a and b, it follows that $\vartheta(a,b) \leqslant \varphi$. Finally, since $\vartheta(a,b)$ is a congruence we have that

$$(a,b) \in \vartheta(a,b) \Rightarrow (\forall n)\ \big(f^n(a), f^n(b)\big) \in \vartheta(a,b),$$

from which we deduce that, for each n,

$$\vartheta\big(f^n(a), f^n(b)\big) \leqslant \vartheta(a,b)$$

and hence that

$$\varphi = \bigvee_{n\geqslant 0} \vartheta_{\text{lat}}\big(f^n(a), f^n(b)\big) \leqslant \bigvee_{n\geqslant 0} \vartheta\big(f^n(a), f^n(b)\big) \leqslant \vartheta(a,b).$$

Thus $\varphi = \vartheta(a,b)$ as asserted. ◇

Corollary *If* $(L,f) \in \mathbf{K}_{p,q}$ *then* $\vartheta(a,b) = \bigvee_{n=0}^{2p+q-1} \vartheta_{\text{lat}}\big(f^n(a), f^n(b)\big)$. ◇

A class $\mathbf{K}$ of algebras is said to enjoy the *(principal) congruence extension property* if, for all $A, B \in \mathbf{K}$ with A a subalgebra of B, every (principal) congruence ϑ on A is the restriction of some congruence φ on B (this being denoted by $\varphi|_A = \vartheta$). In fact, as was shown by A. Day **[66]**, in an equational class of algebras these properties are equivalent; indeed, they are each equivalent to the condition

for all subalgebras A of B and all $a, b \in A$, $\vartheta_A(a,b) = \vartheta_B(a,b)|_A$.

Using this fact, J. Berman **[28]** established the following result.

Theorem 2.2 *The class* **O** *enjoys the congruence extension property.*

Proof Let $A, B \in \mathbf{O}$ with A a subalgebra of B. For $a, b \in A$ let $\lambda = \vartheta_A(a,b)$ and $\mu = \vartheta_B(a,b)$. Then, by the above result of Day, it suffices to prove that $\lambda = \mu|_A$. Now by Theorem 2.1 we have

$$\lambda = \bigvee_{n\geqslant 0} \vartheta_{\text{lat}}(f^n(a), f^n(b)) \in \text{Con}_{\text{lat}} A,$$

$$\mu = \bigvee_{n\geqslant 0} \vartheta_{\text{lat}}(f^n(a), f^n(b)) \in \text{Con}_{\text{lat}} B.$$

Denote the lattice congruences that appear in the right hand side of these equalities by λ_n and μ_n respectively. Clearly, A is a sublattice of B and so, since the class of distributive lattices has the congruence extension property, we have $\lambda_n = \mu_n|_A$. Now since λ is a lattice congruence it has an extension $\overline{\lambda} \in \text{Con}_{\text{lat}} B$ such that $\overline{\lambda}|_A = \lambda$. It follows that $\overline{\lambda} \geqslant \mu_n$ for all n and so

$$\lambda = \overline{\lambda}|_A \geqslant \Big(\bigvee_{n\geqslant 0} \mu_n\Big)\Big|_A \geqslant \lambda,$$

whence $\lambda = \mu|_A$. ◇

Consider now, for any $a \in L$, the relation ϑ_a defined by

$$(x,y) \in \vartheta_a \iff x \wedge a = y \wedge a \ \text{ and } \ x \vee f(a) = y \vee f(a).$$

Theorem 2.3 *For every* $a \in L$,

(1) ϑ_a *is a principal lattice congruence; moreover,* $[a]\vartheta_a = [a, a \vee f(a)]$ *and* $\vartheta_a = \omega \Leftrightarrow a \geqslant f(a)$.

(2) *If* $a \leqslant f^2(a)$ *then* ϑ_a *is a principal congruence.*

(3) *If* $a = f^2(a)$ *and* $a \parallel f(a)$ *then* ϑ_a *and* $\vartheta_{f(a)}$ *are non-trivial congruences with* $\vartheta_a \wedge \vartheta_{f(a)} = \omega$.

Proof (1) By its very definition,

$$\vartheta_a = \vartheta_{\text{lat}}(a,1) \wedge \vartheta_{\text{lat}}(0, f(a)) = \vartheta_{\text{lat}}(a \wedge f(a), f(a)).$$

That $[a]\vartheta_a = [a, a \vee f(a)]$ follows immediately from the definition of ϑ_a ; and that $\vartheta_a = \omega \Leftrightarrow a \geqslant f(a)$ is immediate from the above.

(2) If $a \leqslant f^2(a)$ and $(x,y) \in \vartheta_a$ then $(f(x), f(y)) \in \vartheta_a$; for, from $x \wedge a = y \wedge a$ we obtain $f(x) \vee f(a) = f(y) \vee f(a)$ and from $x \vee f(a) = y \vee f(a)$ we obtain $f(x) \wedge f^2(a) = f(y) \wedge f^2(a)$, hence $f(x) \wedge a = f(y) \wedge a$.

(3) If $a = f^2(a)$ and $a \parallel f(a)$ then by (1) we have $\vartheta_{f(a)} = \vartheta_{\text{lat}}(a \wedge f(a), a)$ and consequently $\vartheta_a \wedge \vartheta_{f(a)} = \omega$. It also follows by (1) that both ϑ_a and $\vartheta_{f(a)}$ are non-trivial. ◇

Theorem 2.4 *For an Ockham algebra* $(L;f)$,

(1) $\{a \in L \mid a \leqslant f^2(a)\}$ *is a sublattice of* L *which contains* 0 *and* 1;

(2) $a \leqslant b \Rightarrow \vartheta_a \geqslant \vartheta_b$;

(3) $\vartheta_a \wedge \vartheta_b = \vartheta_{a \vee b}$.

Proof (1) follows from the fact that $f^2(a \wedge b) = f^2(a) \wedge f^2(b)$ and $f^2(a \vee b) = f^2(a) \vee f^2(b)$.

(2) follows from the fact that $x \wedge b = y \wedge b$ implies $x \wedge a = y \wedge a$, and $x \vee f(b) = y \vee f(b)$ implies $x \vee f(a) = y \vee f(a)$.

(3) If $(x, y) \in \vartheta_a \wedge \vartheta_b$ then

$$x \wedge a = y \wedge a, \quad x \wedge b = y \wedge b, \quad x \vee f(a) = y \vee f(a), \quad x \vee f(b) = y \vee f(b),$$

which implies that

$$x \wedge (a \vee b) = y \wedge (a \vee b), \quad x \vee f(a \vee b) = y \vee f(a \vee b).$$

Consequently, $\vartheta_a \wedge \vartheta_b \leqslant \vartheta_{a \vee b}$. The converse inequality follows from (2). ◊

Theorems 2.3 and 2.4 show that in **MS** the meet of two principal congruences each of the form ϑ_a is again a principal congruence. Unfortunately, this property does not hold for arbitrary principal congruences, even when the Ockham algebra belongs to a 'small' class such as that of de Morgan algebras. The interested reader may consult M. E. Adams [**17**].

For an Ockham algebra $(L;f)$ consider now, for every $n \in \mathbb{N}$, the relation Φ_n on L defined by

$$(x, y) \in \Phi_n \iff f^n(x) = f^n(y).$$

It is clear that Φ_n is a congruence on L. Moreover, the subset

$$f^n(L) = \{f^n(x) \mid x \in L\}$$

is a subalgebra of L.

We now consider some basic results concerning these congruences and subalgebras. Of especial importance in this is the congruence

$$\Phi_\omega = \bigvee_{i \geqslant 0} \Phi_i,$$

a practical description of which is as follows.

Theorem 2.5 *If* $(L;f) \in \mathbf{O}$ *then* $(x, y) \in \Phi_\omega$ *if and only if* $f^n(x) = f^n(y)$ *for some* n. *Moreover, if* L *is non-trivial then* $\Phi_\omega < \iota$.

Proof If $(x, y) \in \Phi_\omega$ then there exist $t_0, \dots, t_k$ and $\Phi_{i_1}, \dots, \Phi_{i_k}$ such that

$$x = t_0 \, \Phi_{i_1} \, t_1 \, \Phi_{i_2} \, t_2 \, \dots \, t_{k-1} \, \Phi_{i_k} \, t_k = y.$$

Denote the greatest of these Φ_i by Φ_n. Then we have $x\,\Phi_n\,y$, i.e. $f^n(x) = f^n(y)$. Conversely, if $f^n(x) = f^n(y)$ then clearly $(x,y) \in \Phi_\omega$.

Finally, if L is non-trivial, $f^n(0) \neq f^n(1)$ for all n gives $\Phi_\omega \neq \iota$. ◇

For every non-trivial Ockham algebra $(L;f)$ it is clear that we have

$$\omega = \Phi_0 \leqslant \Phi_1 \leqslant \Phi_2 \leqslant \cdots \leqslant \Phi_i \leqslant \Phi_{i+1} \leqslant \ldots \leqslant \Phi_\omega < \iota,$$

and, with $\leqslant$ meaning 'is a subalgebra of',

$$\{0,1\} \leqslant \cdots \leqslant f^{i+1}(L) \leqslant f^i(L) \leqslant \cdots \leqslant f(L) \leqslant f^0(L) = L.$$

It is readily seen that $[x]\Phi_i \mapsto f^i(x)$ describes an Ockham algebra isomorphism when i is even and a dual isomorphism when i is odd, a situation which we shall denote by writing $L/\Phi_i \sim f^i(L)$. The following result is therefore clear.

Theorem 2.6 *If* $(L;f) \in \mathbf{K}_{p,q}$ *then, for* $n \leqslant q$,

$$L/\Phi_n \sim f^n(L) \in \mathbf{K}_{p,q-n}. \quad ◇$$

Theorem 2.7 *If* $(L;f) \in \mathbf{K}_{p,q}$ *then*

$$\omega = \Phi_0 \leqslant \Phi_1 \leqslant \cdots \leqslant \Phi_q = \Phi_{q+1} = \cdots = \Phi_\omega.$$

Proof Observe that

$$\begin{aligned}(x,y) \in \Phi_{q+1} &\iff f^{q+1}(x) = f^{q+1}(y)\\ &\iff f^q(x) = f^{2p-1}[f^{q+1}(x)] = f^{2p-1}[f^{q+1}(y)] = f^q(y)\\ &\iff (x,y) \in \Phi_q.\end{aligned}$$

We thus have $\Phi_q = \Phi_{q+1}$. If now $(x,y) \in \Phi_{q+r}$ with $r \geqslant 1$ then $f^{q+r}(x) = f^{q+r}(y)$ gives

$$(f^{r-1}(x), f^{r-1}(y)) \in \Phi_{q+1} = \Phi_q$$

and so $f^{q+r-1}(x) = f^{q+r-1}(y)$, i.e. $(x,y) \in \Phi_{q+r-1}$. Consequently,

$$\Phi_q = \Phi_{q+1} = \Phi_{q+2} = \cdots,$$

and therefore $\Phi_\omega = \bigvee_{i \geqslant 0} \Phi_i = \Phi_q$. ◇

Corollary *The following statements are equivalent:*

(1) $\Phi_\omega = \omega$;

(2) $(\forall i \geqslant 1) \quad \Phi_i = \omega$;

(3) f *is injective.*

Moreover, if $(L;f)$ belongs to the Berman class $\mathbf{K}_{p,q}$ then each of the above is equivalent to $(L;f) \in \mathbf{K}_{p,0}$.

Proof The equivalence of (1), (2), (3) is clear. As for the final statement, if for every $x \in L$ we have $f^{2p+q}(x) = f^q(x)$ then, by (3), we have $f^{2p}(x) = x$ and then $(L;f) \in \mathbf{K}_{p,0}$. Conversely, suppose that $(L;f) \in \mathbf{K}_{p,0}$ and that $\Phi_1 \neq \omega$. Then there exist $x, y \in L$ such that $x \neq y$ and $f(x) = f(y)$. This implies that $f^{2p}(x) = f^{2p}(y)$ and hence the contradiction $x = y$. ◇

Note that the hypothesis that L belong to some Berman class is necessary in the above. As the following example shows, it is possible for f to be injective with $L \notin \mathbf{K}_{n,0}$ for any n.

Example 2.1 On the chain $L = \{-\infty\} \oplus \mathbb{Z} \oplus \{\infty\}$ define f by

$$\begin{array}{rccccccccc} x: & -\infty & \ldots & -2 & -1 & 0 & 1 & 2 & \ldots & \infty \\ f(x): & \infty & \ldots & 3 & 2 & 0 & -2 & -3 & \ldots & -\infty \end{array}$$

Then $(L;f)$ is an Ockham algebra with f injective. But for $m \neq n$ we have $f^m(x) \neq f^n(x)$, so $f^m \neq f^n$ and L does not belong to any Berman class.

In general, a Φ_n-class can contain more than one element of $f^n(L)$, or even none, as the following example shows.

Example 2.2 Let L be the chain $0 < a < b < c < 1$. Define f by

$$\begin{array}{rccccc} x: & 0 & a & b & c & 1 \\ f(x): & 1 & 1 & c & a & 0 \end{array}$$

Then it is readily seen that $(L;f) \in \mathbf{O}$ and that the smallest Berman class that it belongs to is $\mathbf{K}_{1,3}$. Now $f^2(L) = \{0, a, 1\}$, and

$$\begin{array}{ll} [0]\Phi_2 = \{0, a\} & \text{with } [0]\Phi_2 \cap f^2(L) = \{0, a\}; \\ [b]\Phi_2 = \{b\} & \text{with } [b]\Phi_2 \cap f^2(L) = \emptyset. \end{array}$$

Of some interest is the case where every Φ_n-class of L contains exactly one element of $f^n(L)$. Here we shall deal with the subvarieties of $\mathbf{O}$ defined by $x \wedge f^{2p}(x) = x$ (see Chapter 1) and denoted by Ramalho and Sequeira **[82]** by $\mathbf{K}_{p,0}^{\leqslant}$. Clearly,

$$\mathbf{K}_{p,0} \subset \mathbf{K}_{p,0}^{\leqslant} \subset \mathbf{K}_{p,1}.$$

Note that in $\mathbf{K}_{p,0}^{\leqslant}$ the mapping $\varphi : x \mapsto f^{2p}(x)$ is a closure operator, since

$$\varphi(x) \geqslant x; \quad \varphi^2(x) = \varphi(x); \quad x \leqslant y \Rightarrow \varphi(x) \leqslant \varphi(y).$$

This closure is *additive* in the sense that $\varphi(x \vee y) = \varphi(x) \vee \varphi(y)$, and also *multiplicative* in the sense that $\varphi(x \wedge y) = \varphi(x) \wedge \varphi(y)$. By Theorem 5 of

[97], it follows that $\varphi(L)$ is a *semiconvex* subalgebra of L, i.e. if $x \wedge y$ and $x \vee y$ both belong to $\varphi(L)$ then both x and y belong to $\varphi(L)$. As a direct consequence, $\varphi(L)$ contains $Z(L)$, the *centre* of L (i.e. the boolean sublattice of L formed by the complemented elements).

Theorem 2.8 *If $L \in \mathbf{K}_{p,q}$ and $q = 2kp$ then every Φ_q-class contains exactly one element of $f^q(L)$. If $L \in \mathbf{K}^{\leqslant}_{p,0}$ then every Φ_{2p}-class has a greatest element, this being the only element of the class that belongs to $f^{2p}(L)$.*

Proof Since $L \in \mathbf{K}_{p,q}$, for every $x \in L$ we have $(x, f^{2p}(x)) \in \Phi_q$, and therefore $(f^{2p}(x), f^{4p}(x)) \in \Phi_q$ and so $(x, f^{2kp}(x)) \in \Phi_q$, i.e. $(x, f^q(x)) \in \Phi_q$. It follows that $[x]\Phi_q$ contains exactly one element of $f^q(L)$ since $(x,y) \in \Phi_q$ is equivalent to $f^q(x) = f^q(y)$.

Now let $L \in \mathbf{K}^{\leqslant}_{p,0}$. Then $L \in \mathbf{K}_{p,1} \subseteq \mathbf{K}_{p,2p}$. Since f^{2p} is a closure operator, the greatest element of $[x]\Phi_{2p}$ is the only one that belongs to $f^{2p}(L)$. ◇

If an Ockham algebra $(L;f)$ belongs to a Berman class then there is a smallest Berman class to which it belongs; we denote this by $\mathbf{V}_B(L)$.

Theorem 2.9 *For an Ockham algebra $(L;f)$ the following statements are equivalent:*

(1) $\mathbf{V}_B(L) = \mathbf{K}_{p,q}$;

(2) $L \supset f(L) \supset \cdots \supset f^q(L) = f^{q+1}(L) = \cdots$, *and* $\mathbf{V}_B(f^q(L)) = \mathbf{K}_{p,0}$.

Proof (1) $\Rightarrow$ (2) : From $f^q(x) = f^{2p+q}(x) = f^{q+1}[f^{2p-1}(x)] \in f^{q+1}(L)$ it follows that $f^q(L) \subseteq f^{q+1}(L)$, whence $f^q(L) = f^{q+1}(L) = \cdots$. Suppose now, by way of obtaining a contradiction, that for some n with $n \leqslant q$ we have $f^{n-1}(L) = f^n(L)$. Then clearly $f^{q-1}(L) = f^q(L)$. But, by (1), $L \notin \mathbf{K}_{p,q-1}$ and so there exists $x \in L$ with $f^{q-1}(x) \neq f^{2p+q-1}(x)$. But $f^{q-1}(x) \in f^{q-1}(L) = f^q(L)$ and so $f^{q-1}(x) = f^q(y)$ for some $y \in L$. Hence $f^q(y) \neq f^{2p+q}(y)$, and this contradicts the fact that $L \in \mathbf{K}_{p,q}$. This then establishes the chain

$$L \supset f(L) \supset f^2(L) \supset \cdots \supset f^q(L) = f^{q+1}(L) = \cdots .$$

Now since $f^q = f^{2p+q}$ we see that f^{2p} acts as the identity on $f^q(L)$. Hence $f^q(L) \in \mathbf{K}_{p,0}$. That $\mathbf{V}_B(f^q(L)) = \mathbf{K}_{p,0}$ follows from the fact that if $f^q(L) \in \mathbf{K}_{t,0} \subset \mathbf{K}_{p,0}$ then necessarily $t|p$ and, since f^{2t} is then the identity on $f^q(L)$, we have $f^q = f^{2t+q}$ so that $L \in \mathbf{K}_{t,q}$, which contradicts $\mathbf{V}_B(L) = \mathbf{K}_{p,q}$.

(2) $\Rightarrow$ (1) : If $\mathbf{V}_B(f^q(L)) = \mathbf{K}_{p,0}$ then clearly $f^q = f^{2p+q}$ and so $L \in \mathbf{K}_{p,q}$. If $\mathbf{V}_B(L) = \mathbf{K}_{p',q'}$ then from (1) $\Rightarrow$ (2) it follows that $q' = q$ and $p' = p$. ◇

Corollary 1 *If $\mathbf{V}_B(L) = \mathbf{K}_{p,q}$ then, for $1 \leqslant i \leqslant q$, $\mathbf{V}_B(f^i(L)) = \mathbf{K}_{p,q-i}$.* ◇

Corollary 2 *If* $\mathbf{V}_B(L) = \mathbf{K}_{p,q}$ *then*

$$\omega = \Phi_0 < \Phi_1 < \Phi_2 < \cdots < \Phi_q = \Phi_{q+1} = \cdots,$$

and Con L *has length at least* $q+1$. ◇

In a bounded distributive lattice every ideal I is the *kernel* of at least one congruence, i.e. there is a lattice congruence ϑ such that $[0]\vartheta = I$. We also write $I = \operatorname{Ker}\vartheta$. Dually, every filter F is the *cokernel* of a congruence, i.e. there is a lattice congruence ψ such that $[1]\psi = F$. These properties do not carry over to Ockham algebras. For instance, in Example 2.2 the principal ideal $c^{\downarrow}$ is not the kernel of any congruence; for $(0,c) \in \vartheta$ implies $(a,1) \in \vartheta$, whence the contradiction $(0,1) \in \vartheta$. So it is of interest to characterise those ideals of an Ockham algebra that are congruence kernels.

Let I be an ideal of the Ockham algebra $(L;f)$. For each $n \geqslant 0$, define

$$I_{2n} = \left(f^{2n}(I)\right)^{\downarrow}, \qquad I^{2n+1} = \left(f^{2n+1}(I)\right)^{\uparrow}.$$

Then clearly I_{2n} is an ideal and I^{2n+1} is a filter. Finally, let

$$I_{\circ\circ} = \bigvee_{n\geqslant 0} I_{2n}, \qquad I^{\circ} = \bigvee_{n\geqslant 0} I^{2n+1}.$$

We shall now investigate, for a given ideal I, the smallest congruence $\Theta(I)$ on L that identifies the elements of I. We recall that, in a distributive lattice, $\Theta(I)$ is characterised as follows :

$$(x,y) \in \Theta(I) \iff (\exists i \in I)\ x \vee i = y \vee i.$$

With the obvious subscript 'lat' to denote the corresponding smallest lattice congruence, we have the following result.

Theorem 2.10 *For every ideal* I *of* $(L;f) \in \mathbf{O}$,

$$\Theta(I) = \Theta_{\text{lat}}(I^{\circ}) \vee \Theta_{\text{lat}}(I_{\circ\circ}).$$

Proof This follows immediately from Theorem 2.1 on noting that for every subset X of L we have $\Theta(X) = \bigvee\{\vartheta(a,b) \mid a,b \in X\}$. ◇

Theorem 2.11 *For every ideal* I *of* $(L;f) \in \mathbf{O}$,

$$(x,y) \in \Theta(I) \iff (\exists i \in I_{\circ\circ})(\exists j \in I^{\circ}) \quad (x \vee i) \wedge j = (y \vee i) \wedge j.$$

Proof For convenience, denote by Ψ the relation defined on L by the above condition. Since L is distributive, Ψ is a lattice congruence. Suppose that $(x,y) \in \Psi$. Then there exist $i \in I_{\circ\circ}$ and $j \in I^{\circ}$ with $(x \vee i) \wedge j = (y \vee i) \wedge j$, whence

$$x \wedge j = x \wedge (x \vee i) \wedge j = x \wedge (y \vee i) \wedge j$$

and hence $(x, x \wedge (y \vee i)) \in \Theta_{\text{lat}}(I^{\circ})$.

Now we also have

$$(x \wedge (y \vee i)) \vee i = (x \vee i) \wedge (y \vee i) = (x \wedge y) \vee i$$

and so $(x \wedge (y \vee i), x \wedge y) \in \Theta_{\text{lat}}(I_{\circ\circ})$. Thus we see that

$$(x, x \wedge y) \in \Theta_{\text{lat}}(I^{\circ}) \vee \Theta_{\text{lat}}(I_{\circ\circ}).$$

Similarly, we can show that

$$(y, x \wedge y) \in \Theta_{\text{lat}}(I^{\circ}) \vee \Theta_{\text{lat}}(I_{\circ\circ}).$$

Thus, by Theorem 2.10, $(x, y) \in \Theta(I)$.

Conversely, suppose that $(x, y) \in \Theta(I)$. Then, again by Theorem 2.10, we have the finite sequence

$$x \equiv p_1 \equiv p_2 \equiv \cdots \equiv p_n \equiv y$$

where each $\equiv$ is either $\Theta_{\text{lat}}(I^{\circ})$ or $\Theta_{\text{lat}}(I_{\circ\circ})$. Suppose that we have

$$x \ \Theta_{\text{lat}}(I_{\circ\circ}) \ p_1 \ \Theta_{\text{lat}}(I^{\circ}) \ p_2 \ \Theta_{\text{lat}}(I_{\circ\circ}) \ \ldots \ \Theta_{\text{lat}}(I^{\circ}) \ y.$$

Then we have finitely many equalities

$$x \vee i_1 = p_1 \vee i_1,\ p_1 \wedge j_1 = p_2 \wedge j_1,\ p_2 \vee i_2 = p_3 \vee i_2,\ p_3 \wedge j_2 = p_4 \wedge j_2,\ \ldots$$

where each $i_k \in I_{\circ\circ}$ and each $j_k \in I^{\circ}$. Since $I_{\circ\circ}$ is an ideal and I° is a filter, we have $\bigvee i_k \in I_{\circ\circ}$ and $\bigwedge j_k \in I^{\circ}$. Moreover,

$$\begin{aligned}
(x \vee \textstyle\bigvee i_k) \wedge \textstyle\bigwedge j_k &= (p_1 \vee \textstyle\bigvee i_k) \wedge \textstyle\bigwedge j_k \\
&= (p_1 \wedge \textstyle\bigwedge j_k) \vee (\textstyle\bigvee i_k \wedge \textstyle\bigwedge j_k) \\
&= (p_2 \wedge \textstyle\bigwedge j_k) \vee (\textstyle\bigvee i_k \wedge \textstyle\bigwedge j_k) \\
&= (p_2 \vee \textstyle\bigvee i_k) \wedge \textstyle\bigwedge j_k \\
&= \cdots \\
&= (y \vee \textstyle\bigvee i_k) \wedge \textstyle\bigwedge j_k,
\end{aligned}$$

from which we see that $(x, y) \in \Psi$ and hence that Ψ coincides with $\Theta(I)$. ◇

The following observation is immediate.

Lemma 2.1 *If the ideal I is such that $f^2(I) \subseteq I$ then, for every n,*

$$f^{2n}(I) \subseteq I, \qquad f^{2n+1}(I) \subseteq (f(I))^{\uparrow},$$

from which it follows that

$$I_{\circ\circ} = I, \qquad I^{\circ} = (f(I))^{\uparrow}. \quad ◇$$

Theorem 2.12 *An ideal I of an Ockham algebra $(L; f)$ is a congruence kernel if and only if*

(α) $f^2(I) \subseteq I$;

(β) $(\forall x \in L)(\forall j \in (f(I))^{\uparrow}) \qquad x \wedge j \in I \Rightarrow x \in I.$

Proof If I is the kernel of a congruence Θ on L then for every $i \in I$ we have $(f^2(i), 0) \in \Theta$, so that $f^2(i) \in I$ and we have property (α). If $j \in L$ is such that $j \geqslant f(i)$ for some $i \in I$ then $(i, 0) \in \Theta$ gives $(j, 1) \in \Theta$, whence $(x \wedge j, x) \in \Theta$ for every $x \in L$. Property (β) is now immediate.

Conversely, suppose that (α) and (β) hold. By Theorem 2.11, Lemma 2.1, and condition (α),

$$(x, 0) \in \Theta(I) \iff (\exists i \in I)(\exists j \in (f(I))^{\uparrow}) \qquad (x \vee i) \wedge j = i \wedge j.$$

We have $i \wedge j \in I$ and so, by (β), $x \vee i \in I$ and hence $x \in I$. It follows that $\text{Ker}\, \Theta(I) \subseteq I$. Since $I = \text{Ker}\, \Theta_{\text{lat}}(I) \subseteq \text{Ker}\, \Theta(I)$, it follows that $\text{Ker}\, \Theta(I) = I$. ◇

Corollary 1 *If a proper ideal I is a congruence kernel then*

$$I \cap f(I) = \emptyset. \qquad \diamond$$

An element of an ordered set P is said to be a *node* if it is comparable with all the elements of P. A node is *non-trivial* if it differs from 0 and 1. An element a of an Ockham algebra $(L; f)$ is a *fixed point* of f if $a = f(a)$. With these notions, we have :

Corollary 2 *Let $(L; f) \in \mathbf{O}$ and let a be a non-trivial node of L that is also a (necessarily unique) fixed point of f. Then the congruence kernels of L are the proper ideals of $a^{\downarrow}$ that satisfy condition (α) above, and L itself.* ◇

We now proceed with some considerations concerning the structure of the congruence lattice Con L of an Ockham algebra $(L; f)$. Since Con L is a sublattice of $\text{Con}_{\text{lat}} L$, which is known to be distributive, Con L is also distributive. It is also *algebraic*, in the sense that it is complete and *compactly generated* (every element of Con L is the supremum of a set of compact elements, a compact element being an element ϑ such that if $\vartheta \leqslant \sup X$ for some $X \subseteq \text{Con}\, L$ then $\vartheta \leqslant \sup X_1$ for some finite $X_1 \subseteq \text{Con}\, L$).

As far as the structure of Con L is concerned, it is important to observe that every lattice congruence that is contained in Φ_1 is a congruence. It follows that if L is finite then the interval $[\omega, \Phi_1]$ of Con L is boolean. We also point out that Φ_1 is *dually dense* in Con L, in the sense that if $\vartheta \vee \Phi_1 = \iota$ then $\vartheta = \iota$. In fact, for any $\vartheta \in \text{Con}\, L$ we have $(0, 1) \in \vartheta \vee \Phi_1$ if and only if $(0, 1) \in \vartheta$; for

$$0 \;\vartheta\; x_1 \;\Phi_1\; x_2 \;\vartheta\; x_3 \;\Phi_1 \;\ldots\; \Phi_1\; x_{n-2} \;\vartheta\; x_{n-1} \;\Phi_1\; 1$$

implies that

$$1 \,\vartheta\, f(x_1) = f(x_2) \,\vartheta\, f(x_3) = \cdots = f(x_{n-2}) \,\vartheta\, f(x_{n-1}) = 0.$$

We have already seen that the class **O** of Ockham algebras enjoys the congruence extension property. This fact can be useful in the determination of Con L, especially when f has a fixed point. In order to illustrate this, we consider the following notions.

Let A and B be algebras of the same type. Then we say that A is an *extension* of B if B is a subalgebra of A. We say that A is a *strong extension* of B if every congruence on B has at most one extension to A, in which case B is said to be a *strongly large subalgebra* of A. Finally, A is a *perfect extension* of B if every congruence on B has exactly one extension to A, in which case B is a *perfect subalgebra* of A. When this happens, we have Con $A \simeq$ Con B.

In **[100]** it is shown that, for any element a of a modular lattice L, the *cone* C_a generated by a, i.e. the set $a^{\downarrow} \cup a^{\uparrow}$, is a strongly large sublattice. This leads to the following property.

Theorem 2.13 *Let* $(L; f) \in$ **O** *and let* a *be a fixed point of* L. *Then*

$$\text{Con } L \simeq \text{Con } C_a.$$

Proof Clearly, C_a is a subalgebra of L and this subalgebra is strongly large. In fact, since **O** has the congruence extension property, C_a is a perfect subalgebra. Hence we have the isomorphism stated. ◊

The usefulness of Theorem 2.13 lies in the fact that it enables us to work with C_a instead of L, and we can benefit from this in two ways. Firstly, the size of C_a can be considerably less than that of L. Secondly, C_a will in general belong to a subvariety of **O** that is smaller than that of L; for instance, it is clear that $(C_a; f)$ always satisfies the axiom $x \wedge f(x) \leqslant y \vee f(y)$ even if $(L; f)$ does not.

Example 2.3 On the lattice L with Hasse diagram

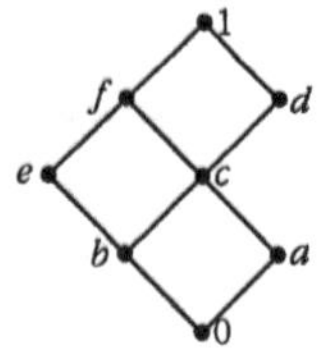

define f by

$$\begin{array}{r l} x : & 0\ a\ b\ c\ d\ e\ f\ 1 \\ f(x) : & 1\ d\ f\ c\ a\ e\ b\ 0 \end{array}$$

Then $(L;f)$ is an Ockham algebra that belongs to $\mathbf{K}_{1,0}$ and has two fixed points, namely c and e. The cone C_e generated by e is a five-element chain. As a subalgebra of $(L;f)$ it also belongs to $\mathbf{K}_{1,0}$ and satisfies the supplementary relation $x \wedge f(x) \leqslant y \vee f(y)$, which L itself does not (consider the elements e and c). It is easily seen that Con C_e is a four-element boolean lattice and therefore so also is Con L. Note also that by Theorem 2.12 the only ideals of L that are congruence kernels are $0^{\downarrow}$, $a^{\downarrow}$, $b^{\downarrow}$, and $1^{\downarrow}$.

Example 2.4 (*The pineapple*) Consider the ordered set L with Hasse diagram

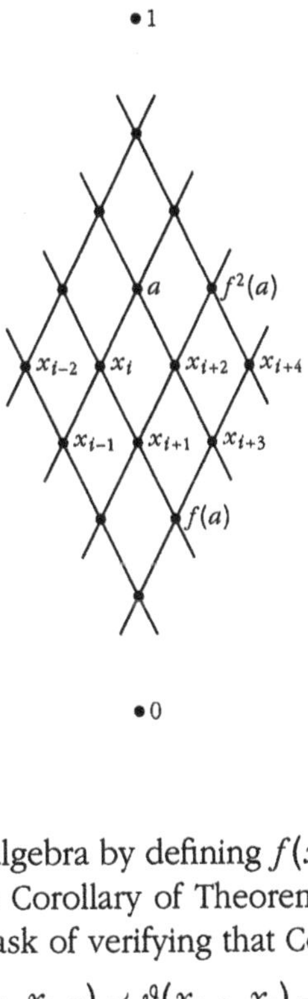

and made into an Ockham algebra by defining $f(x_i) = x_{i+1}$ for each i. Note that f is injective so, by the Corollary of Theorem 2.7, $\omega = \Phi_1 = \cdots = \Phi_\omega$. We leave to the reader the task of verifying that Con L is the chain

$$\omega < \cdots \prec \vartheta(x_{i+1}, x_{i+2}) \prec \vartheta(x_{i+1}, x_i) \prec \cdots < \Psi \prec \iota$$

where Ψ has classes $\{0\}, \{1\}, L \setminus \{0,1\}$. We shall return to this particular example later.

Concerning the basic congruences in a general Ockham algebra $(L;f)$ we also have the following results.

Theorem 2.14 *If $a, b \in L$ with $a \leqslant b$ and $f(a) = f(b)$ then $\vartheta(a,b)$ has a complement in $[\omega, \Phi_1]$.*

Proof By Theorem 2.1 we have

$$\vartheta(a,b) = \bigvee_{n \geqslant 0} \vartheta_{\text{lat}}\big(f^n(a), f^n(b)\big).$$

Since $f(a) = f(b)$ by hypothesis, it follows that $\vartheta(a,b) = \vartheta_{\text{lat}}(a,b) \in \operatorname{Con} L$ with, clearly, $\vartheta(a,b) \leqslant \Phi_1$. Write $\vartheta(a,b) = \alpha$ and observe that since α is a principal lattice congruence it has a complement in $\operatorname{Con}_{\text{lat}}(L)$, namely

$$\beta = \vartheta_{\text{lat}}(0,a) \vee \vartheta_{\text{lat}}(b,1).$$

Consider the lattice congruence $\alpha' = \beta \wedge \Phi_1$. Since every lattice congruence contained in Φ_1 is a congruence, we have $\alpha' \in \operatorname{Con} L$. Now

$$\begin{aligned} \alpha \vee \alpha' = \vartheta(a,b) \vee (\beta \wedge \Phi_1) &= \big(\vartheta_{\text{lat}}(a,b) \vee \beta\big) \wedge \big(\vartheta_{\text{lat}}(a,b) \vee \Phi_1\big) \\ &= \iota \wedge \Phi_1 \\ &= \Phi_1, \end{aligned}$$

and

$$\alpha \wedge \alpha' = \vartheta_{\text{lat}}(a,b) \wedge \beta \wedge \Phi_1 = \omega \wedge \Phi_1 = \omega.$$

It follows that α' is the complement of α in $[\omega, \Phi_1]$. ◇

If α is a congruence on an Ockham algebra L then an α-class $[a]\alpha$ will be called *locally finite* if, whenever $x, y \in [a]\alpha$ with $x \leqslant y$, the interval $[x,y]$ is finite.

The following two results are due to Jie Fang [**67**].

Theorem 2.15 *Let $a, b \in L$ be such that*

$$a \prec b, \quad (a,b) \notin \Phi_n, \quad (a,b) \in \Phi_{n+1}.$$

Then $\Phi_n \vee \vartheta(a,b)$ is an atom of $[\Phi_n, \Phi_{n+1}]$. Moreover, if every Φ_{n+1}-class is locally finite then every atom of $[\Phi_n, \Phi_{n+1}]$ is obtained in this way.

Proof Since $f^{n+1}(a) = f^{n+1}(b)$ we have $\big(f(a), f(b)\big) \in \Phi_n$. It follows that $\vartheta\big(f(a), f(b)\big) \leqslant \Phi_n$ and consequently, by Theorem 2.1,

$$(1) \qquad \Phi_n \vee \vartheta(a,b) = \Phi_n \vee \vartheta_{\text{lat}}(a,b) \vee \vartheta\big(f(a), f(b)\big) = \Phi_n \vee \vartheta_{\text{lat}}(a,b).$$

Clearly, we have $\Phi_n < \Phi_n \vee \vartheta(a,b) \leqslant \Phi_{n+1}$. Suppose that $\varphi \in \operatorname{Con} L$ is such that $\Phi_n \leqslant \varphi < \Phi_n \vee \vartheta(a,b)$. Then we note that

$$(2) \qquad \varphi \wedge \vartheta_{\text{lat}}(a,b) = \omega.$$

In fact if $(x,y) \in \varphi \wedge \vartheta_{\text{lat}}(a,b)$ then we have

$$(\star) \qquad x \wedge a = y \wedge a, \quad x \vee b = y \vee b, \quad (x,y) \in \varphi.$$

Writing $s = (x \vee a) \wedge b$ and $t = (y \vee a) \wedge b$ we see that $(s,t) \in \varphi$ and therefore, since $a \prec b$ by hypothesis, we have $\{s,t\} \subseteq \{a,b\}$. Now if

$s \neq t$ then one of s, t must be a and the other must be b, whence $(a, b) \in \varphi$. This gives the contradiction $\vartheta(a, b) \leqslant \varphi$. Hence we must have $s = t$, i.e. $(x \vee a) \wedge b = (y \vee a) \wedge b$. But from $(\star)$ we have $x \vee a \vee b = y \vee a \vee b$ and so, by the distributivity of L, $x \vee a = y \vee a$. Again by $(\star)$ and the distributivity of L, we obtain $x = y$ and hence $\varphi \wedge \vartheta_{\text{lat}}(a, b) = \omega$.

It now follows from (1) and (2) that

$$\varphi = \varphi \wedge (\Phi_n \vee \vartheta(a, b)) = \varphi \wedge (\Phi_n \vee \vartheta_{\text{lat}}(a, b)) = \varphi \wedge \Phi_n$$

and therefore $\varphi \leqslant \Phi_n$, whence $\varphi = \Phi_n$. Hence $\Phi_n \vee \vartheta(a, b)$ is an atom of $[\Phi_n, \Phi_{n+1}]$.

Finally, let α be an atom of $[\Phi_n, \Phi_{n+1}]$. Then there exist $a, b \in L$ such that $a < b$, $(a, b) \notin \Phi_n$, and $(a, b) \in \alpha \leqslant \Phi_{n+1}$. If every Φ_{n+1}-class is locally finite there exist $p, q \in [a, b]$ such that $p \prec q$, $(p, q) \notin \Phi_n$, $(p, q) \in \alpha$. For such p, q we have $\Phi_n < \Phi_n \vee \vartheta(p, q) \leqslant \alpha$, whence $\alpha = \Phi_n \vee \vartheta(p, q)$. ◇

Corollary *If $a, b \in L$ are such that $a \prec b$ and $f(a) = f(b)$ then $\vartheta(a, b)$ is an atom of* Con L.

Proof Take $n = 0$ in the above. Then $\vartheta(a, b)$ is an atom of $[\omega, \Phi_1]$ and hence an atom of Con L. ◇

Theorem 2.16 *If the congruence Φ_ω has locally finite classes then every non-trivial interval $[\Phi_n, \Phi_{n+1}]$ of* Con L *is a complete atomic boolean lattice.*

Proof For every $\varphi \in \text{Con } L$ we have $\varphi = \bigvee\{\vartheta(a, b) \mid (a, b) \in \varphi\}$. Thus, if $\Phi_n \leqslant \varphi$ then we have

$$\varphi = \Phi_n \vee \bigvee\{\vartheta(a, b) \mid (a, b) \notin \Phi_n,\ (a, b) \in \varphi\}.$$

If now $\varphi \in [\Phi_n, \Phi_{n+1}]$ then, since the Φ_{n+1}-classes are locally finite, we have

$$\varphi = \Phi_n \vee \bigvee\{\vartheta(p, q) \mid (p, q) \notin \Phi_n,\ (p, q) \in \varphi,\ p \prec q\}.$$

Now for such p, q we have, by Theorem 2.15 , that $\Phi_n \vee \vartheta(p, q)$ is an atom of $[\Phi_n, \Phi_{n+1}]$. Consequently, $[\Phi_n, \Phi_{n+1}]$ is a complete atomistic lattice in which the infinite distributive law $x \wedge \bigvee_i y_i = \bigvee_i (x \wedge y_i)$ holds, and so is boolean. ◇

Example 2.5 Let L be the chain $0 < a < b < c < d < e < 1$ made into an Ockham algebra by defining f as follows :

x	0	a	b	c	d	e	1
$f(x)$	1	1	1	1	b	b	0

There are 21 congruences on $(L; f)$, depicted as follows :

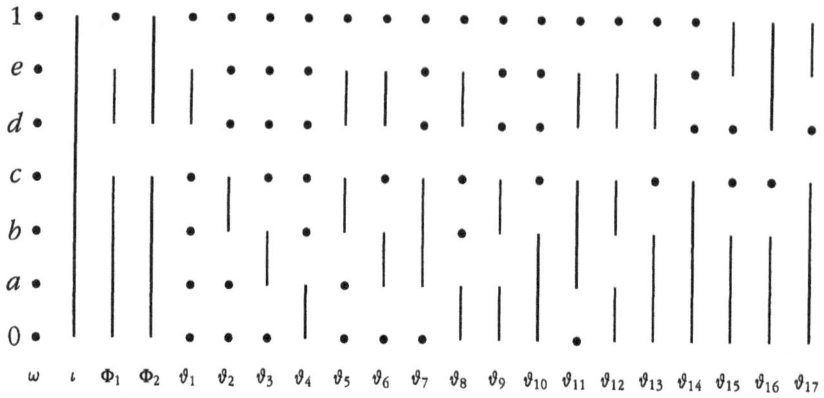

The lattice of congruences of L is

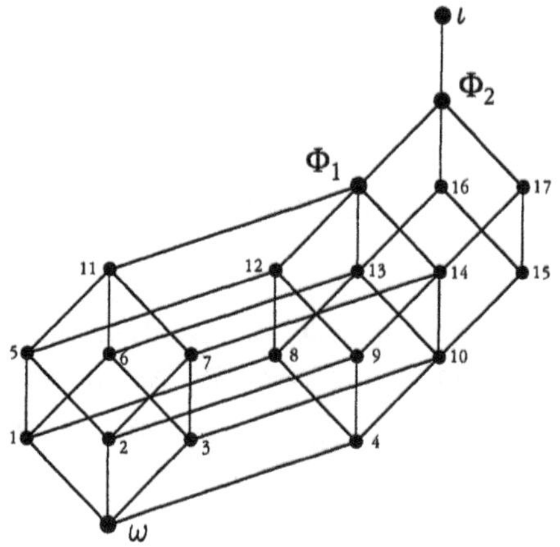

Note that $[\omega, \Phi_1] \simeq \mathbf{2}^4$ and $[\Phi_1, \Phi_2] \simeq \mathbf{2}$.

The condition that every Φ_ω-class be locally finite cannot be removed from Theorem 2.16, as the following example shows.

Example 2.6 Consider the infinite chain C given by

$$0 < x_1 < x_2 < \cdots < \alpha < \cdots < y_2 < y_1 < 1$$

made into an Ockham algebra by defining

$$f(0) = 1, \quad f(1) = 0, \quad (\forall i)\ f(x_i) = f(y_i) = f(\alpha) = \alpha.$$

Here we have $\Phi_\omega = \Phi_1$ with classes $\{0\}$, $\{1\}$, and $C \setminus \{0, 1\}$. The Φ_ω-class $C \setminus \{0, 1\}$ is not locally finite. Consider now the partition

$$\{\{0\},\ \{x_i \mid i \geqslant 1\},\ \{\alpha\},\ \{y_i \mid i \geqslant 1\},\ \{1\}\}.$$

This defines a congruence in $[\omega, \Phi_1]$ which has no complement in $[\omega, \Phi_1]$. So in this case $[\omega, \Phi_1]$ is not boolean.

The following interesting result was obtained by J. Vaz de Carvalho **[105]**.

Theorem 2.17 *If* $L \in \mathbf{K}_{n,0}$ *is finite then* Con L *is boolean. Moreover, if the length of* L *is* m *then* Con L *has at most* m *atoms, and has precisely* m *atoms if and only if* L *itself is boolean.*

Proof We first show that if $a \prec b$ then $\vartheta(a,b)$ is an atom of Con L. Suppose that $\vartheta \in \operatorname{Con} L$ is such that $\omega \leqslant \vartheta < \vartheta(a,b)$. By Theorem 2.1 we have

$$\vartheta(a,b) = \bigvee_{k=0}^{2n-1} \vartheta_{\mathrm{lat}}\big(f^k(a), f^k(b)\big).$$

Since in this case f is a dual automorphism and $a \prec b$, if k is even we have $f^k(a) \prec f^k(b)$, and if k is odd then $f^k(b) \prec f^k(a)$. In what follows we shall suppose that k is even; a similar argument holds when k is odd. We now show that

$$\vartheta \wedge \vartheta_{\mathrm{lat}}\big(f^k(a), f^k(b)\big) = \omega.$$

In fact, if $(x,y) \in \vartheta \wedge \vartheta_{\mathrm{lat}}\big(f^k(a), f^k(b)\big)$ then

(1) $x \wedge f^k(a) = y \wedge f^k(a)$;

(2) $x \vee f^k(b) = y \vee f^k(b)$;

(3) $(x,y) \in \vartheta$.

Consider the elements

$$u = \big(x \vee f^k(a)\big) \wedge f^k(b), \quad v = \big(y \vee f^k(a)\big) \wedge f^k(b).$$

We have $u, v \in \{f^k(a), f^k(b)\}$ and $(u,v) \in \vartheta$. Suppose that $u \neq v$. Then one of these elements is $f^k(a)$, the other is $f^k(b)$, and $\big(f^k(a), f^k(b)\big) \in \vartheta$. Since ϑ is a congruence it follows that

$$(a,b) = \big(f^{2n-k}[f^k(a)],\ f^{2n-k}[f^k(b)]\big) \in \vartheta,$$

whence $\vartheta(a,b) \leqslant \vartheta$, a contradiction. Thus we have $u = v$. This, together with condition (2) and the distributivity of L, gives $x \vee f^k(a) = y \vee f^k(a)$; and this together with condition (1) gives likewise $x = y$.

Using this observation, we now have

$$\vartheta = \vartheta \wedge \vartheta(a,b) = \bigvee_{k=0}^{2n-1} \big(\vartheta \wedge \vartheta_{\mathrm{lat}}\big(f^k(a), f^k(b)\big)\big) = \omega$$

and consequently $\vartheta(a,b)$ is an atom of Con L. Now take a maximal chain

$$0 = z_0 \prec z_1 \prec \cdots \prec z_m = 1$$

in L. Then we have

$$\iota = \bigvee_{i=0}^{m-1} \vartheta(z_i, z_{i+1})$$

and so $\mathrm{Con}\, L$ is a bounded distributive lattice whose greatest element is a join of at most m distinct atoms, whence it is boolean.

If now L is boolean with m atoms then it is well known that $\mathrm{Con}\, L$ is also boolean with m atoms. Conversely, suppose that $L \in \mathbf{K}_{n,0}$ is of length m and that $\mathrm{Con}\, L$ is boolean with precisely m atoms. Observe first that if $x \in L \setminus \{0,1\}$ then $x \parallel f(x)$. In fact, suppose that $x < f(x)$ and consider a maximal chain passing through both, of the form

$$0 = z_0 \prec \cdots \prec x = z_{i+1} \prec \cdots \prec f(x) = z_t \prec f(z_i) = z_{t+1} \prec \cdots \prec z_m = 1.$$

Since $\mathrm{Con}\, L$ has precisely m atoms, all the congruences $\vartheta(z_j, z_{j+1})$ are distinct. But $\big(f(z_i), f(x)\big) \in \vartheta(z_i, x)$ and $\vartheta(z_i, x)$ is an atom of $\mathrm{Con}\, L$, so we have the contradiction $\vartheta(z_i, x) = \vartheta\big(f(x), f(z_i)\big)$. Similar arguments show that $x > f(x)$ and $x = f(x)$ give contradictions.

Since for every $x \in L$ we have $f\big(x \vee f(x)\big) \leqslant x \vee f(x)$, the above observation gives $x \vee f(x) = 1$; and from $x \wedge f(x) \leqslant f\big(x \wedge f(x)\big)$ we obtain $x \wedge f(x) = 0$. Thus x and $f(x)$ are complementary, whence L is boolean. ◇

Theorem 2.18 *Let L be a finite Ockham algebra for which $\mathbf{V}_B(L) = \mathbf{K}_{p,q}$. Then each summand of*

$$[\omega, \Phi_1] \overline{\oplus} [\Phi_1, \Phi_2] \overline{\oplus} \cdots \overline{\oplus} [\Phi_q, \iota]$$

is boolean.

Proof By Theorem 2.7 we have $\Phi_\omega = \Phi_q$ and so, by Theorem 2.6,

$$L/\Phi_q \sim f^q(L) \in \mathbf{K}_{p,0}$$

whence, by Theorem 2.17, $\mathrm{Con}\, L/\Phi_q$ is boolean; and by a standard result in universal algebra **[13]**,

$$\mathrm{Con}\, L/\Phi_q \simeq [\Phi_q, \iota].$$

The result now follows by Theorem 2.16. ◇

3 Subdirectly irreducible algebras

An algebra L is said to be *subdirectly irreducible* if it has a smallest non-trivial congruence; i.e. a congruence α such that $\vartheta \geqslant \alpha$ for all $\vartheta \in \text{Con } L$ with $\vartheta \neq \omega$. Such a congruence α is called the *monolith* of Con L. The importance of such algebras is shown in a classic theorem of Birkhoff **[2]** which states that in an equational class of algebras every algebra can be embedded in a direct product of subdirectly irreducible algebras. An immediate consequence of the above definition is that if L is subdirectly irreducible then in Con L the trivial congruence ω is $\wedge$-irreducible. A particularly important case of a subdirectly irreducible algebra is a *simple algebra*, namely one for which the lattice of congruences is the two-element chain $\{\omega, \iota\}$.

The existence of infinite subdirectly irreducible Ockham algebras can be seen from the following two examples. These are due to Berman, who also showed that every class $\mathbf{K}_{p,q}$ has only finitely many subdirectly irreducible algebras all of which are finite [**28**, Theorems 7,8].

Example 3.1 Let $n \in \mathbb{N}$ and consider the set

$$L_n = \{0,1\} \cup \{a_i \mid -n \leqslant i \leqslant n\},$$

totally ordered by

$$0 < a_{-n} < a_{-n+1} < \cdots < a_{-1} < a_0 < a_1 < \cdots < a_{n-1} < a_n < 1.$$

Define $f : L_n \to L_n$ by $f(0) = 1, f(1) = 0, f(a_0) = a_0$, and

$$(\forall m \geqslant 1) \qquad f(a_m) = a_{-m+1},\ f(a_{-m}) = a_m.$$

It is readily seen that $(L_n; f)$ is an Ockham algebra and that it belongs to the Berman class $\mathbf{K}_{1,2n}$. From the definition of f it follows that any congruence ϑ on L_n that identifies an adjacent pair a_i, a_{i+1} (i.e. any congruence different from ω) must identify a_0, a_1. Consequently, Con L_n has a smallest non-trivial element, namely the principal congruence $\vartheta(a_0, a_1)$. Hence L_n is subdirectly irreducible. The algebra L_0 is simple.

Example 3.2 (*The centripetal see-saw*) With L_n and f as in Example 3.1, let

$$L_\infty = \bigcup_{n \geqslant 0} L_n = \{0,1\} \cup \{a_i \mid i \in \mathbb{Z}\}.$$

Then $(L_\infty; f)$ is an infinite totally ordered subdirectly irreducible Ockham algebra, with monolith $\vartheta(a_0, a_1)$.

Associated with the notion of a subdirectly irreducible algebra is that of a *finitely subdirectly irreducible* algebra, this being defined as an algebra in which the intersection of two non-trivial principal congruences is non-trivial. Clearly, every subdirectly irreducible algebra is finitely subdirectly irreducible. The following is an example of a finitely subdirectly irreducible Ockham algebra that is not subdirectly irreducible.

Example 3.3 (*The centrifugal see-saw*) Let L be the chain

$$0 < \cdots < a_3 < a_1 < a_0 < a_2 < a_4 < \cdots < 1$$

and define $f : L \to L$ by

$$f(0) = 1, \quad f(1) = 0, \quad (i \geqslant 0)\ f(a_i) = a_{i+1}.$$

Then $(L; f)$ is an Ockham algebra. We leave as an exercise for the reader the task of showing that Con L is the lattice

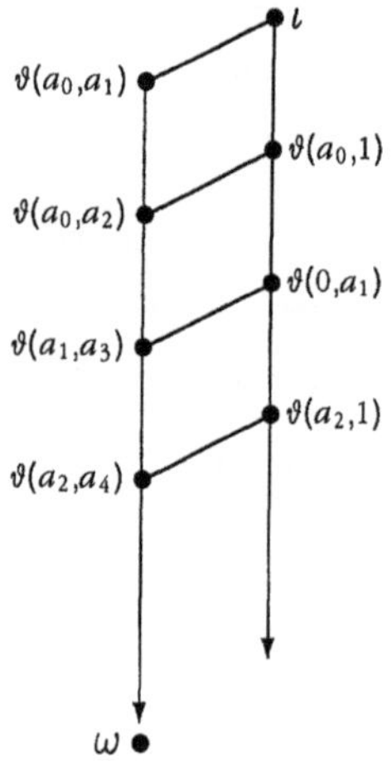

Clearly, L is finitely subdirectly irreducible but not subdirectly irreducible.

Example 3.4 The pineapple (Example 2.4) is finitely subdirectly irreducible but not subdirectly irreducible.

Given an Ockham algebra $(L; f)$, consider now for each $i \geqslant 1$ the subset

$$T_i(L) = \{x \in L \mid f^i(x) = x\}.$$

In particular, $T_1(L)$ is the set of fixed points of f. Of course, it can happen that $T_1(L)$ is empty. More generally, every subset $T_{2n+1}(L)$ is either empty or is an antichain; for if $x, y \in T_{2n+1}(L)$ with $x \leqslant y$ then $x = f^{2n+1}(x) \geqslant f^{2n+1}(y) = y$ so $x = y$. In contrast, $T_2(L)$ is never empty, for it clearly contains 0 and 1. It is readily seen that every subset $T_{2n}(L)$ is a subalgebra of L; in fact, $T_{2n}(L)$ is the largest $\mathbf{K}_{n,0}$-subalgebra of L. Clearly, we have the chains

$$T_{2i}(L) \subseteq T_{4i}(L) \subseteq \cdots \subseteq T_{2^n i}(L) \subseteq T_{2^{n+1} i}(L) \subseteq \cdots .$$

Consider now the subset

$$T(L) = \bigcup_{n \geqslant 1} T_{2n}(L),$$

i.e. the set of $x \in L$ for which there exists an even positive integer m_x such that $f^{m_x}(x) = x$. Given $x, y \in T(L)$, let $m = \text{lcm}\{m_x, m_y\}$. Then, m being even, we have

$$f^m(x \vee y) = f^m(x) \vee f^m(y) = x \vee y,$$

and similarly $f^m(x \wedge y) = x \wedge y$. Since $x \in T(L)$ clearly implies $f(x) \in T(L)$, it follows that $T(L)$ is also a subalgebra of L. Note that $T(L) \subseteq S(L)$.

Consider also the subset $K(L)$ defined by

$$K(L) = \{0, 1\} \cup T_1(L).$$

We shall call this the *core* of L. This is not in general a subalgebra of L.

Theorem 3.1 *If the core of $(L; f)$ is a subalgebra then f has at most two fixed points.*

Proof Suppose that $K(L)$ is a subalgebra and that f has at least two fixed points a, b. Then $a \vee b \in K(L)$ and $a \wedge b \in K(L)$. Since distinct fixed points cannot be comparable, it follows that we must have $a \vee b = 1$ and $a \wedge b = 0$. Thus any two fixed points are complementary. The distributivity of L now shows that there can be at most two fixed points. ◇

It is clear that $K(L) \subseteq T_2(L)$. In general, we have strict inclusion; for example, the 4-element chain **4** made into a de Morgan algebra is fixed point free and $K(\mathbf{4}) = \{0, 1\}$, $T_2(\mathbf{4}) = \mathbf{4}$.

It is natural to consider the particular case where $K(L) = T_2(L)$. An important situation in which this happens arises from the following.

Theorem 3.2 *Let $(L; f)$ be a finitely subdirectly irreducible Ockham algebra. If $a, b \in L$ are such that $f(a) = b$ and $f(b) = a$ then either $a = b$ or $\{a, b\} = \{0, 1\}$.*

Proof If $\{a, b\} \neq \{0, 1\}$ then we have $0 < a < 1$ or $0 < b < 1$. If $a \wedge b = 0$ (in which case $1 = f(a) \vee f(b) = b \vee a$) then, by Theorem 2.1,

$$\vartheta(0, a) \wedge \vartheta(0, b) = [\vartheta_{\text{lat}}(0, a) \vee \vartheta_{\text{lat}}(b, 1)] \wedge [\vartheta_{\text{lat}}(0, b) \vee \vartheta_{\text{lat}}(a, 1)].$$

Since $\text{Con}_{\text{lat}}\, L$ is distributive, we can expand the right hand side. Observing that $\vartheta_{\text{lat}}(0, a) \wedge \vartheta_{\text{lat}}(0, b) = \omega$, that $\vartheta_{\text{lat}}(0, a) \wedge \vartheta_{\text{lat}}(a, 1) = \vartheta_{\text{lat}}(a, a) = \omega$, and using the hypothesis that $a \wedge b = 0$, we see that the right hand side reduces to ω. This contradicts the hypothesis that L is finitely subdirectly irreducible. Consequently we must have $a \wedge b \neq 0$ in which case, by Theorem 2.1,

$$\begin{aligned}\vartheta(0, a \wedge b) \wedge \vartheta(a \wedge b, a \vee b) &= [\vartheta_{\text{lat}}(0, a \wedge b) \vee \vartheta_{\text{lat}}(a \vee b, 1)] \wedge \vartheta_{\text{lat}}(a \wedge b, a \vee b) \\ &= \vartheta_{\text{lat}}(a \wedge b, a \wedge b) \vee \vartheta_{\text{lat}}(a \vee b, a \vee b) \\ &= \omega.\end{aligned}$$

It follows that $\vartheta(a \wedge b, a \vee b) = \omega$, whence $a \wedge b = a \vee b$ and hence $a = b$. ◇

Corollary 1 *If L is finitely subdirectly irreducible then $K(L) = T_2(L)$.*

Proof If $a \in T_2(L)$ then $f[f(a)] = a$. Taking $b = f(a)$ we deduce that either $a = f(a)$ or $\{a, f(a)\} = \{0, 1\}$, i.e. $a \in T_1(L) \cup \{0, 1\} = K(L)$. ◇

Corollary 2 *If $(L; f)$ is finitely subdirectly irreducible then f has at most two fixed points.*

Proof This follows from Corollary 1 and Theorem 3.1. ◇

Consider now the particular case where L is a finitely subdirectly irreducible de Morgan algebra. By the above, we have

$$L = T_2(L) = K(L) = \{0, 1\} \cup T_1(L)$$

where, by Theorem 3.1, $T_1(L)$ is either empty or is an antichain of at most two elements. We can therefore state the following result, which was first obtained by Kalman **[72]**.

Theorem 3.3 *In the class* **M** *of de Morgan algebras there are only three (finitely) subdirectly irreducible algebras, each of which is simple, namely the algebras*

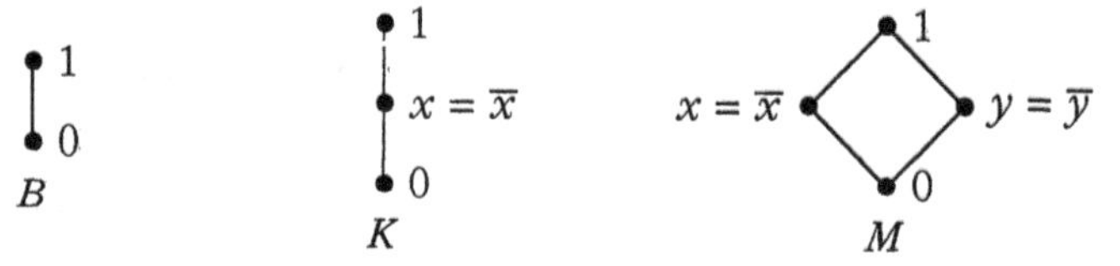

Proof By the above, the algebras shown are the only candidates. It is readily seen that for these we have

$$\text{Con } B = \text{Con } K = \text{Con } M \simeq \mathbf{2},$$

so these algebras are indeed subdirectly irreducible; in fact, they are simple. ◇

Theorem 3.4 *If $L \in \mathbf{O}$ is such that $K(L) = T_2(L)$ and if $a, b \in T(L)$ are such that $a < b$ then $\vartheta(a,b) = \iota$.*

Proof For every $x \in T(L)$ let m_x be the least even positive integer such that $f^{m_x}(x) = x$. With $n_x = \frac{1}{2}m_x$, consider the elements

$$\alpha(x) = \bigwedge_{i=0}^{n_x-1} f^{2i}(x), \qquad \beta(x) = \bigvee_{i=0}^{n_x-1} f^{2i}(x).$$

Observe that $f^2[\alpha(x)] = \alpha(x)$ and $f^2[\beta(x)] = \beta(x)$, so $\alpha(x), \beta(x) \in T_2(L) = K(L)$. Now let $a, b \in T(L)$ be such that $a < b$. Consider the sublattice M that is generated by

$$\{f^{2i}(a), f^{2j}(b) \mid 0 \leqslant i \leqslant n_a,\ 0 \leqslant j \leqslant n_b\}.$$

Clearly, M is finite with smallest element $\alpha(a)$ and greatest element $\beta(b)$. Let p be an atom of M and consider the interval $B = [\alpha(a), \beta(p)]$ in M. Since every atom of M is of the form $\bigwedge_{i \neq j} f^{2i}(a)$ for some j, it follows that $f^2(p)$ is also an atom of M. Consequently, B is boolean; for it is a finite distributive lattice whose greatest element is a join of atoms.

Observe that $\alpha(a) < \beta(p)$ and so, since both belong to $K(L)$, we have that $\alpha(a)$ is either 0 or a fixed point, and $\beta(p)$ is either 1 or a fixed point.

Clearly, $a \wedge \beta(p)$ and $b \wedge \beta(p)$ belong to B, and

$$\big(a \wedge \beta(p), b \wedge \beta(p)\big) \in \vartheta(a,b).$$

If $a \wedge \beta(p) < b \wedge \beta(p)$, let c be an atom of B with $c \nleqslant a \wedge \beta(p)$ and $c \leqslant b \wedge \beta(p)$. Then we have

$$\big(\alpha(a), c\big) = \big(a \wedge \beta(p) \wedge c, b \wedge \beta(p) \wedge c\big) \in \vartheta(a,b).$$

It follows that $\big(\alpha(a), \beta(c)\big) \in \vartheta(a,b)$. Since $\alpha(a), \beta(c) \in K(L)$ with $\alpha(a) < \beta(c)$, we deduce that $(0,1) \in \vartheta(a,b)$ and that therefore $\vartheta(a,b) = \iota$.

If now $a \wedge \beta(p) = b \wedge \beta(p)$ let $a_1 = a \vee \beta(p) < b \vee \beta(p) = b_1$. Then $(a_1, b_1) \in \vartheta(a,b)$. Moreover, we cannot have $\beta(p) = 1$, so $\beta(p)$ must be a fixed point. Then $\beta(b) = 1$; for otherwise $\beta(b) = \beta(p)$ gives the contradiction

$$a = a \wedge \beta(b) = a \wedge \beta(p) = b \wedge \beta(p) = b \wedge \beta(b) = b.$$

Considering therefore the interval $[\beta(p), 1]$ in M and a coatom q such that $q \geqslant a \vee \beta(p)$ and $q \ngeqslant b \vee \beta(p)$, we see in a dual manner that $\vartheta(a,b) = \iota$. ◇

Theorem 3.5 *For an Ockham algebra L the following are equivalent:*

(1) $K(L) = T_2(L)$;

(2) *the subalgebra $T(L)$ is simple;*

(3) *all de Morgan subalgebras of L are simple.*

Proof (1) $\Rightarrow$ (2) : If (1) holds then by Theorem 3.4 every non-trivial principal congruence on $T(L)$ coincides with ι. Since every congruence is the supremum of the principal congruences that it contains, it follows that $T(L)$ is simple.

(2) $\Leftrightarrow$ (3) : $T_2(L)$ is the largest de Morgan subalgebra of L.

(3) $\Rightarrow$ (1) : If (3) holds then $T_2(L)$ is simple. But $T_2(L) \in \mathbf{K}_{1,0} = \mathbf{M}$; and by Theorem 3.3 there are only three non-isomorphic simple de Morgan algebras, in each of which $K(L) = T_2(L)$. Since $T_2(L)$ and L have the same core, (1) follows. ◇

Corollary *For an Ockham algebra L we have $K(L) = T(L)$ if and only if $T(L)$ is a simple de Morgan algebra.*

Proof If $K(L) = T(L)$ then $K(L) = T_2(L) = T(L)$ and so $T(L)$ is simple de Morgan. Conversely, if $T(L)$ is simple de Morgan then $K(L) = T_2(L)$; and $T(L) = T_2(L)$ since $T_2(L)$ is the largest de Morgan subalgebra of L. ◇

In order to examine more closely the (finitely) subdirectly irreducible Ockham algebras, we shall concentrate on a class that contains all the Berman classes, namely the subclass $\mathbf{K}_\omega$ of $\mathbf{O}$ defined by

$$(L;f) \in \mathbf{K}_\omega \iff (\forall x \in L)(\exists m,n \in \mathbb{N})(m \neq 0) \quad f^{m+n}(x) = f^n(x).$$

Without loss of generality, we may assume that both m and n are even; for $f^{m+n}(x) = f^n(x)$ implies

$$f^{2m+2n}(x) = f^{m+n}[f^n(x)] = f^n[f^{m+n}(x)] = f^{2n}(x).$$

By definition, $\mathbf{K}_\omega$ is closed under the formation of subalgebras, homomorphic images, and arbitrary direct powers. But it is not closed under arbitrary direct products, as can be seen by taking an algebra $L_q \in \mathbf{K}_{p,q}$ for each $q \geqslant 0$ and considering $L_0 \times L_1 \times L_2 \times \cdots$. However, the following result, due to Jie Fang [67], enables us to claim that $\mathbf{K}_\omega$ is closed under finite direct products, and therefore forms a *generalised variety* in the sense of Ash [23].

Theorem 3.6 *Let $(L_1;f)$ and $(L_2;f)$ be Ockham algebras. If $x_1 \in L_1$ and $x_2 \in L_2$ are such that there exist natural numbers m_1, n_1 and m_2, n_2 with*

$$f^{m_1+n_1}(x_1) = f^{n_1}(x_1), \quad f^{m_2+n_2}(x_2) = f^{n_2}(x_2),$$

then there are natural numbers m, n such that, in the algebra $L_1 \times L_2$,

$$f^{m+n}(x_1, x_2) = f^n(x_1, x_2).$$

Proof Observe first that $f^{p+q}(x) = f^q(x)$ implies $f^{kp+q}(x) = f^q(x)$ for all $k \in \mathbb{N}_0$. Now let $m = \mathrm{lcm}\{m_1, m_2\}$ and $n = \mathrm{lcm}\{n_1, n_2\}$. Then $m = m_1 r$, $n = n_1 s$ and

$$\begin{aligned} f^{m+n}(x_1) &= f^{m_1 r + n_1 + (s-1)n_1}(x_1) \\ &= f^{n_1 + sn_1 - n_1}(x_1) \\ &= f^n(x_1). \end{aligned}$$

Similarly, $f^{m+n}(x_2) = f^n(x_2)$. The result now follows. ◊

A class $\mathbf{C}$ of algebras is said to be *locally finite* if every finitely generated member of $\mathbf{C}$ is finite. In particular, it is well known that the class $\mathbf{D}_{01}$ of bounded distributive lattices is locally finite.

Theorem 3.7 *The generalised variety* $\mathbf{K}_\omega$ *is locally finite.*

Proof Suppose that $L \in \mathbf{K}_\omega$ is $\mathbf{O}$-generated by $\{x_1, \dots, x_k\}$. Then there are natural numbers m_i, n_i with $m_i \neq 0$ such that $f^{m_i+n_i}(x_i) = f^{n_i}(x_i)$ for $i = 1, \dots, k$. By Theorem 3.6 and induction, there exist m, n $(m \neq 0)$ such that

$$(i = 1, \dots, k) \qquad f^{m+n}(x_i) = f^n(x_i) \;\; [= f^{2m+n}(x_i)].$$

It follows that L belongs to $\mathbf{K}_{m,n}$ and is $\mathbf{D}_{01}$-generated by

$$\{f^j(x_i) \mid 0 \leqslant j < 2m + n, \;\; 1 \leqslant i \leqslant k\}.$$

The result now follows from the fact that $\mathbf{D}_{01}$ is locally finite. ◊

Our objective now is to show that for every algebra in $\mathbf{K}_\omega$ (and therefore every algebra that belongs to a Berman class) the properties of being finitely subdirectly irreducible and subdirectly irreducible are equivalent.

Theorem 3.8 *If an Ockham algebra* L *is finitely subdirectly irreducible then every* Φ_1*-class in* L *contains at most two elements.*

Proof Suppose that a Φ_1-class contains at least three elements. Then it contains a 3-element chain $x < y < z$ with $f(x) = f(y) = f(z)$. Then, by Theorem 2.1, we have $\vartheta(x, y) = \vartheta_{\mathrm{lat}}(x, y)$ and $\vartheta(y, z) = \vartheta_{\mathrm{lat}}(y, z)$, whence we have the contradiction $\vartheta(x, y) \wedge \vartheta(y, z) = \omega$. ◊

Theorem 3.9 *If* $L \in \mathbf{K}_\omega$ *then the following statements are equivalent:*

(1) L *is finitely subdirectly irreducible;*

(2) L *is subdirectly irreducible.*

Proof (1) ⇒ (2) : Since $L \in \mathbf{K}_\omega$, for every $x \in L$ we have $f^{m+n}(x) = f^n(x)$ for some $m, n \in \mathbb{N}$ with $m \neq 0$ and even. If $\Phi_1 = \omega$ then f is injective and $x = f^m(x)$, whence $x \in T(L)$. Thus $L = T(L)$ and it follows by Corollary 1 of

Theorem 3.2 and Theorem 3.5 that L is simple, hence subdirectly irreducible. On the other hand, if $\Phi_1 \neq \omega$ then by Theorem 3.8 there is a two-element Φ_1-class $\{a, b\}$ and, by the Corollary of Theorem 2.15, $\vartheta(a, b)$ is an atom in the interval $[\omega, \Phi_1]$ of Con L. If now $\alpha \in$ Con L with $\alpha \neq \omega$ then, since α is the supremum of the non-trivial principal congruences which it contains and since Con L satisfies the infinite distributive law $\beta \wedge \bigvee_i \gamma_i = \bigvee_i (\beta \wedge \gamma_i)$, it follows by the hypothesis that L is finitely subdirectly irreducible that $\vartheta(a, b) \wedge \alpha \neq \omega$. Since $\vartheta(a, b)$ is an atom in Con L it follows that $\vartheta(a, b) \wedge \alpha = \vartheta(a, b)$ and hence $\vartheta(a, b) \leqslant \alpha$. Thus $\vartheta(a, b)$ is the smallest non-trivial congruence on L, so L is subdirectly irreducible.

(2) $\Rightarrow$ (1) : This is clear. ◇

Corollary *If an Ockham algebra L is finitely subdirectly irreducible but not subdirectly irreducible then necessarily $L \notin \mathbf{K}_\omega$ and f is injective.*

Proof That $L \notin \mathbf{K}_\omega$ follows from the above. Suppose that f were not injective. Then by Theorem 3.8 the interval $[\omega, \Phi_1]$ contains an atom $\vartheta(a, b)$. As shown above, this implies that L is subdirectly irreducible, a contradiction. ◇

The $\mathbf{K}_\omega$-analogue of Theorem 2.6 is the following.

Theorem 3.10 *If $L \in \mathbf{K}_\omega$ then $L/\Phi_\omega \simeq T(L)$.*

Proof Observe first that $\Phi_\omega|_{T(L)} = \omega$. In fact, let $x, y \in T(L)$ be such that $(x, y) \in \Phi_\omega$. Then $x = f^p(x)$, $y = f^q(y)$ for some p, q; and $f^n(x) = f^n(y)$ for some n. Let $r = \text{lcm}\{p, q, n\}$; then we have $x = f^r(x) = f^r(y) = y$.

Now in the quotient algebra $(L/\Phi_\omega; \widehat{f})$ we have $\widehat{f}([x]\Phi_\omega) = [f(x)]\Phi_\omega$. Consider the morphism $\vartheta : T(L) \to L/\Phi_\omega$ given by $\vartheta(x) = [x]\Phi_\omega$. By the above observation, ϑ is injective. To see that ϑ is also surjective, observe that, since $L \in \mathbf{K}_\omega$ by hypothesis, for every $y \in L$ there exist m, n such that $f^n(y) = f^{m+n}(y)$. It follows that there exists $p \geqslant n$ such that

$$f^{p+n}(y) = f^n(y) = f^{2p+n}(y).$$

The first of these equalities gives $(y, f^p(y)) \in \Phi_n \leqslant \Phi_\omega$; and the second gives $f^n(y) \in T_{2p}(L) \subseteq T(L)$ from which it follows that $f^p(y) \in T(L)$ since $n \leqslant p$. Thus we see that

$$\vartheta : f^p(y) \mapsto [f^p(y)]\Phi_\omega = [y]\Phi_\omega$$

and so ϑ is also surjective. ◇

Our objective now is to determine precisely when, for $L \in \mathbf{K}_\omega$, the interval $[\Phi_\omega, \iota]$ of Con L is boolean. For this purpose, we require the following result of J. Vaz de Carvalho [**106**].

Theorem 3.11 *If $L \in \mathbf{K}_{n,0}$ then L is a strong extension of $T_2(L)$.*

Proof Given $\vartheta \in \operatorname{Con} T_2(L)$, let ϑ_1, ϑ_2 be extensions of ϑ to L. We may assume without loss of generality that $\vartheta_1 \leqslant \vartheta_2$.

Observe first that if $a \in T_2(L)$ then $[a]\vartheta_1 = [a]\vartheta_2$. In fact, for every $x \in L$ define

$$x^\star = \bigvee_{k=0}^{n-1} f^{2k}(x), \quad x_\star = \bigwedge_{k=0}^{n-1} f^{2k}(x).$$

If $(x,a) \in \vartheta_2$ then clearly $(x^\star, a) \in \vartheta_2$ and $(x_\star, a) \in \vartheta_2$. Since $x^\star, x_\star \in T_2(L)$ and $\vartheta_1|_{T_2(L)} = \vartheta = \vartheta_2|_{T_2(L)}$ we have $(x^\star, a) \in \vartheta_1$ and $(x_\star, a) \in \vartheta_1$. Since $x_\star \leqslant x \leqslant x^\star$ we deduce that $(x,a) \in \vartheta_1$. Thus $[a]\vartheta_2 \subseteq [a]\vartheta_1$ whence we have equality.

Suppose now that L is finite and $(x,y) \in \vartheta_2$. To show that $(x,y) \in \vartheta_1$ it suffices, since L/ϑ_1 is a finite distributive lattice, to show that $[x]\vartheta_1$ and $[y]\vartheta_1$ contain the same set of $\vee$-irreducible elements. Suppose then that $[c]\vartheta_1$ is $\vee$-irreducible in L/ϑ_1 and such that $[c]\vartheta_1 \leqslant [x]\vartheta_1$, i.e. such that $(c \wedge x, c) \in \vartheta_1$. Let m be the smallest positive integer such that $[c]\vartheta_1 = [f^{2m}(c)]\vartheta_1$, noting that $m \leqslant n$. Define

$$\widehat{c} = \begin{cases} f^2(c) \vee \cdots \vee f^{2m-2}(c) & \text{if } m > 1; \\ 0 & \text{if } m = 1. \end{cases}$$

Then $[c^\star]\vartheta_1 = [c \vee \widehat{c}]\vartheta_1$; and $(c, c \wedge x) \in \vartheta_1$ gives

$$\big(c \vee \widehat{c},\ (c \wedge x) \vee \widehat{c}\big) \in \vartheta_1 \leqslant \vartheta_2.$$

But $\big((c \wedge x) \vee \widehat{c},\ (c \wedge y) \vee \widehat{c}\big) \in \vartheta_2$ and so $\big(c \vee \widehat{c},\ (c \wedge y) \vee \widehat{c}\big) \in \vartheta_2$. Since $\big(c^\star,\ c \vee \widehat{c}\big) \in \vartheta_1 \leqslant \vartheta_2$ we have that $\big(c^\star,\ (c \wedge y) \vee \widehat{c}\big) \in \vartheta_2$. Since $c^\star \in T_2(L)$ it follows by the observation above that $\big(c^\star,\ (c \wedge y) \vee \widehat{c}\big) \in \vartheta_1$ and therefore $\big(c \vee \widehat{c},\ (c \wedge y) \vee \widehat{c}\big) \in \vartheta_1$. Thus$\big(c,\ (c \wedge y) \vee (c \wedge \widehat{c})\big) \in \vartheta_1$ and so

$$[c]\vartheta_1 = [c \wedge y]\vartheta_1 \vee [c \wedge \widehat{c}]\vartheta_1.$$

If $m = 1$ then $\widehat{c} = 0$ and we obtain $[c]\vartheta_1 \leqslant [y]\vartheta_1$. If $m > 1$ then

$$[c]\vartheta_1 = [c \wedge y]\vartheta_1 \vee [c \wedge f^2(c)]\vartheta_1 \vee \cdots \vee [c \wedge f^{2m-2}(c)]\vartheta_1$$

and so, since $[c]\vartheta_1$ is $\vee$-irreducible in L/ϑ_1, either $[c]\vartheta_1 \leqslant [y]\vartheta_1$ or there exists k with $1 \leqslant k \leqslant m-1$ such that $[c]\vartheta_1 \leqslant [f^{2k}(c)]\vartheta_1$. Since f is injective, the latter gives the contradiction $[c]\vartheta_1 = [f^{2k}(c)]\vartheta_1$ with $1 \leqslant k \leqslant m-1$. Hence we have $[c]\vartheta_1 \leqslant [y]\vartheta_1$. Similarly, every $\vee$-irreducible contained in $[y]\vartheta_1$ is contained in $[x]\vartheta_1$ whence $[x]\vartheta_1 = [y]\vartheta_1$ and consequently $\vartheta_1 = \vartheta_2$.

Suppose now that L is arbitrary and that $(x,y) \in \vartheta_2$. Let A be the subalgebra of L generated by $\{x,y\}$. Since $\mathbf{K}_{n,0}$ is locally finite by Theorem 3.7, A is finite with $T_2(A) = A \cap T_2(L)$. By the above, we deduce from

$\vartheta_1|_{T_2(A)} = \vartheta_2|_{T_2(A)}$ that $\vartheta_1|_A = \vartheta_2|_A$. Since $x,y \in A$ we then have $(x,y) \in \vartheta_1$. Hence $\vartheta_1 = \vartheta_2$ as required. ◇

Theorem 3.12 *If $L \in \mathbf{O}$ then $T(L)$ is a strong extension of $T_2(L)$.*

Proof Let $\varphi \in \mathrm{Con}\, T_2(L)$ and let $\varphi_1, \varphi_2 \in \mathrm{Con}\, T(L)$ be extensions of φ. Suppose that $x,y \in T(L)$ are such that $(x,y) \in \varphi_1$. If A is the subalgebra of $T(L)$ generated by $\{x,y\}$ then $A \in \mathbf{K}_{p,0}$ for some p. Since $\varphi_1|_{T_2(L)} = \varphi = \varphi_2|_{T_2(L)}$ and $T_2(A) \subseteq T_2(L)$ we have $\varphi_1|_{T_2(A)} = \varphi_2|_{T_2(A)}$. By Theorem 3.11, A is a strong extension of $T_2(A)$. Hence $\varphi_1|_A = \varphi_2|_A$ and consequently $(x,y) \in \varphi_2$. Thus we have $\varphi_1 \leqslant \varphi_2$. Similarly, $\varphi_2 \leqslant \varphi_1$ and so $\varphi_1 = \varphi_2$ as required. ◇

Corollary $\mathrm{Con}\, T(L) \simeq \mathrm{Con}\, T_2(L)$.

Proof Immediate from the above and the congruence extension property. ◇

We can now use the above results to determine precisely when $[\Phi_\omega, \iota]$ is boolean. For this purpose we use the fact, established by Sankappanavar **[84]** using techniques from universal algebra, that *if M is a de Morgan algebra then* Con M *is boolean if and only if M is finite.*

Theorem 3.13 *If $L \in \mathbf{K}_\omega$ then $[\Phi_\omega, \iota]$ is boolean if and only if the de Morgan subalgebra $T_2(L)$ is finite.*

Proof By Theorem 3.10 and the Corollary to Theorem 3.12 we have

$$[\Phi_\omega, \iota] \simeq \mathrm{Con}\, L/\Phi_\omega \simeq \mathrm{Con}\, T(L) \simeq \mathrm{Con}\, T_2(L).$$

The result follows now from the above result of Sankappanavar. ◇

Definition We shall say that Con L has *comonolith* α whenever

$$\mathrm{Con}\, L = [\omega, \alpha] \oplus \{\iota\}.$$

The comonolith is therefore the only coatom in Con L, when such exists.

Theorem 3.14 *If $L \in \mathbf{K}_\omega$ then the following statements are equivalent:*

(1) $K(L) = T_2(L)$;

(2) Con L *has comonolith* Φ_ω;

(3) Φ_ω *is maximal in* Con L.

Proof (1) $\Rightarrow$ (2) : Since $(0,1) \notin \Phi_\omega$ we always have $\Phi_\omega < \iota$. Let $a,b \in L$ be such that $a < b$ and $(a,b) \notin \Phi_\omega$, so that $f^n(a) \neq f^n(b)$ for all $n \in \mathbb{N}$. We first show that $\vartheta(a,b) = \iota$. Since $L \in \mathbf{K}_\omega$ we see by Theorem 3.6 that there exist $m \geqslant 1$, $n \geqslant 0$ such that

$$f^{m+n}(a) = f^n(a), \quad f^{m+n}(b) = f^n(b).$$

Let $c = f^n(a)$ and $d = f^n(b)$. Then we have $c, d \in T(L)$ with $c < d$ or $c > d$. It follows by (1) and Theorem 3.4 that $\vartheta(c,d) = \iota$. Consequently, by Theorem 2.1,

$$\vartheta(a,b) = \bigvee_{k=0}^{n-1} \vartheta_{\text{lat}}\big(f^k(a), f^k(b)\big) \vee \vartheta(c,d) = \iota.$$

Suppose now that $\varphi \in \text{Con } L$ is such that $\varphi \neq \iota$. Since $\varphi = \bigvee_{(a,b)\in\varphi} \vartheta(a,b)$ we have

$$(a,b) \in \varphi \Rightarrow \vartheta(a,b) \neq \iota.$$

But by the above observation

$$\vartheta(a,b) \neq \iota \Rightarrow (a,b) \in \Phi_\omega.$$

Hence $\varphi \leqslant \Phi_\omega$ and consequently

$$\text{Con } L = [\omega, \Phi_\omega] \oplus \{\iota\}.$$

(2) $\Rightarrow$ (3) : This is clear.

(3) $\Rightarrow$ (1) : If (3) holds then, by Theorem 3.10, $T(L)$ is simple, whence (1) follows by Theorem 3.5. ◇

Corollary 1 *If $L \in \mathbf{K}_\omega$ then the following statements are equivalent :*

(1) $K(L) = T(L)$;

(2) *L/Φ_ω is a simple de Morgan algebra.*

Proof This is immediate from the above and the Corollary to Theorem 3.5. ◇

Corollary 2 *If $L \in \mathbf{K}_\omega$ then the following statements are equivalent :*

(1) *L is simple;*

(2) *$K(L) = T_2(L)$ and f is injective.*

Proof (1) $\Rightarrow$ (2) : If L is simple then Con $L = \{\omega, \iota\}$ and so $\Phi_\omega = \omega$ since $(0,1) \notin \Phi_\omega$. Consequently, $\Phi_1 = \omega$ whence f is injective; and Φ_ω is maximal, so $K(L) = T_2(L)$.

(2) $\Rightarrow$ (1) : If (2) holds then clearly $\Phi_\omega = \omega$ and, by the above, Con L reduces to $\{\omega, \iota\}$. ◇

Corollary 3 *The simple algebras in $\mathbf{K}_{p,q}$ are precisely those algebras L in $\mathbf{K}_{p,0}$ for which $K(L) = T_2(L)$.* ◇

Note that Corollary 2 above does not extend beyond $\mathbf{K}_\omega$. This fact is illustrated by the pineapple (Example 2.4).

We now proceed to characterise the (finitely) subdirectly irreducible algebras in $\mathbf{K}_\omega$. For this purpose, we require the following results.

Theorem 3.15 *Let A be an algebra that belongs to a class that has the congruence extension property. If A is subdirectly irreducible with monolith φ then every subalgebra B with $\varphi|_B \neq \omega$ is also subdirectly irreducible and has monolith $\varphi|_B$.*

Proof Every congruence $\vartheta^\star$ on B with $\vartheta^\star \neq \omega$ extends to a congruence ϑ on A such that $\vartheta \neq \omega$, and therefore $\vartheta \geqslant \varphi$. It follows that $\vartheta^\star \geqslant \varphi|_B \neq \omega$, and hence that B is subdirectly irreducible with monolith $\varphi|_B$. ◊

Theorem 3.16 *If $L \in \mathbf{O}$ is subdirectly irreducible and if f is not injective then the monolith of L is Φ_1.*

Proof If f is not injective then $\Phi_1 \neq \omega$. By Theorem 3.8 every Φ_1-class has at most two elements. Let α be the monolith of L. Then $\omega \prec \alpha \leqslant \Phi_1$ and so α has a two-element class, say $\{a, b\}$ with $a \prec b$. Since every lattice congruence contained in Φ_1 is a congruence, it follows that $\alpha = \vartheta(a, b) = \vartheta_{\text{lat}}(a, b)$. Since α is then a principal lattice congruence it has a complement β in $\text{Con}_{\text{lat}} L$. Now $\beta \wedge \Phi_1$ is a lattice congruence contained in Φ_1 and so $\beta \wedge \Phi_1 \in \text{Con}\, L$. Since L is subdirectly irreducible by hypothesis, it follows that either $\beta \wedge \Phi_1 \geqslant \alpha$ or $\beta \wedge \Phi_1 = \omega$. The former is excluded since it gives

$$\alpha = \beta \wedge \Phi_1 \wedge \alpha = \beta \wedge \alpha = \omega.$$

Thus we must have $\beta \wedge \Phi_1 = \omega$. But $\iota = \beta \vee \alpha$ and $\Phi_1 \geqslant \alpha$ give $\iota = \beta \vee \Phi_1$. Hence β is the complement of Φ_1 in $\text{Con}_{\text{lat}} L$ and therefore $\Phi_1 = \alpha$. ◊

Theorem 3.17 *$L \in \mathbf{K}_\omega$ is subdirectly irreducible if and only if* Con L *reduces to the chain*

$$\omega = \Phi_0 \preceq \Phi_1 \preceq \cdots \preceq \Phi_\omega \prec \iota.$$

More precisely,

(1) *if L belongs to a Berman class and $\mathbf{V}_B(L) = \mathbf{K}_{p,q}$ then L is subdirectly irreducible if and only if* Con L *reduces to the finite chain*

$$\omega = \Phi_0 \prec \Phi_1 \prec \cdots \prec \Phi_q = \Phi_\omega \prec \iota;$$

(2) *if L belongs properly to $\mathbf{K}_\omega$ then L is subdirectly irreducible if and only if* Con L *reduces to the infinite chain*

$$\omega = \Phi_0 \prec \Phi_1 \prec \cdots < \Phi_\omega \prec \iota.$$

Proof $\Rightarrow$: Suppose that L is subdirectly irreducible. By Corollary 1 of Theorem 3.2 and Theorem 3.14, we have

$$\text{Con}\, L = [\omega, \Phi_\omega] \oplus \{\iota\}.$$

We now show that the subalgebra $f(L)$ is also subdirectly irreducible. In fact, suppose first that $\Phi_1 = \omega$. Then we have

$$\text{Con}\, f(L) \simeq \text{Con}\, L/\Phi_1 = \text{Con}\, L,$$

whence $f(L)$ is subdirectly irreducible. Suppose now that $\Phi_1 \neq \omega$ and let $\Phi_1^\star$ be the restriction of Φ_1 to $f(L)$. Since

$$(f(x), f(y)) \in \Phi_1^\star \iff f^2(x) = f^2(y) \iff (x, y) \in \Phi_2,$$

it follows that

$$\Phi_1^\star = \omega \iff \Phi_1 = \Phi_2 = \cdots = \Phi_\omega.$$

Thus, if $\Phi_1^\star = \omega$ then Con L reduces to the three-element chain

$$\omega \prec \Phi_1 = \cdots = \Phi_\omega \prec \iota.$$

It follows by the congruence extension property that $f(L)$ is also subdirectly irreducible. On the other hand, if $\Phi_1^\star \neq \omega$ then exactly the same conclusion follows by Theorem 3.15, the monolith of Con $f(L) \simeq$ Con L/Φ_1 being $\Phi_1^\star$ by Theorem 3.16.

In conclusion, $f(L)$ is subdirectly irreducible, whence so are all $f^n(L)$ since for every i we have that $f^i(L) = f[f^{i-1}(L)]$. We thus have

$$\text{Con}\, L = \{\omega\} \oplus [\Phi_1, \iota] \quad \text{with} \quad [\Phi_1, \iota] \simeq \text{Con}\, L/\Phi_1 \simeq \text{Con}\, f(L).$$

Similarly,

$$\text{Con}\, L = \{\omega\} \oplus \{\Phi_1\} \oplus [\Phi_2, \iota] \quad \text{with} \quad [\Phi_2, \iota] \simeq \text{Con}\, L/\Phi_2 \simeq \text{Con}\, f^2(L).$$

We conclude from this that if L belongs to a Berman class and $\mathbf{V}_B(L) = \mathbf{K}_{p,q}$ then, by Theorem 2.7, Con L is the finite chain

$$\omega = \Phi_0 \prec \Phi_1 \prec \Phi_2 \prec \cdots \prec \Phi_q \prec \iota.$$

If on the other hand L belongs properly to $\mathbf{K}_\omega$ then since there are infinitely many Φ_i, with Φ_{i+1} covering Φ_i and the supremum of the Φ_i being Φ_ω, we conclude that Con L is the infinite chain

$$\omega = \Phi_0 \prec \Phi_1 \prec \Phi_2 \prec \cdots < \Phi_\omega \prec \iota.$$

$\Leftarrow$: This is clear. ◇

Corollary 1 *If $L \in \mathbf{K}_\omega$ is subdirectly irreducible then so is every subalgebra of L.* ◇

Corollary 2 *If $L \in \mathbf{K}_\omega$ then L is simple if and only if L is subdirectly irreducible and f is injective.* ◇

Example 3.5 (*The sink*) Consider the lattice L with Hasse diagram

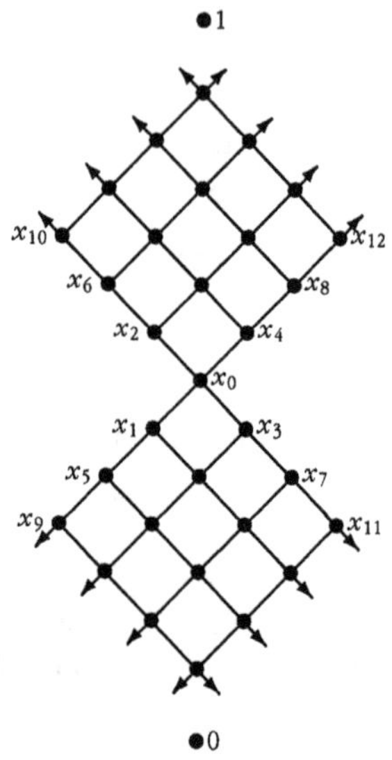

made into an Ockham algebra by defining

$$f(0) = 1,\ f(1) = 0,\ f(x_0) = x_0, \quad (\forall i \geqslant 1)\ f(x_i) = x_{i-1}$$

and extending to the whole of L. It is readily seen that $L \in \mathbf{K}_\omega$. Moreover, every congruence that identifies any adjacent pair also identifies x_0 and x_1. Consequently the smallest non-trivial congruence is $\Phi_1 = \vartheta(x_1, x_0)$, and L is subdirectly irreducible. The congruence Φ_ω has classes $\{0\}, \{1\}, L \setminus \{0, 1\}$.

The reader can easily verify that $\mathrm{Con}\, L$ is the infinite chain

$$\omega \prec \Phi_1 \prec \Phi_2 \prec \cdots < \Phi_\omega \prec \iota$$

where

$$\Phi_i = \begin{cases} \vartheta(x_0, x_i) & \text{if } i \text{ is even;} \\ \vartheta(x_i, x_0) & \text{if } i \text{ is odd.} \end{cases}$$

Every congruence on L is determined by the class containing x_0.

Note that Examples 3.2 and 3.5 show that, unlike $\mathbf{K}_{p,q}$, the generalised variety $\mathbf{K}_\omega$ has infinite subdirectly irreducible algebras.

Example 3.6 Let F be the finite/cofinite algebra on $\mathbb{N}_0$. In the lattice $L = F \overline{\oplus} F^{\mathrm{op}}$ depicted as follows

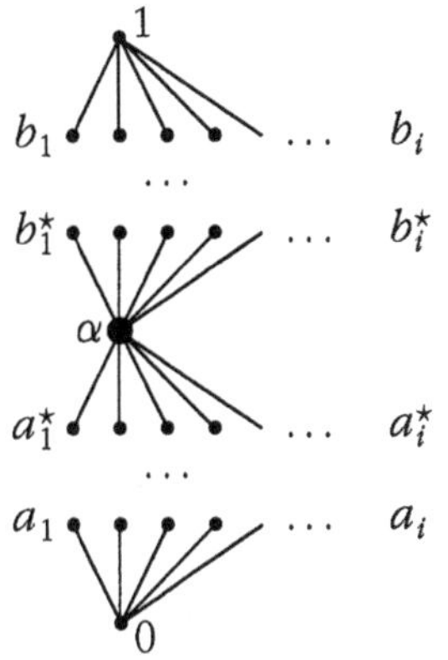

where $a_i^\star$ is the complement in F of the atom a_i and $b_i^\star$ is the complement in F^{op} of the coatom b_i, define

$$\begin{cases} f(0) = 1, \ f(1) = 0, \ f(\alpha) = \alpha; \\ f(a_i) = b_i, \ f(b_i) = a_{i-1}, \ f(b_1) = 0; \\ f(a_i^\star) = b_i^\star, \ f(b_i^\star) = a_{i-1}^\star, \ f(b_1^\star) = \alpha. \end{cases}$$

Since every element of L can be expressed either as a finite join of the a_i or the $b_i^\star$, or a finite meet of the $a_i^\star$ or the b_i, we can extend f to a dual endomorphism on L. Observe that $f^{2i}(b_i) = 1$ so $f^{2i+1}(a_i) = 1$; and $f^{2i}(b_i^\star) = \alpha$ so $f^{2i+1}(a_i^\star) = \alpha$. It follows that $L \in \mathbf{K}_\omega$. It is readily seen that any congruence on L that identifies a pair $x, y \in F$ identifies $0, a_i$ for some i, whence it identifies $0, a_1$ and hence $b_1, 1$; and any congruence on L that identifies a pair $x, y \in F^{\mathrm{op}}$ identifies $b_i, 1$ for some i, whence it identifies $b_1, 1$. Hence L is subdirectly irreducible with monolith $\Phi_1 = \vartheta(b_1, 1)$. As for Φ_ω, we have $L/\Phi_\omega \simeq \{0, \alpha, 1\} = T(L)$.

4 Duality theory

Duality theory in the context of distributive lattice-ordered algebras has rather a long history. It was in 1933 that G. Birkhoff established his famous representation theorem for finite distributive lattices and, about three years later, M. H. Stone developed a representation theory for arbitrary boolean algebras, using topological methods. In the early seventies, H. A. Priestley provided an ingenious common generalisation of both these theories, thus allowing questions of a lattice-theoretic nature 'to be translated into the language of ordered topological spaces' and usually be resolved more easily, the dual space generally being simpler and more tractable than the algebra itself. Of course, it is impossible to give here a complete survey of the theory of duality; for this we refer the reader to **[78]**, **[79]**, and **[63]**. Here we shall simply introduce the concepts and results that we shall require.

A set X which carries a topology τ and an order relation $\leqslant$ is called an *ordered topological space*. Such a space $(X;\tau,\leqslant)$ is said to be *totally order-disconnected* if

(TOD) given $x, y \in X$ with $x \nleqslant y$ there exists a *clopen*(= closed and open) down-set U such that $y \in U$ and $x \notin U$.

Clearly, every totally order-disconnected X is *totally disconnected* in the sense that

(TD) given $x, y \in X$ with $x \neq y$ there exists a *clopen* subset U such that $x \in U$ and $y \notin U$.

Every totally disconnected space X is *Hausdorff* in the sense that

(H) given $x, y \in X$ with $x \neq y$ there exist *open* sets U_1, U_2 such that $x \in U_1$, $y \in U_2$, and $U_1 \cap U_2 = \emptyset$.

In summary, therefore, we have (TOD) $\Rightarrow$ (TD) $\Rightarrow$ (H).

A *compact* totally order-disconnected space is called a *Priestley space*. The family of clopen down-sets of a Priestley space X will be denoted by $\mathcal{O}(X)$.

Let L be a distributive lattice and let $I_p(L)$ be the set of prime ideals of L. The *dual space* or *prime ideal space* of L is $(X;\tau,\subseteq)$ where $X = I_p(L)$ and τ has as a base the sets $\{x \in I_p(L) \mid x \ni a\}$ and $\{x \in I_p(L) \mid x \not\ni a\}$ for every $a \in L$. The fundamental result concerning this is the following :

$(X;\tau,\subseteq)$ is a Priestley space and $L \simeq \mathcal{O}(X)$ via $a \mapsto \{x \in X \mid x \not\ni a\}$. Conversely, if P is a Priestley space then $\mathcal{O}(P)$ is a distributive lattice and $P \simeq \big(I_p(\mathcal{O}(P));\tau,\subseteq\big)$.

In the language of category theory, if we denote by $\mathbf{D}_{01}$ the category of bounded distributive lattices and $0,1$-preserving lattice homomorphisms, and by **P** the category of Priestley spaces and continuous order-preserving maps, then the above isomorphisms give a dual equivalence between $\mathbf{D}_{01}$ and **P**.

The power of duality theory is particularly evident in the study of congruence relations. If $L \in \mathbf{D}$ and $X = \big(I_p(\mathcal{O}(P));\tau,\subseteq\big)$ is the dual space of L then for every closed subset Q of X the relation ϑ_Q defined on $\mathcal{O}(X)$ by

$$(A,B) \in \vartheta_Q \iff A \cap Q = B \cap Q$$

is a congruence. Note that $A \cap Q = B \cap Q$ is equivalent to $A \Delta B \subseteq Q'$ where Δ means 'symmetric difference' and Q' is the set-theoretic complement of Q in X. A direct consequence of this is that, for $L \in \mathbf{D}$,

Con L *is dually isomorphic to the lattice of closed subsets of* X.

At this point we would draw the reader's attention to the '**D**–**P** dictionary' which ends Priestley's survey paper **[79]** and which summarises some of the most commonly used dual equivalents.

Since Ockham algebras are bounded distributive lattices they are dually equivalent to a suitable subcategory of **P**. By an *Ockham space* we shall mean a Priestley space endowed with a continuous order-reversing map g. Let **Q** be the category whose objects are the Ockham spaces and whose morphisms are those order-preserving maps that commute with g. Then the category **O** of Ockham algebras is dually equivalent to the category **Q**.

We shall usually abbreviate $(X;\tau,g)$ to simply $(X;g)$. Whenever possible, we shall use letters $a,b,c,\ldots$ for elements of the Ockham algebra and letters $p,q,r,\ldots$ for elements of the Ockham space.

The following important results were first established by Urquhart **[94]**.

Theorem 4.1 *If $(X;g)$ is an Ockham space then $\big(\mathcal{O}(X);f\big)$ is an Ockham algebra where*

$$(\forall A \in \mathcal{O}(X)) \qquad f(A) = X \setminus g^{-1}(A).$$

Conversely, if $(L;f)$ is an Ockham algebra then $\big(I_p(L);g\big)$ is an Ockham space where

$$(\forall x \in I_p(L)) \qquad g(x) = \{a \in L \mid f(a) \notin x\}.$$

Moreover, these constructions give a dual equivalence. ♢

We now list some further concepts and terminology that we shall require. A subset Q of an Ockham space $(X;g)$ is called a *g-subset* if it is g-invariant, in the sense that

$$x \in Q \Rightarrow g(x) \in Q.$$

For every $Q \subseteq X$ we denote by $g^{\omega}(Q)$ the smallest g-subset that contains Q, namely

$$g^{\omega}(Q) = \{g^{n}(x) \mid n \geqslant 0,\ x \in Q\}.$$

The lattice of all closed g-subsets of X will be denoted by $G(X)$. A g-subset Q of the form $g^{\omega}(\{x\})$, which we shall write in the simpler form $g^{\omega}\{x\}$, will be called *monogenic*. Clearly, the monogenic g-subsets are the join-irreducible elements of $G(X)$.

Theorem 4.2 *Let $(X;g)$ be the dual space of $(L;f) \in \mathbf{O}$. If $Q \in G(X)$ then the equivalence relation ϑ_Q defined on $\mathcal{O}(X)$ by*

$$(A,B) \in \vartheta_Q \iff A \cap Q = B \cap Q$$

is a congruence.

Moreover, the lattice Con L *is dually isomorphic to* $G(X)$. ◇

For each $Q \in G(X)$ the congruence ϑ_Q will be called the *congruence associated with* Q. Clearly, Con L is boolean if and only if $G(X)$ is boolean. More generally, the congruence ϑ_Q is complemented if and only if $X \setminus Q$ belongs to $G(X)$, i.e. $X \setminus Q$ is a closed g-subset, whence Q is open. We thus have the following

Corollary *If $Q \in G(X)$ then the congruence ϑ_Q is complemented if and only if $X \setminus Q \in G(X)$, a condition that requires Q to be clopen.* ◇

Now in this respect the finite case is worthy of consideration, so we shall illustrate matters by revisiting Theorem 2.17, the proof of which is purely algebraic, and provide a very short proof of a statement that is more general than the theorem in question. For this purpose, and anticipating Chapter 5, we introduce a classification of the subvarieties of $\mathbf{O}$ which is somewhat sharper than that used by Berman. It is also due to Urquhart and is closely related to the dual space.

For $m > n \geqslant 0$ denote by $\mathbf{P}_{m,n}$ the subclass of $\mathbf{O}$ that is formed by those algebras whose dual space satisfies $g^m = g^n$. It will be established in Chapter 5 that when $m - n$ is even we have $L \in \mathbf{P}_{m,n}$ if and only if $f^m = f^n$ (in other words, $\mathbf{P}_{2p+n,n} = \mathbf{K}_{p,n}$); and when $m - n$ is odd we have $L \in \mathbf{P}_{m,n}$ if and only if $f^m(a)$ and $f^n(a)$ are complementary for every $a \in L$.

Theorem 4.3 *Let $(L;f)$ be a finite Ockham algebra and let $(X;g)$ be its dual space. Then the following statements are equivalent:*

(1) $L \in \mathbf{P}_{n,0}$ *for some* n;

(2) g *is surjective*;

(3) g *is injective*;

(4) Con L *is boolean with k atoms, k being the number of monogenic g-subsets of X, which attains its maximum value precisely when L is boolean.*

Proof (1) $\Rightarrow$ (2) : If $L \in \mathbf{P}_{n,0}$ then $g^n = \mathrm{id}_X$.

(2) $\Leftrightarrow$ (3) : Since L is finite, so is X.

(3) $\Rightarrow$ (4) : Let x_1 be an arbitrary element of X. Since X is finite and g is injective, there exists n_1 such that $g^{n_1}(x_1) = x_1$. If $g^{\omega}\{x_1\} \neq X$ we can choose $x_2 \in X \setminus g^{\omega}\{x_1\}$; then there exists n_2 such that $g^{n_2}(x_2) = x_2$, and so on. Since X is finite, this procedure terminates after finitely many steps. In this way we obtain k monogenic g-subsets

$$Q_1 = g^{\omega}\{x_1\}, \quad Q_2 = g^{\omega}\{x_2\}, \quad \dots, \quad Q_k = g^{\omega}\{x_k\}$$

of respective cardinalities $n_1, n_2, \dots, n_k$. Moreover, since g is injective, these g-subsets are pairwise disjoint. It follows that $G(X) \simeq \mathbf{2}^k$ and so Con L is a boolean lattice with k atoms. Note that in this case the maximum value of k is obtained when every monogenic g-subset reduces to a singleton. This is so precisely when $g(x) = x$ for every $x \in X$, in which case $L \in \mathbf{P}_{1,0} = \mathbf{B}$, the class of boolean algebras.

(4) $\Rightarrow$ (2) : If Con L is boolean then so is $G(X)$. In this case $g(X)$ and its complement are both g-subsets. This implies that $X \setminus g(X) = \emptyset$, whence $X = g(X)$ and so g is surjective.

(3) $\Rightarrow$ (1) : If g is injective then, as we have shown above, $G(X)$ is the disjoint union of k monogenic g-subsets $Q_1, \dots, Q_k$ of respective cardinalities $n_1, \dots, n_k$. If we consider $n = \mathrm{lcm}\{n_1, \dots, n_k\}$ then we see that $g^n(x) = x$ for every $x \in X$. Consequently, $L \in \mathbf{P}_{n,0}$. ◇

Example 4.1 Let $(X;g)$ be the Ockham space

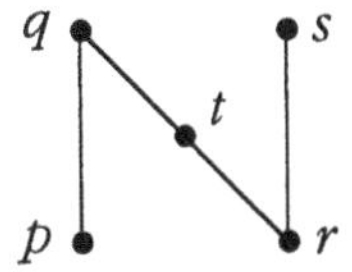

x	p	q	r	s	t
$g(x)$	s	r	q	p	t

The dual algebra $(L;f)$ belongs to $\mathbf{P}_{2,0}$ and so is a de Morgan algebra. There are 3 monogenic g-subsets, namely

$$Q_1 = \{p, s\}, \quad Q_2 = \{q, r\}, \quad Q_3 = \{t\}.$$

Hence Con $L \simeq \mathbf{2}^3$. The algebra $(L; f)$ is described as follows :

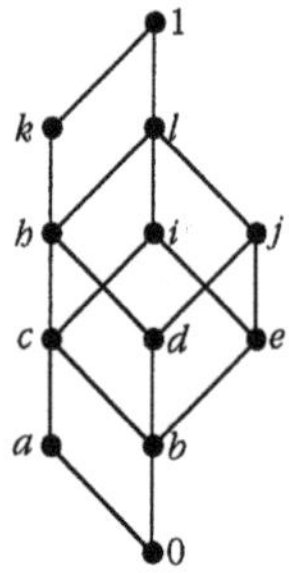

α	0	a	b	c	d	e	h	i	j	k	l	1
$f(\alpha)$	1	k	l	h	i	j	c	d	e	a	b	0

The three coatoms of Con L are ϑ_{Q_1}, ϑ_{Q_2}, and ϑ_{Q_3}, descriptions of which are obtained by Theorem 4.2.

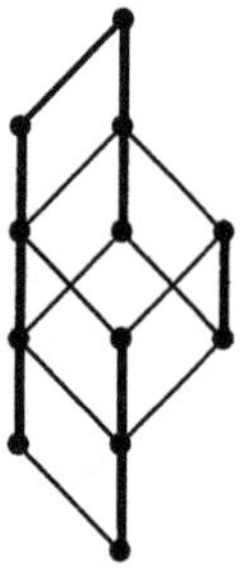

ϑ_{Q_1}

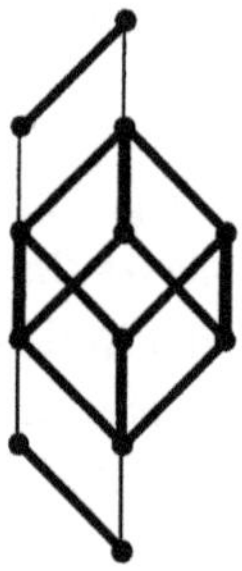

ϑ_{Q_2}

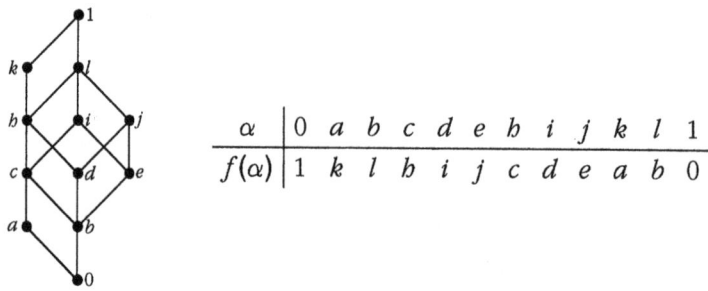

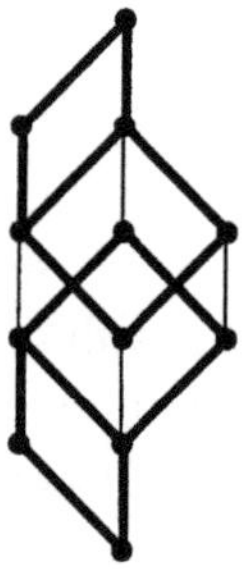

ϑ_{Q_3}

The other three non-trivial congruences are easily described. Note also that $\vartheta_{Q_1} = \vartheta(a,k)$, $\vartheta_{Q_2} = \vartheta(b,l)$, and $\vartheta_{Q_3} = \vartheta(0,i)$.

Example 4.2 Consider again Example 2.5. The dual space X is the 6-element chain $p < q < r < s < t < u$ with g defined by

x	p	q	r	s	t	u
$g(x)$	u	u	s	s	s	s

Here the monogenic g-subsets are

$$\{s\},\ \{s,u\},\ \{s,t\},\ \{r,s\},\ \{p,s,u\},\ \{q,s,u\}.$$

The Hasse diagram of $(G(X))^{\mathrm{op}}$ is depicted below and, compared with the diagram for Con L given in Chapter 2, shows the correspondence $Q \leftrightarrow \vartheta_Q$ obtained in Theorem 4.2. To simplify the notation in the diagram, we write $\{q,s,u\}$ as qsu, etc..

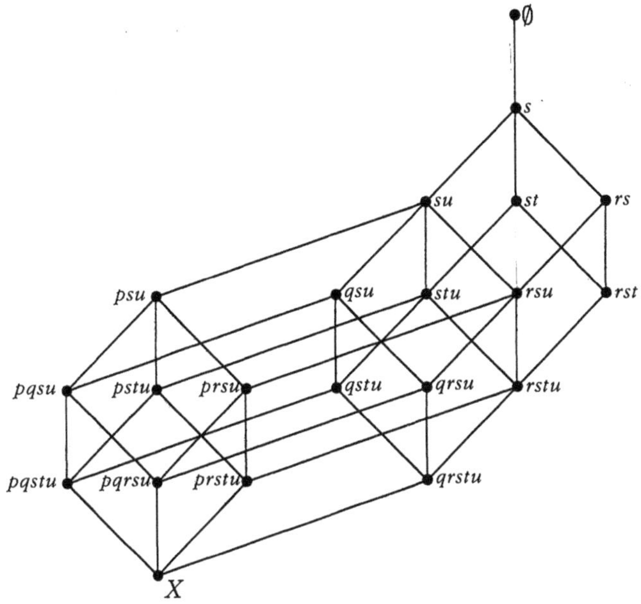

Example 4.3 Let X be an ordered set (the order may be discrete) of cardinality n. Let p be an element of X and let $g : X \to X$ be the constant map given by $g(x) = p$ for every $x \in X$. Clearly, $(X; g)$ is an Ockham space. The non-empty g-subsets of X are all the subsets of X that contain p. It follows that

$$G(X) \simeq \mathbf{1} \oplus \mathbf{2}^{n-1}.$$

If $(L; f)$ is the dual algebra then

$$\operatorname{Con} L \simeq \mathbf{2}^{n-1} \oplus \mathbf{1},$$

the coatom being the congruence associated with $\{p\}$. Since $g^2 = g$ we have $(L; f) \in \mathbf{P}_{2,1}$.

Let A be a down-set of X. If $p \in A$ then we have $g^{-1}(A) = X$ and $f(A) = \emptyset$. If $p \notin A$ then $g^{-1}(A) = \emptyset$ and $f(A) = X$. Let d be the element of L that represents $p^{\downarrow}$. Then it follows that we have

$$(\forall \alpha \in d^{\uparrow}) \quad f(\alpha) = 0; \qquad (\forall \alpha \notin d^{\uparrow}) \quad f(\alpha) = 1.$$

In particular, if $p \leqslant x$ for every $x \in X$ then $f(\alpha) = 0$ for every $\alpha \in L \setminus \{0\}$, which implies that 0 is meet-irreducible. In this case, L is a Stone algebra. If, on the contrary, we have $p \geqslant x$ for every $x \in X$ then $f(\alpha) = 1$ for every $x \in L \setminus \{1\}$, whence 1 is join-irreducible and L is a dual Stone algebra.

In Chapter 2 we introduced the congruences Φ_n defined for every $n \in \mathbb{N}$ by

$$(x,y) \in \Phi_n \iff f^n(x) = f^n(y).$$

In the dual space $(X;g)$ we can consider for every $n \in \mathbb{N}$ the g-subset

$$g^n(X) = \{g^n(x) \mid x \in X\}.$$

We can also consider the g-subset

$$g^\omega(X) = \bigcap_{n \geqslant 0} g^n(X).$$

Clearly, all g^n are continuous maps and, since X is a compact Hausdorff space, all $g^n(X)$ are compact, hence closed. The g-subset $g^\omega(X)$ enjoys the same property. Obviously, these closed g-subsets are related by

$$X \supseteq g(X) \supseteq g^2(X) \supseteq \cdots \supseteq g^\omega(X) \supseteq \{0,1\}.$$

The following result **[55]** elucidates the correspondence between the congruences Φ_n and the closed g-subsets $g^n(X)$.

Theorem 4.4 *Let $(L;f) \in \mathbf{O}$ and let $(X;g)$ be its dual space. Then for every $n \geqslant 0$ the congruence Φ_n is associated with the closed g-subset $g^n(X)$, and the congruence Φ_ω is associated with $g^\omega(X)$.*

Proof Clearly, Φ_0 is associated with $g^0(X) = X$. As for Φ_1, we have the following equivalences which show that Φ_1 is associated with $g(X)$:

$$\begin{aligned}(a,b) \in \Phi_1 &\iff f(a) = f(b)\\ &\iff X \setminus g^{-1}(A) = X \setminus g^{-1}(B)\\ &\iff g^{-1}(A) = g^{-1}(B)\\ &\iff A \cap g(X) = B \cap g(X)\\ &\iff (A,B) \in \vartheta_{g(X)}.\end{aligned}$$

Taking into account the fact that, for every n, $f^{n+1}(L)$ is a subalgebra of $f^n(L)$, and using an easy inductive argument, we can show that Φ_n is associated with $g^n(X)$.

Finally, $\Phi_\omega = \bigvee_{n \geqslant 0} \Phi_n$ is associated with $\bigcap_{n \geqslant 0} g^n(X) = g^\omega(X)$. ◇

Example 4.4 Let $\mathbb{N}_\infty$ consist of $\mathbb{N}_0$ with an additional point ∞ adjoined. It is known (see **[9]** for the details) that $\mathbb{N}_\infty$ becomes a Priestley space if one takes as a sub-basis for the topology the subsets U such that

$$U \not\ni \infty \quad \text{or} \quad (U \ni \infty \text{ and } U' \text{ is finite}).$$

Then a subset V is closed if

$$V \ni \infty \quad \text{or} \quad (V \not\ni \infty \text{ and } V \text{ is finite}).$$

The clopen subsets of $\mathbb{N}_\infty$ are therefore the finite sets that do not contain ∞, together with their complements.

Now define g by setting $g(n) = n + 1$ for every $n \in \mathbb{N}$, and $g(\infty) = \infty$. Since the order on $\mathbb{N}_\infty$ is discrete, g is trivially order-reversing. It is easily verified that g is continuous. It follows that $(\mathbb{N}_\infty; g)$ is an Ockham space which we can represent as follows :

The g-subsets are $\emptyset$, $\{\infty\}$ and, for every n,

$$\{n, n+1, n+2, \ldots\} \quad \text{and} \quad \{n, n+1, n+2, \ldots, \infty\} = g^{n-1}(\mathbb{N}_\infty).$$

The closed g-subsets are $\emptyset$, $g^n(\mathbb{N}_\infty)$ for all $n \geqslant 0$, and $g^\omega(\mathbb{N}_\infty) = \{\infty\}$. Now

$$g^0(\mathbb{N}_\infty) = \mathbb{N}_\infty \supset g(\mathbb{N}_\infty) \supset g^2(\mathbb{N}_\infty) \supset \cdots \supset g^\omega(\mathbb{N}_\infty) \supset \emptyset,$$

and so, if $(L; f)$ is the dual algebra, Con L reduces to the chain

$$\omega \prec \Phi_1 \prec \Phi_2 \prec \cdots < \Phi_\omega \prec \iota$$

whence L is subdirectly irreducible.

Now (again see **[9]**) it is known that $\mathbb{N}_\infty$ is the 1-point compactification of a countable discrete space. As such, it is a boolean space and is (homeomorphic to) the prime ideal space of the finite-cofinite algebra FC($\mathbb{N}$). Thus L is an atomic boolean lattice that is not complete. We shall now show that $(L; f)$ does not belong to any Berman class, but does belong to $\mathbf{K}_\omega$. For this purpose, let U be a finite subset of $\mathbb{N}_\infty$ that does not contain ∞ and let z be its numerically greatest element. Since

$$f^{2n}(U) = \{x \in \mathbb{N}_\infty \mid g^{2n}(x) \in U\},$$

we see that $f^{2n}(U)$ is the subset of $\mathbb{N}_\infty$ obtained from U by translating U to the left through an amplitude of $2n$ and deleting the resulting elements that fall outside $\mathbb{N}_\infty$. From this observation, we have that if $z = 2n$ then $f^{2n}(U) = \emptyset$ and therefore $f^{2n+2}(U) = f^{2n}(U)$; and that if $z = 2n + 1$ then $f^{2n}(U) = \{1\}$ and $f^{2n+1}(U) = \mathbb{N}_\infty$, and therefore $f^{2n+3}(U) = f^{2n+1}(U)$.

The above example can be modified slightly. Consider the effect of taking g to be the constant map given by

$$(\forall n \in \mathbb{N}) \quad g(n) = g(\infty) = \infty.$$

Here the closed g-subsets are $\emptyset$ and all the subsets of $\mathbb{N}_\infty$ that contain ∞. Consequently we have

$$\text{Con } L \simeq \mathbf{2}^{\mathbb{N}} \oplus \{\iota\},$$

the comonolith being Φ_1. In this case, $(L;f) \in \mathbf{P}_{2,1}$ and is not subdirectly irreducible.

Another variation on the same theme consists of ordering $\mathbb{N}_\infty$ as the chain $\mathbb{N}$ with ∞ as a greatest element, and taking g to be the same constant map as in the above. Here the clopen down-sets are ∞, $\mathbb{N}_\infty$, and $n^\downarrow$ for every n. Consequently, L is isomorphic to $\mathbb{N} \oplus \mathbf{1}$. The closed g-subsets are $\emptyset$ and all the subsets of $\mathbb{N}_\infty$ that contain ∞, so that

$$\text{Con } L \simeq \mathbf{2}^{\mathbb{N}} \oplus \{\iota\},$$

the comonolith being Φ_1. Since for every $a \in L \setminus \{1\}$ we have $f(a) = 1$, it follows that $(L;f)$ is a Stone algebra.

Our objective now is to characterise those closed g-subsets that correspond to principal congruences under the duality. The following theorem was inspired by Lemmas 2 and 3 of **[17]**.

We shall say that $Q \in G(X)$ is *maximally disjoint* from $R \subseteq X$ if $Q \cap R = \emptyset$ and $T \cap R \neq \emptyset$ for every $T \in G(X)$ with $T \supset Q$.

Theorem 4.5 *A closed g-subset Q represents a principal congruence if and only if there is a convex clopen subset R from which Q is maximally disjoint.*

Proof $\Rightarrow$: Let $Q \in G(X)$ represent the principal congruence $\vartheta(a,b)$. If the elements a and b correspond to the clopen decreasing subsets A and B of X (with $A \subseteq B$) then, since $(a,b) \in \vartheta(a,b)$, we have

$$Q \subseteq (A \Delta B)' = (B \setminus A)'.$$

Even more, Q is the greatest closed g-subset contained in $(B \setminus A)'$. Clearly, $R = B \setminus A$ is a convex clopen subset, and Q is maximally disjoint from R.

$\Leftarrow$: Conversely, suppose that Q is maximally disjoint from the convex clopen subset R. Since R is closed, $R^\downarrow$ is a closed down-set; and since R is open, $R^\downarrow \setminus R$ is also a closed down-set. Since X is a Priestley space, there is a clopen down-set A such that $R^\downarrow \setminus R \subseteq A$ and $R \cap A = \emptyset$. Consider the set $B = A \cup R$. Since R is convex and clopen, B is also a clopen down-set. Let a and b be the elements of L that are represented in X by A and B respectively, and let $P \in G(X)$ represent the principal congruence $\vartheta(a,b)$. By the first part, P is maximally disjoint from $B \setminus A = R$. The fact that P and Q are both closed now gives $P = Q$, whence Q represents a principal congruence. $\diamondsuit$

Remarks Note that, in the above,

(1) the convex clopen subset R is not necessarily unique;

(2) the principal congruence $\vartheta(a,b)$ is represented by the closed g-subset that is maximally disjoint from $B \setminus A$;

(3) if the complement of a g-closed subset Q is convex and clopen then Q represents a principal congruence. It follows that if L is finite and $\ell(X) \leqslant 1$ then all congruences of L are principal.

(4) if L is finite and the g-subset Q is covered by a unique g-subset P then Q represents a principal congruence (for the convex subset R take $\{x\}$ in $P \setminus Q$).

Example 4.5 Consider once more the pineapple lattice made into an Ockham algebra as in Example 2.4. In this, all prime ideals of L, except $\{0\}$, are non-principal. They form two chains connecting $\{0\}$ to $J = L \setminus \{1\}$. We shall denote by I_{2n} (resp. I_{2n-1}) the prime ideal that separates x_{2n} and x_{2n+1} (resp. x_{2n-1} and x_{2n}). The dual space X is then the following ordered set.

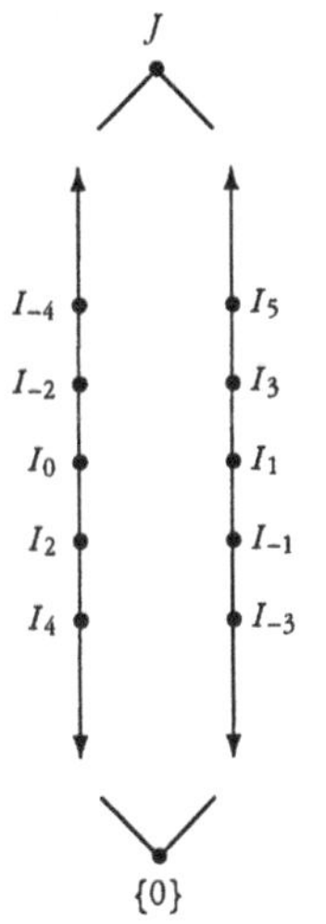

This becomes a Priestley space if we take as a sub-basis for the topology τ the subsets $I_n^{\uparrow} \setminus \{J\}$, $I_n^{\downarrow} \setminus \{\{0\}\}$, $I_{2m}^{\uparrow} \cup I_{2n+1}^{\uparrow}$, and $I_{2m}^{\downarrow} \cup I_{2n+1}^{\downarrow}$ where $m, n \in \mathbb{Z}$. The clopen down-sets of X are $\emptyset$, X, and $I_k^{\downarrow} \cup I_\ell^{\downarrow}$ for $k \in 2\mathbb{Z}$ and $\ell \in 2\mathbb{Z}+1$. Then we define g as follows :

$$(\forall n \in \mathbb{Z})\ g(I_n) = I_{n-1},\ g(J) = \{0\},\ g(\{0\}) = J.$$

$(X; g, \tau)$ is an Ockham space and it is easy to verify that the dual algebra is that described in Example 2.4. The open g-subsets other than $\emptyset$ and X are generated by any element of $\{I_n \mid n \in \mathbb{Z}\}$. The closed g-subsets are obtained by adding to the preceding sets the set $\{\{0\}, J\}$. Ordered by inclusion, this collection forms an infinite chain with $\{\{0\}, J\}$ covering $\emptyset$.

The congruence $\vartheta(x_{2n-1}, x_{2n})$ is represented by the closed g-subset that is maximally disjoint from $\{I_{2n-1}\}$, i.e. the closed g-subset generated by I_{2n-2}. Similarly, we see that $\vartheta(x_{2n+1}, x_{2n})$ is represented by the closed g-subset generated by I_{2n-1}.

The only non-principal congruence Ψ corresponds to $\{\{0\}, J\}$, and Con L is the chain

$$\omega < \cdots \prec \vartheta(x_3, x_2) \prec \vartheta(x_1, x_2) \prec \vartheta(x_1, x_0) \prec \cdots < \Psi \prec \iota.$$

Example 4.6 (*The diamond*)

(1) *The finite case.* Consider the lattice L, which is the direct product of two 6-element chains, described by the Hasse diagram

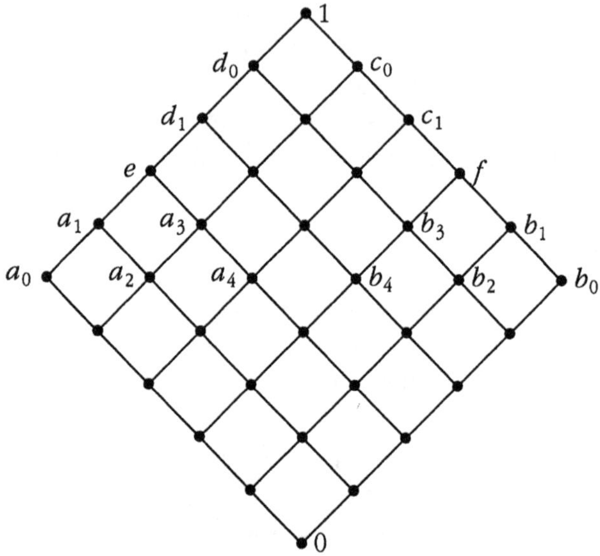

and made into an Ockham algebra by defining

$$f(a_0) = a_0, \quad f(b_0) = b_0, \quad (1 \leqslant i \leqslant 4) \ f(a_i) = a_{i-1}, \ f(b_i) = b_{i-1}$$

then extending to the whole of L. The dual space X is as follows, in which the arrows indicate the action of g :

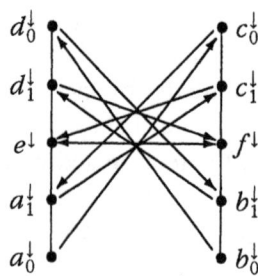

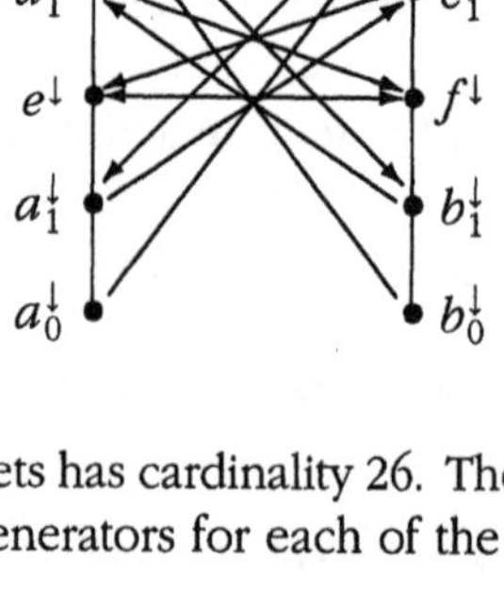

The lattice $G(X)$ of g-subsets has cardinality 26. The Hasse diagram of its dual $(G(X))^{\mathrm{op}}$ and a table of generators for each of the g-subsets are as follows :

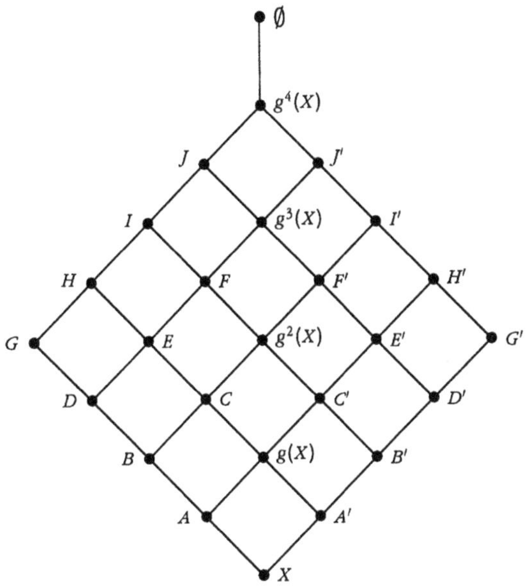

X	$\{a_0^{\downarrow}, b_0^{\downarrow}\}$	A	$\{a_0^{\downarrow}, d_0^{\downarrow}\}$	F	$\{a_1^{\downarrow}, d_1^{\downarrow}\}$	A'	$\{b_0^{\downarrow}, c_0^{\downarrow}\}$	F'	$\{b_1^{\downarrow}, c_1^{\downarrow}\}$
$g(X)$	$\{c_0^{\downarrow}, d_0^{\downarrow}\}$	B	$\{a_0^{\downarrow}, b_1^{\downarrow}\}$	G	$\{a_0^{\downarrow}\}$	B'	$\{b_0^{\downarrow}, a_1^{\downarrow}\}$	G'	$\{b_0^{\downarrow}\}$
$g^2(X)$	$\{a_1^{\downarrow}, b_1^{\downarrow}\}$	C	$\{c_0^{\downarrow}, b_1^{\downarrow}\}$	H	$\{c_0^{\downarrow}\}$	C'	$\{d_0^{\downarrow}, a_1^{\downarrow}\}$	H'	$\{d_0^{\downarrow}\}$
$g^3(X)$	$\{c_1^{\downarrow}, d_1^{\downarrow}\}$	D	$\{a_0^{\downarrow}, d_1^{\downarrow}\}$	I	$\{a_1^{\downarrow}\}$	D'	$\{b_0^{\downarrow}, c_1^{\downarrow}\}$	I'	$\{b_1^{\downarrow}\}$
$g^4(X)$	$\{e^{\downarrow}\}$	E	$\{c_0^{\downarrow}, d_1^{\downarrow}\}$	J	$\{c_1^{\downarrow}\}$	E'	$\{d_0^{\downarrow}, c_1^{\downarrow}\}$	J'	$\{d_1^{\downarrow}\}$

The Hasse diagram for Con L is the same as that above with X replaced by ω; $\emptyset$ replaced by ι; each $g^i(X)$ replaced by Φ_i; and each g-subset K replaced by the corresponding congruence ϑ_K.

Using Theorem 4.5, we can verify that there are 8 non-principal congruences, namely $\vartheta_C, \vartheta_{C'}, \vartheta_F, \vartheta_{F'}, \vartheta_H, \vartheta_{H'}, \vartheta_J, \vartheta_{J'}$. Moreover, we have

$$\begin{array}{lll}
\Phi_1 = \vartheta(0, a_1 \wedge b_1) & \Phi_2 = \vartheta(a_2 \vee b_2, 1) & \Phi_3 = \vartheta(0, a_3 \wedge b_3) \\
\Phi_4 = \vartheta(a_4 \vee b_4, 1) & \vartheta_A = \vartheta(b_0, b_1) & \vartheta_{A'} = \vartheta(a_0, a_1) \\
\vartheta_B = \vartheta(b_2, b_1) & \vartheta_{B'} = \vartheta(a_2, a_1) & \vartheta_D = \vartheta(b_2, b_3) \\
\vartheta_{D'} = \vartheta(a_2, a_3) & \vartheta_G = \vartheta(b_4, b_3) & \vartheta_{G'} = \vartheta(a_4, a_3) \\
\vartheta_E = \vartheta(0, a_1 \wedge b_3) & \vartheta_{E'} = \vartheta(0, a_3 \wedge b_1) & \vartheta_I = \vartheta(a_2 \vee b_4, 1) \\
\vartheta_{I'} = \vartheta(a_4 \vee b_2, 1) & &
\end{array}$$

(2) *The infinite case.* Consider the lattice L which is the direct product of two chains isomorphic to $\mathbb{N} \oplus \mathbb{N}^{op}$, described by the Hasse diagram

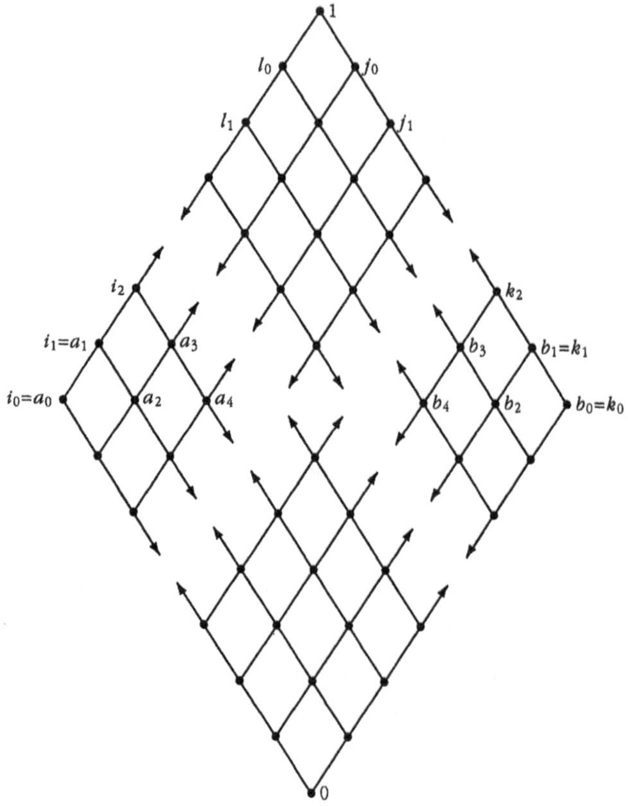

and made into an Ockham algebra by defining

$$f(a_0) = a_0, \; f(b_0) = b_0, \quad (\forall i \geqslant 1) \; f(a_i) = a_{i-1}, \; f(b_i) = b_{i-1},$$

then extending to the whole of L.

All prime ideals of L are principal except for

$$r = \bigcup_{n \in \mathbb{N}} i_n^{\downarrow} \quad \text{and} \quad s = \bigcup_{n \in \mathbb{N}} k_n^{\downarrow}.$$

They form two chains C_1 and C_2 as in the following diagram. The set $X = C_1 \cup C_2$ becomes a Priestley space if we take as a sub-basis for the topology τ the subsets $\{x \mid x > p\}$ and $\{x \mid x < p\}$ for every $p \in X$. Note that the sets $r^{\uparrow}$, $s^{\uparrow}$, $r^{\downarrow}$, $s^{\downarrow}$ are not open.

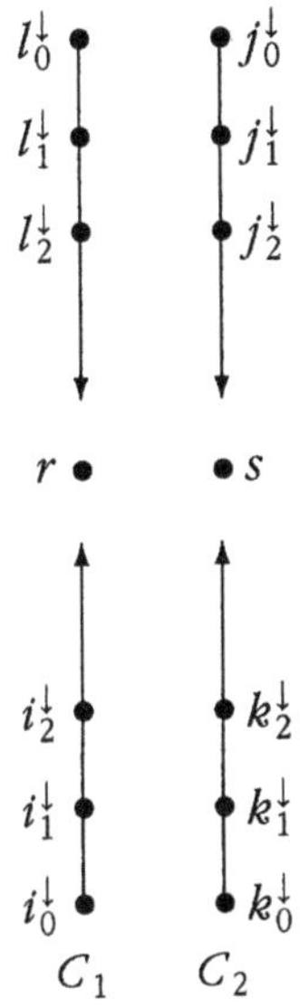

The clopen down-sets of X are $\emptyset$, X, $\alpha^\downarrow$, $\beta^\downarrow$, and $\alpha^\downarrow \cup \beta^\downarrow$ where $\alpha \in C_1 \setminus \{r\}$ and $\beta \in C_2 \setminus \{s\}$. The action of g on X is described by

$$i_0^\downarrow \to j_0^\downarrow \to i_1^\downarrow \to j_1^\downarrow \to i_2^\downarrow \to \cdots \quad (1)$$
$$k_0^\downarrow \to l_0^\downarrow \to k_1^\downarrow \to l_1^\downarrow \to k_2^\downarrow \to \cdots \quad (2)$$
$$r \leftrightarrow s$$

The g-subsets of X are easily determined : they are $\emptyset$; those generated by any element γ of the sequence (1); those generated by any element δ of the sequence (2); those generated by any pair $\{\gamma, \delta\}$; and finally all of the previous with $\{r, s\}$ adjoined. Only $\emptyset$ and the last mentioned g-subsets are closed, so $(G(X))^{\mathrm{op}}$ is given by

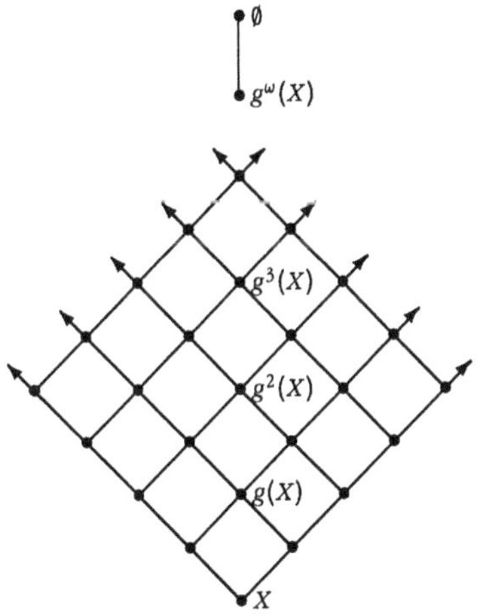

The diagram for Con L is the same with the label change $X \to \omega$, $g^i(X) \to \Phi_i$, $g^\omega(X) \to \Phi_\omega$, and $\emptyset \to \iota$. Note that

$$g^\omega(X) = \bigcap_{i \geqslant 0} g^i(X) = \{r, s\}$$

gives the comonolith Φ_ω which is not principal, and that L/Φ_ω is isomorphic to the simple 4-element de Morgan algebra.

Example 4.7 Here is an example of a small Ockham algebra with only one non-principal congruence.

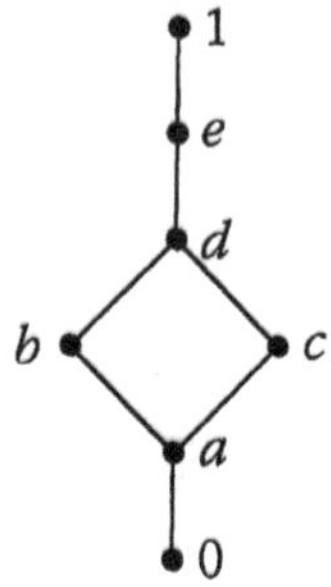

L

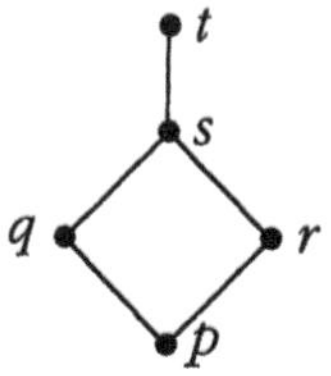

X

n	0	a	b	c	d	e	1
$f(n)$	1	d	b	c	a	0	0

x	p	q	r	s	t
$g(x)$	s	r	q	p	p

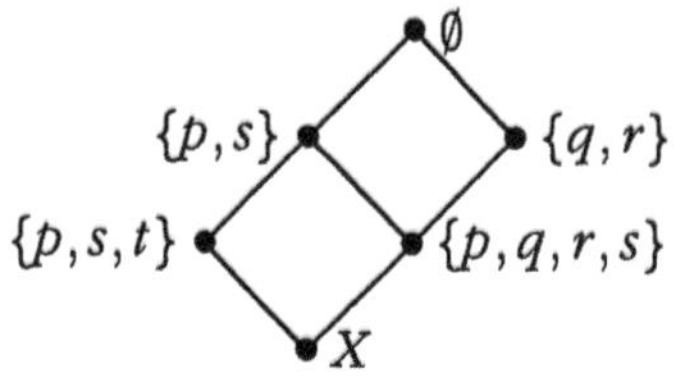

$(G(X))^{op}$

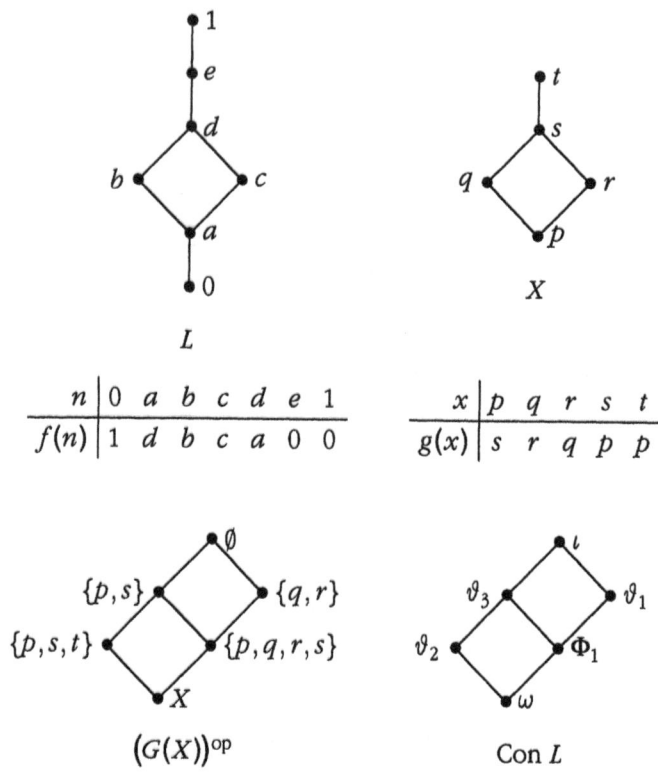

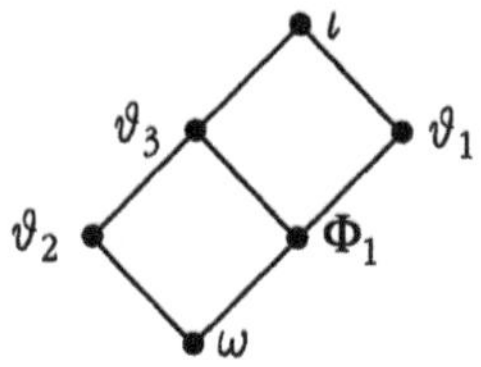

Con L

It is easily seen that $\Phi_1 = \vartheta(e,1)$ corresponds to $\{p,q,r,s\}$, $\vartheta_1 = \vartheta(0,a)$ corresponds to $\{q,r\}$, and $\vartheta_2 = \vartheta(a,d)$ corresponds to $\{p,s,t\}$. Only

$$\vartheta_3 = \vartheta_{\{p,s\}} \equiv \{\{0\},\{a,b,c,d\},\{e,1\}\}$$

is not principal. In fact, the greatest convex sets disjoint from $\{p,s\}$ are $\{q,r\}$ and $\{t\}$; and $\{p,s\}$ is not maximally disjoint from either.

We now consider subdirectly irreducible and simple algebras from the duality point of view. For this purpose, we require the following fundamental results of Urquhart [**94**].

Lemma 4.1 *Let* $(X;g)$ *be an Ockham space. Then a subset* Y *of* X *is a* g-*subset if and only if* $Y \subseteq g^{-1}(Y)$.

Proof If Y is a g-subset then $g(Y) \subseteq Y$ gives $Y \subseteq g^{-1}[g(Y)] \subseteq g^{-1}(Y)$. Conversely, $Y \subseteq g^{-1}(Y)$ gives $g(Y) \subseteq g[g^{-1}(Y)] \subseteq Y$. ◇

Lemma 4.2 *If Y is a g-subset of X and $Y \subseteq Q$ then $Y \subseteq g^{-1}(Q)$.*

Proof $Y \subseteq Q$ gives $g^{-1}(Y) \subseteq g^{-1}(Q)$ and the result follows by Lemma 4.1. ◇

Lemma 4.3 *Let $(X; g)$ be an Ockham space. If Y is a g-subset of X then so is its closure $\overline{Y}$.*

Proof Since $\{A \mid A \in \mathcal{O}(X)\} \cup \{A' \mid A \in \mathcal{O}(X)\}$ is a sub-base of X, the closure $\overline{Y}$ of any g-subset Y is

$$\overline{Y} = \bigcap \{A' \cup B \mid A, B \in \mathcal{O}(X),\ \ Y \subseteq A' \cup B\}.$$

Now for $A' \cup B \in \overline{Y}$ we have

$$g^{-1}(A' \cup B) = g^{-1}(A') \cup g^{-1}(B) = f(A) \cup [f(B)]'$$

and, by Lemma 4.2, $Y \subseteq g^{-1}(A' \cup B)$. It follows that $g^{-1}(A' \cup B) \in \overline{Y}$. Thus $g(\overline{Y}) \subseteq \overline{Y}$ and so $\overline{Y}$ is g-closed. ◇

Corollary $\overline{g^{\omega}(Y)}$ *is the least closed g-subset that contains Y.* ◇

Theorem 4.6 *Let $(X; g)$ be the dual space of $(L; f) \in \mathbf{O}$ and let*

$$Q = \overline{\left\{x \in X \mid \overline{g^{\omega}\{x\}} \neq X\right\}}.$$

Then $(L; f)$ is subdirectly irreducible if and only if $Q \neq X$.

Proof $\Leftarrow$: Suppose that $Q \neq X$ and let X be such that $\overline{g^{\omega}\{x\}} \neq X$. Since $g^{\omega}\{g(x)\} \subseteq g^{\omega}\{x\}$, we have $\overline{g^{\omega}\{g(x)\}} \subseteq \overline{g^{\omega}\{x\}} \subset X$ and, by Lemma 4.3, $Q \in G(X)$. Let $R \in G(X) \setminus \{X\}$. If $x \in R$ then $\overline{g^{\omega}\{x\}} \subseteq R$ and $x \in Q$. Thus $R \subseteq Q$ and Q is maximal in $G(X) \setminus \{X\}$. Consequently, by Theorem 4.2, L is subdirectly irreducible.

$\Rightarrow$: Suppose now that L is subdirectly irreducible and denote by Y the maximum element of $G(X) \setminus \{X\}$. If $x \in Y$ then $\overline{g^{\omega}\{x\}} \subseteq Y$ and $\overline{g^{\omega}\{x\}} \neq X$. Conversely, if $\overline{g^{\omega}\{x\}} \neq X$ then $\overline{g^{\omega}\{x\}} \subseteq Y$, which gives $x \in Y$. We thus have $Y = Q$. ◇

Corollary 1 *If $(L; f) \in \mathbf{O}$ is finite then $(L; f)$ is subdirectly irreducible if and only if $g^{\omega}\{x\} = X$ for some $x \in X$.*

Proof The topology on X being discrete, L is subdirectly irreducible if and only if

$$\{x \mid g^{\omega}\{x\} \neq X\} \neq X;$$

i.e. if and only if there exists $x \in X$ such that $g^{\omega}\{x\} = X$. ◇

Corollary 2 *If $(L; f) \in \mathbf{O}$ is finite then $(L; f)$ is simple if and only if $g^{\omega}\{x\} = X$ for every $x \in X$.*

Proof For every $x \in X$ we have $g^{\omega}\{x\} \in G(X) \setminus \{\emptyset\}$, and $(L; f)$ is simple if and only if $G(X)$ has only two elements, namely $\emptyset$ and X. ◇

By Corollary 1 above, the dual space $(X; g)$ of a finite subdirectly irreducible algebra $L \in \mathbf{O}$ can be represented as follows (in which the order on X is ignored and the action of g is indicated by the arrows) :

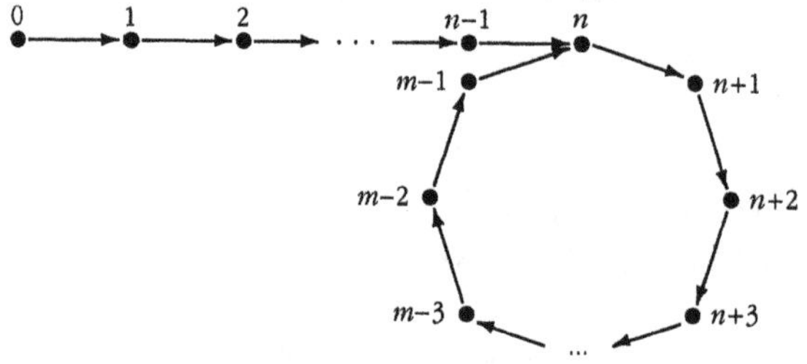

We suppose that $g^m(0) = g^n(0)$. The element 0 then generates X under g; such an element is called an *end* of X. The subsets $\{0, 1, \ldots, n-1\}$ and $\{n, n+1, \ldots, m-1\}$ are called the *tail* and the *loop* of X respectively. Such an Ockham space with the discrete order is usually denoted by m_n. The dual algebra of m_n is denoted by $L_{m,n}$. Clearly, the end of X is unique except in the case where X reduces to a loop; i.e., by Corollary 2 above, precisely when L is simple.

It is clear that the non-empty g-subsets of X are the sets

$$\{k, k+1, \ldots, m-1\} \quad (0 \leqslant k \leqslant n).$$

The following is now a consequence of Theorem 4.2.

Corollary 3 Con $L_{m,n}$ *is an $(n+2)$-element chain.* ◇

Any order on $\{0, 1, \ldots, m-1\}$ with respect to which g is order-reversing yields the dual space of a subdirectly irreducible Ockham algebra. Conversely, all dual spaces of finite subdirectly irreducible Ockham algebras arise in this way. Now the identity map on $\{0, 1, \ldots, m-1\}$ is an Ockham space morphism of m_n onto X and, by the duality, corresponds to an injective Ockham algebra morphism. We therefore have :

Corollary 4 *A finite Ockham algebra is subdirectly irreducible if and only if it is isomorphic to a subalgebra of $L_{m,n}$ for some m and n.* ◇

We shall now apply the preceding results to describe the subdirectly irreducible algebras in the class $\mathbf{P}_{3,1} = \mathbf{K}_{1,1}$. These were determined in 1980 by Sankappanavar but not published until 1985 **[85]**. During this time they were determined independently by Beazer **[24]** using both algebraic and topological approaches. Of course, all MS-algebras belong to $\mathbf{P}_{3,1}$; the subdirectly irreducible MS-algebras were determined by the authors in 1983 **[33]**.

Example 4.8 The subdirectly irreducible algebras in $\mathbf{P}_{3,1}$ are the subalgebras of $L_{3,1}$, i.e. the algebra whose dual space $X = \{p, q, r\}$ has discrete order and on which g acts as follows :

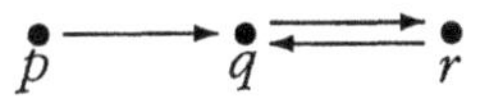

The Hasse diagram of $L_{3,1}$ is the boolean lattice

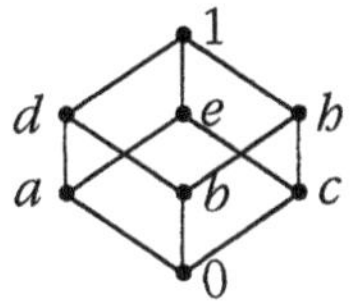

with the operation f defined by

x	0	a	b	c	d	e	h	1
$f(x)$	1	1	b	e	b	e	0	0

We could of course seek to find all the subalgebras of $L_{3,1}$. Instead, we shall adopt the tactic of determining them from their dual spaces. For this purpose we have to determine all the orders on $\{p, q, r\}$ with respect to which g is order-reversing. The enumeration is made simpler by the observation that for such an order we must have $p < q \Rightarrow r < q$ and dually, whereas the relations $p < r$, $q < r$ and their duals have no consequences.

There are in all 19 non-trivial subdirectly irreducible algebras in $\mathbf{P}_{3,1}$ (including its subclasses $\mathbf{P}_{1,0}, \mathbf{P}_{2,0}, \mathbf{P}_{2,1}$). These are as shown on pages 70 and 71, together with their duals. Note that B, M, K, S_1, B_1 are the only subdirectly irreducible algebras in $\mathbf{P}_{3,1}$ that are self-dual. In these diagrams we use the notation ⊙ to denote g-fixed points in X and f-fixed points in L. By Corollary 3 of Theorem 4.6, we know that Con L has 2 elements if $L \in \mathbf{P}_{1,0}$ or $L \in \mathbf{P}_{2,0}$, and 3 elements if $L \in \mathbf{P}_{2,1}$ or $L \in \mathbf{P}_{3,1}$. In the latter case we indicate the classes of the only non-trivial congruence Φ_1.

Ockham spaces	Ockham algebras		
	$\mathbf{P}_{1,0}$		B
	$\mathbf{P}_{2,0}$	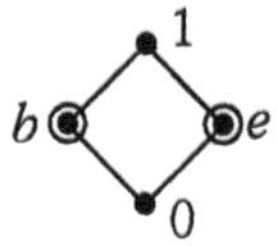	M
			K
	$\mathbf{P}_{2,1}$		S
		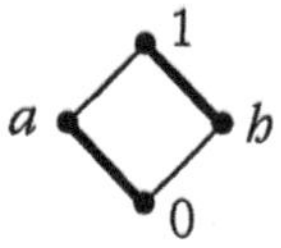	S_1
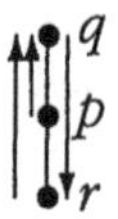	$\mathbf{P}_{3,1}$		K_1
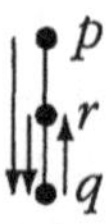			K_2

$\mathbf{P}_{3,1}$ *continued*

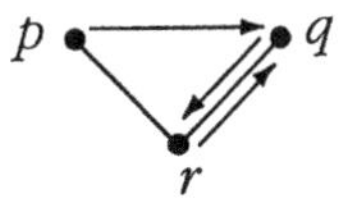

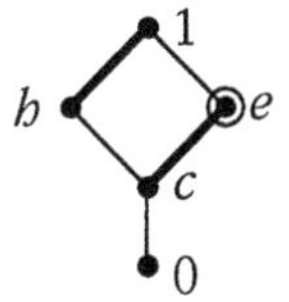

K_3

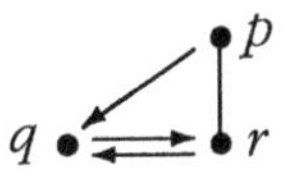

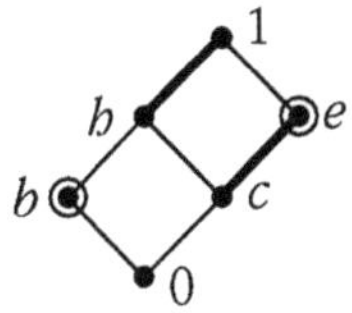

M_1

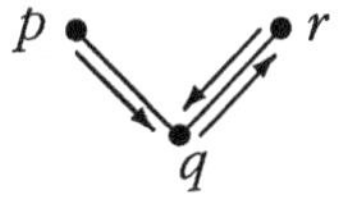

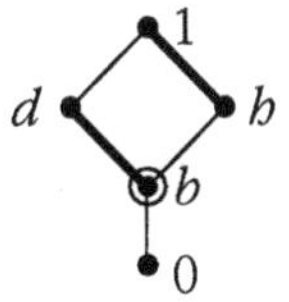

L

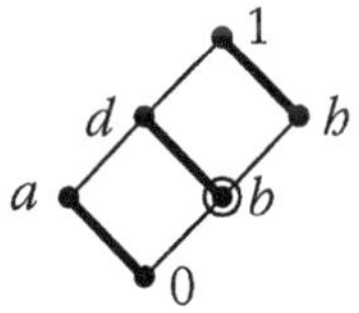

N

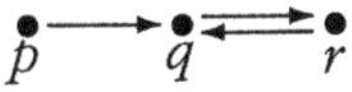

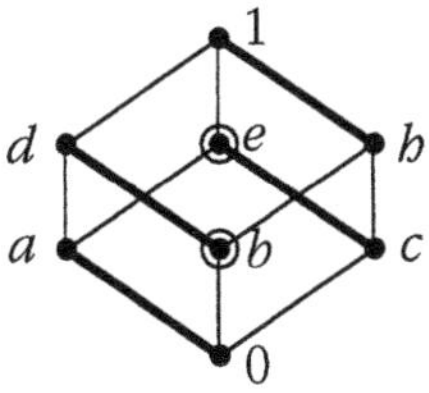

B_1

Ockham spaces	*Ockham algebras*

We close this chapter with some observations on the decomposability of an Ockham algebra [**41**]. For this purpose, we recall that an algebra is said to be *directly decomposable* if it is isomorphic to a direct product of two non-trivial algebras.

Theorem 4.7 *Let L be an Ockham algebra. If $a \in Z(L)$ with $f(a) = a'$ then $a^{\downarrow}$ and $a'^{\downarrow}$ inherit from L the structure of Ockham algebras.*

Proof For every $x \in a^{\downarrow}$ define $f_a(x) = f(x) \wedge a$. Then we have $f_a(0) = a$ and $f_a(a) = 0$. If $u, v \in a^{\downarrow}$ then

$$f_a(u \wedge v) = \big(f(u) \vee f(v)\big) \wedge a = f_a(u) \vee f_a(v),$$

and dually $f_a(u \vee v) = f_a(u) \wedge f_a(v)$. Hence $(a^{\downarrow}; f_a)$ is an Ockham algebra.

Since $f(a) \wedge f^2(a) = 0$ and $f(a) \vee f^2(a) = 1$, we have that $f(a') = f^2(a) = \big(f(a)\big)' = a'' = a$. It follows by the first part that $a'^{\downarrow}$ is also an Ockham algebra. ◇

Theorem 4.8 *An Ockham algebra L is directly decomposable if and only if there exists $a \in Z(L) \setminus \{0,1\}$ such that $f(a) = a'$.*

Proof $\Leftarrow$: Suppose that there exists $a \in Z(L) \setminus \{0,1\}$ such that $f(a) = a'$. By Theorem 4.7, $a^{\downarrow}$ and $a'^{\downarrow}$ are Ockham algebras. Define $h : L \to a^{\downarrow} \times a'^{\downarrow}$ by

$$h(x) = (x \wedge a,\, x \wedge a').$$

It is well known that h is a lattice isomorphism, and since

$$\begin{aligned} f(h(x)) &= \big(f(x \wedge a) \wedge a,\, f(x \wedge a') \wedge a'\big) \\ &= \big([f(x) \vee f(a)] \wedge a,\, [f(x) \vee f(a')] \wedge a'\big) \\ &= \big(f(x) \wedge a,\, f(x) \wedge a'\big) \\ &= h\big(f(x)\big), \end{aligned}$$

it follows that h is an Ockham isomorphism.

$\Rightarrow$: Suppose that $L = U \times V$ with $U, V \neq L$. Then (see, for example, [**1**]) there exists $u, v \in Z(L) \setminus \{0,1\}$ with $U = u^{\downarrow}$ and $V = v^{\downarrow}$. Since $f(u,0) = \big(f_u(u), f_v(0)\big) = (0, v)$, the result follows. ◇

The direct decomposability of an Ockham algebra is sharply reflected in its dual space. We already know that a direct product $L \times M$ of distributive lattices corresponds to a disjoint union $A \mathbin{\dot\cup} B$ in the dual space.

Theorem 4.9 *Let $(L; f)$ be an Ockham algebra and let $(X; g)$ be its dual space. Then L is directly decomposable if and only if X is a disjoint union of two g-subsets.*

Proof Two complementary elements a and a' of L are represented in the dual space X by two clopen down-sets A, B such that $A \cap B = \emptyset$ and $A \cup B = X$. If $f(a) = a'$, i.e. $f(A) = B$, it follows that $X \setminus g^{-1}(A) = B$, hence $g^{-1}(A) = A$. Since $f(a) = a'$ implies $f(a') = a$, we also have that $B = g^{-1}(B)$. Hence, by Lemma 4.1, A and B are g-subsets. ◇

Example 4.9 If we are given the Priestley space

we can make it into the Ockham space of a directly decomposable Ockham algebra by defining g in one of the following ways :

	p	q	r	s	0	1	a	b	c	d	e	f	g	h	i	j	
X_1	q	p	r	s	1	0	j	i	h	g	f	e	d	c	b	a	L_1
X_2	q	q	r	s	1	0	1	i	h	h	f	e	e	c	b	b	L_2
X'_2	p	p	r	s	1	0	i	i	h	f	f	e	c	c	b	0	L'_2
X_3	q	p	s	r	1	0	j	i	e	d	c	h	g	f	b	a	L_3
X_4	q	q	s	r	1	0	1	i	e	e	c	h	h	f	b	b	L_4
X'_4	p	p	s	r	1	0	i	i	e	c	c	h	f	f	b	0	L'_4
X_5	q	q	r	r	1	0	1	i	1	1	i	b	b	0	b	b	L_5
X'_5	p	p	r	r	1	0	i	i	1	i	i	b	0	0	b	0	L'_5
X_6	q	p	r	r	1	0	j	i	1	j	i	b	a	0	b	a	L_6
X_7	q	p	q	s	1	0	j	c	h	g	0	1	j	c	h	g	L_7
X'_7	q	p	p	s	1	0	d	c	h	a	0	1	d	c	h	a	L'_7
X_8	q	q	q	s	1	0	1	c	h	h	0	1	1	c	h	h	L_8
X'_8	p	p	p	s	1	0	c	c	h	0	0	1	c	c	h	0	L'_8
X_9	r	r	p	s	1	0	e	e	h	h	h	i	c	c	f	0	L_9
X'_9	r	r	q	s	1	0	1	e	h	h	h	i	i	c	f	f	L'_9
X_{10}	r	r	r	s	1	0	1	1	h	h	h	c	c	c	0	0	L_{10}
X_{11}	q	q	p	s	1	0	e	c	h	b	0	1	e	c	h	b	L_{11}
X'_{11}	p	p	q	s	1	0	i	c	h	f	0	1	i	c	h	f	L'_{11}

This yields 18 non-isomorphic directly decomposable Ockham algebras, each of which has the Hasse diagram

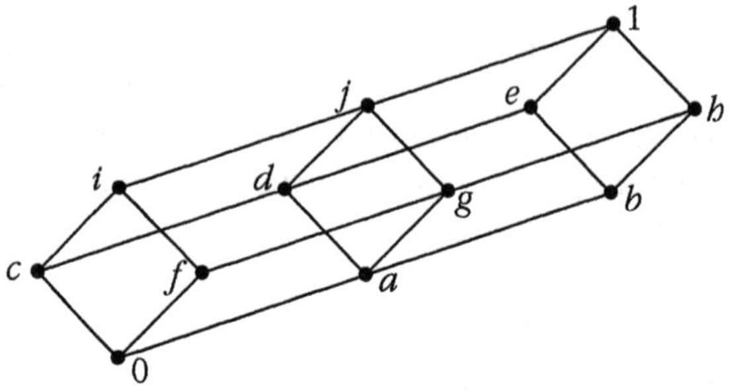

with the corresponding Ockham operation defined as in the table.

5 The lattice of subvarieties

A subclass of a variety **V** which is also a variety is called a *subvariety* of **V**. The subvarieties of **V** form a lattice which we denote by $\Lambda(\mathbf{V})$. In this, the meet $\mathbf{A} \wedge \mathbf{B}$ of two subvarieties $\mathbf{A}, \mathbf{B}$ is their intersection, and the join $\mathbf{A} \vee \mathbf{B}$ is the equational closure of their union, i.e. the smallest subvariety of **V** that contains $\mathbf{A} \cup \mathbf{B}$. A celebrated theorem of B. Jónsson **[15]** states that if **V** is a variety every algebra of which has a distributive congruence lattice then $\Lambda(\mathbf{V})$ is distributive. Applying this to the variety **O** of Ockham algebras, we thus have that $\Lambda(\mathbf{O})$ is distributive. The following three remarkable properties of $\Lambda(\mathbf{O})$ were established by A. Urquhart **[95]** :

(1) $\Lambda(\mathbf{O})$ is uncountable;

(2) every subvariety of **O** is generated by its finite members;

(3) a subvariety of **O** has finite height in $\Lambda(\mathbf{O})$ if and only if it is generated by a finite algebra.

We shall denote by $\Lambda_f(\mathbf{O})$ the set of subvarieties of **O** that are generated by finite algebras. In fact, $\Lambda_f(\mathbf{O})$ is the interesting part of $\Lambda(\mathbf{O})$ since it contains 'at the bottom' the well-known subvarieties

B of boolean algebras;
K of Kleene algebras;
M of de Morgan algebras;
S of Stone algebras, and $\overline{\mathbf{S}}$ of dual Stone algebras;
$\mathbf{P}_{3,1} = \mathbf{K}_{1,1}$ of Ockham algebras with de Morgan skeletons;
$\mathbf{M}_1 = \mathbf{MS}$ of MS-algebras, and $\overline{\mathbf{M}}_1 = \overline{\mathbf{MS}}$ of dual MS-algebras.

Here and throughout what follows we ignore the trivial subvariety.

As a framework for $\Lambda_f(\mathbf{O})$ we may choose either the Berman classes $\mathbf{K}_{p,q}$ $(p \geqslant 1, q \geqslant 0)$ or the classes $\mathbf{P}_{m,n}$ $(m > n \geqslant 0)$. Every finite Ockham algebra is in $\mathbf{K}_{p,q}$ for some p, q and in $\mathbf{P}_{m,n}$ for some m, n. Every $\mathbf{K}_{p,q}$ is a $\mathbf{P}_{m,n}$ (precisely, $\mathbf{K}_{p,q} = \mathbf{P}_{2p+q,q}$) but not conversely.

Since every $\mathbf{P}_{m,n}$ is generated by a single subdirectly irreducible algebra, it is join-irreducible in $\Lambda(\mathbf{O})$.

By comparing the tails and loops in the dual spaces of subdirectly irreducible algebras it is clear that we have

$$\mathbf{P}_{m,n} \subseteq \mathbf{P}_{m',n'} \iff m \leqslant m',\ n \leqslant n',\ m-n \mid m'-n'.$$

In particular, it follows immediately from this that

$$\mathbf{K}_{p,q} \subseteq \mathbf{K}_{p',q'} \iff q \leqslant q',\ p \mid p',$$

a fact that was mentioned in Chapter 2.

If we opt for the $\mathbf{P}_{m,n}$ then the framework at the bottom of $\Lambda_f(\mathbf{O})$ can be depicted as follows :

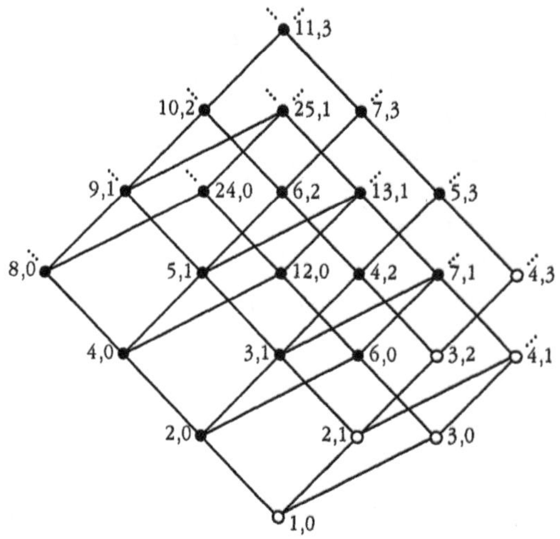

Here the label m,n means $\mathbf{P}_{m,n}$, and the solid circles indicate those that are Berman classes. The $\mathbf{P}_{m,n}$ that are not Berman classes are those for which $m-n$ is odd.

All the classes that we shall consider are defined by identities. For this purpose, we shall generally adopt Urquhart's terminology. Note that we shall very often write $\sim a$ for $f(a)$, particularly where axioms are concerned. A *term* is a polynomial built from variables $a, b, c, \ldots$ and the constants $0, 1$ by means of the operations $\wedge, \vee$ and $\sim$. A term is *atomic* if it is of the form $\sim^n a$ where a is a variable. An atomic term is *even* or *odd* according to whether the exponent n is even or odd. An *inequality* is an expression of the form $A \leqslant B$ where A and B are terms. Two inequalities are *equivalent* if they determine the same equational class in $\mathbf{O}$. An inequality is *basic* if it is of the form $A \leqslant B$ with A a meet of atomic terms and B a join of atomic terms. For example,

$$a \wedge \sim^2 b \leqslant \sim a \vee \sim^3 b \vee c$$

is a basic inequality, as is also

$$\sim a \wedge \sim^2 a = 0.$$

The occurrence of a variable a in an inequality is *positive* if a occurs in an even term of A or in an odd term of B; and the occurrence of a is *negative* if a occurs in an odd term of A or in an even term of B. A basic inequality is *simple* if each variable has precisely one positive and one negative occurrence.

Whereas, for example, the inequality

$$a \wedge b \wedge {\sim}b \wedge {\sim}^2 d \leqslant {\sim}^2 a \vee {\sim}c \vee {\sim}^2 c \vee d$$

is simple, the inequality

$$a \wedge {\sim}^2 b \leqslant {\sim}a \vee {\sim}^2 a \vee b$$

is not.

Clearly, for an inequality to be simple it is necessary that it contain an even number of atomic terms.

The importance of the basic inequalities is emphasised by the following result.

Theorem 5.1 *In* **O** *every inequality is equivalent to a conjunction of basic inequalities.*

Proof Let $A \leqslant B$ be an inequality. By using the equalities

$${\sim}(a \wedge b) = {\sim}a \vee {\sim}b, \qquad {\sim}(a \vee b) = {\sim}a \wedge {\sim}b$$

we can convert A and B into equivalent terms A' and B' in which no lattice operation occurs within the action of any ${\sim}$. By the distributive laws, A' can be put in disjunctive normal form, and likewise B' can be put in conjunctive normal form. The new inequality obtained in this way is equivalent to a conjunction of basic inequalities. ◇

Example 5.1 The inequality

$${\sim}^2(a \vee {\sim}b) \leqslant (a \wedge {\sim}(b \wedge c)) \vee {\sim}a$$

can be written

$${\sim}^2 a \vee {\sim}^3 b \leqslant (a \vee {\sim}a) \wedge ({\sim}a \vee {\sim}b \vee {\sim}\, c)$$

and is equivalent to the conjunction of the four basic inequalities

$${\sim}^2 a \leqslant a \vee {\sim}a, \quad {\sim}^2 a \leqslant {\sim}a \vee {\sim}b \vee {\sim}c,$$
$${\sim}^3 b \leqslant a \vee {\sim}a, \quad {\sim}^3 b \leqslant {\sim}a \vee {\sim}b \vee {\sim}c.$$

The special rôle played by the simple inequalities is illustrated by the following nice property.

Theorem 5.2 [Urquhart] *Let E be a finite subset of $\omega \times \omega$, say*

$$E = \{(p_1, q_1), (p_2, q_2), \ldots, (p_n, q_n)\},$$

and let

$$A = \{\sim^{p_i} x_i \mid p_i \ \textit{even}\} \cup \{\sim^{q_i} x_i \mid q_i \ \textit{odd}\};$$
$$B = \{\sim^{p_i} x_i \mid p_i \ \textit{odd}\} \cup \{\sim^{q_i} x_i \mid q_i \ \textit{even}\},$$

where x_i and x_j are distinct variables for $(p_i, q_i) \neq (p_j, q_j)$.

Consider the basic inequality

$$(\star) \qquad \bigwedge A \leqslant \bigvee B$$

and the universally quantified disjunction

$$(\star\star) \qquad (\forall x) \ \bigvee_i \{g^{p_i}(x) \geqslant g^{q_i}(x) \mid (p_i, q_i) \in E\}.$$

Then $(L; \sim) \in \mathbf{O}$ satisfies $(\star)$ if and only if $(X; g) \in \mathbf{X}$ satisfies $(\star\star)$.

Proof The inequality $(\star)$ fails in L if and only if there are clopen down-sets A_i such that

$$\bigcap_{p_i \text{ even}} \sim^{p_i} A_i \cap \bigcap_{q_i \text{ odd}} \sim^{q_i} A_i \not\subseteq \bigcup_{p_i \text{ odd}} \sim^{p_i} A_i \cup \bigcup_{q_i \text{ even}} \sim^{q_i} A_i,$$

which is the case if and only if there exists $x \in X$ such that

$$x \in \sim^{p_i} A_i \iff p_i \text{ is even},$$
$$x \in \sim^{q_i} A_i \iff q_i \text{ is odd}.$$

Since we have that

$$x \in \sim^{r_i} A_i \iff \begin{cases} g^{r_i}(x) \in A_i & \text{if } r_i \text{ is even;} \\ g^{r_i}(x) \notin A_i & \text{if } r_i \text{ is odd,} \end{cases}$$

the above conditions are equivalent to the existence of down-sets A_i and an element x such that

$$g^{p_i}(x) \in A_i, \quad g^{q_i}(x) \notin A_i.$$

This in turn is equivalent to the existence of an element x with the property that $g^{p_i}(x) \not\geqslant g^{q_i}(x)$ for every i, which is the case if and only if $(\star\star)$ fails in X. ◇

We note that if $A = \emptyset$ then $\bigwedge A = 1$, and if $B = \emptyset$ then $\bigvee B = 0$. Clearly, a basic inequality is simple if and only if it is of the form $(\star)$ in Theorem 5.2.

An important consequence of this is the following, which we have mentioned without proof in Chapter 4.

Corollary *Let $m > n \geqslant 0$ and $(L; \sim) \in \mathbf{O}$. Then*

(1) *when $m - n$ is odd we have*

$$L \in \mathbf{P}_{m,n} \iff (\forall a \in L) \ \sim^m a \wedge \sim^n a = 0, \ \sim^m a \vee \sim^n a = 1;$$

(2) *when $m-n$ is even we have*

$$L \in \mathbf{P}_{m,n} \iff (\forall a \in L)\ \sim^m a = \sim^n a.$$

Proof (1) Consider the equalities $\sim^m a \wedge \sim^n a = 0$ and $\sim^m a \vee \sim^n a = 1$. Without loss of generality we may assume that m is even and n is odd. By Urquhart's theorem the first of these is equivalent to $g^m \geqslant g^n$, and the second is equivalent to $g^n \geqslant g^m$. Their conjunction is equivalent to $g^m = g^n$, i.e. to $L \in \mathbf{P}_{m,n}$.

(2) The equality $\sim^m a = \sim^n a$ is the conjunction of the inequalities $\sim^m a \leqslant \sim^n a$ and $\sim^n a \leqslant \sim^m a$. For m and n of the same parity, these are equivalent by Urquhart's theorem to $g^m \geqslant g^n$ and $g^n \geqslant g^m$, i.e. to $g^m = g^n$. ◇

Urquhart's theorem is extremely powerful. In order to make its impact more transparent, we consider some examples.

Example 5.2 Let $E = \{(0,2),(1,2),(2,0)\}$. We have

$$A = \{\sim^0 a_1, \sim^2 a_3\},$$
$$B = \{\sim^2 a_1, \sim^2 a_2, \sim^0 a_3, \sim^1 a_2\}.$$

Writing a, b, c for a_1, a_2, a_3 respectively, the corresponding simple inequality in the language of $\mathbf{O}$ is

$$(\star) \qquad a \wedge \sim^2 c \leqslant \sim^2 a \vee \sim b \vee \sim^2 b \vee c,$$

and is equivalent in X to the universally quantified disjunction

$$(\star\star) \qquad (\forall x)\ \ x \geqslant g^2(x) \bigvee g(x) \geqslant g^2(x) \bigvee g^2(x) \geqslant x.$$

Example 5.3 Let $E = \{(2,1)\}$. Then $A = \{\sim^2 a, \sim a\}$ and $B = \emptyset$. We have

$$(\star) \qquad \sim^2 a \wedge \sim a = 0,$$

which is equivalent to

$$(\star\star) \qquad (\forall x)\ \ g^2(x) \geqslant g(x).$$

Urquhart's theorem applies only to simple inequalities. Fortunately, however, as shown by Urquhart [**95**, Theorem 2.4], every basic inequality is equivalent to a simple inequality. Hence, if we are given a basic inequality we are able to find its equivalent in the dual space.

Example 5.4 Consider the generalised Kleene identity

$$(K) \qquad a \wedge \sim a \wedge \sim^2 a \leqslant b \vee \sim b \vee \sim^2 b.$$

This is not simple, but is easily seen to be equivalent to

$$(K') \qquad a \wedge {\sim}a \wedge {\sim}c \wedge {\sim}^2 c \leqslant b \vee {\sim}b \vee {\sim}d \vee {\sim}^2 d$$

which is simple and corresponds to the set

$$E = \{(0,1),(1,0),(2,1),(1,2)\}.$$

In fact, it is clear on taking $c = a$ and $d = b$ that (K') implies (K). To obtain the reverse implication, write $a \vee c$ for a and $b \wedge d$ for b in (K) to obtain

$$(a \vee c) \wedge {\sim}a \wedge {\sim}c \wedge ({\sim}^2 a \vee {\sim}^2 c) \leqslant (b \wedge d) \vee {\sim}b \vee {\sim}d \vee ({\sim}^2 b \wedge {\sim}^2 d),$$

which clearly implies (K').

The corresponding universally quantified disjunction in X is

$$(\forall x) \qquad x \geqslant g(x) \bigvee g(x) \geqslant x \bigvee g^2(x) \geqslant g(x) \bigvee g(x) \geqslant g^2(x).$$

Since $x \nparallel g(x)$ implies $g(x) \nparallel g^2(x)$, the equivalent of (K) is

$$(\forall x) \qquad g(x) \nparallel g^2(x).$$

It follows from this that (K) can be simplified, and indeed be replaced by

$$(K'') \qquad {\sim}a \wedge {\sim}^2 a \leqslant {\sim}b \vee {\sim}^2 b.$$

Every equality can be considered as a conjunction of two inequalities, so the procedure can be applied to them also.

Example 5.5 The equality

$$a \vee {\sim}b \vee {\sim}^2 b = {\sim}^2 a \vee {\sim}b \vee {\sim}^2 b$$

is the conjunction of the inequalities

$$a \vee {\sim}b \vee {\sim}^2 b \leqslant {\sim}^2 a \vee {\sim}b \vee {\sim}^2 b \iff a \leqslant {\sim}^2 a \vee {\sim}b \vee {\sim}^2 b \qquad (1)$$

and

$$a \vee {\sim}b \vee {\sim}^2 b \geqslant {\sim}^2 a \vee {\sim}b \vee {\sim}^2 b \iff {\sim}^2 a \leqslant a \vee {\sim}b \vee {\sim}^2 b \qquad (2)$$

Both (1) and (2) are simple and are respectively equivalent to

$$(\forall x) \quad x \geqslant g^2(x) \bigvee g(x) \geqslant g^2(x);$$
$$(\forall x) \quad g^2(x) \geqslant x \bigvee g(x) \geqslant g^2(x).$$

The conjunction of these two disjunctions is

$$(\forall x) \quad x = g^2(x) \bigvee g(x) \geqslant g^2(x).$$

Remark For both visual and typographical reasons, we shall henceforth commit an abuse of notation. Specifically, instead of writing the universally quantified disjunction

$$(\forall x)\ \bigvee_i \{g^{p_i}(x) \geqslant g^{q_i}(x) \mid (p_i, q_i) \in E\}$$

we shall write simply

$$\bigvee_i \{g^{p_i} \geqslant g^{q_i} \mid (p_i, q_i) \in E\}.$$

Thus, for instance, in Example 5.5 we can write

$$\begin{aligned} &(\forall x) \quad x \geqslant g^2(x) \ \bigvee\ g(x) \geqslant g^2(x); \\ &(\forall x) \quad g^2(x) \geqslant x \ \bigvee\ g(x) \geqslant g^2(x). \end{aligned}$$

in the simpler form

$$\begin{aligned} g^0 \geqslant g^2 \ \bigvee\ g \geqslant g^2; \\ g^2 \geqslant g^0 \ \bigvee\ g \geqslant g^2, \end{aligned}$$

the conjunction of which is

$$g^0 = g^2 \ \bigvee\ g \geqslant g^2.$$

In what follows, the reader should bear in mind this abuse of notation.

The 'translation' given by Urquhart's theorem can be done in the opposite direction, in the sense that if we are given the g-inequalities we can find the corresponding $\sim$-inequality.

Example 5.6 Consider (with the above caveat on notation) the disjunction

$$g^0 = g^2 \ \bigvee\ g^0 \geqslant g.$$

This can be written in the form

$$(g^0 \geqslant g^2 \ \bigvee\ g^0 \geqslant g) \wedge (g^2 \geqslant g^0 \ \bigvee\ g^0 \geqslant g),$$

and gives the simple inequalities

$$\begin{aligned} a \wedge b \wedge {\sim}b \leqslant {\sim}^2 a &\iff a \wedge b \wedge {\sim}b \leqslant {\sim}^2 a \wedge b \wedge {\sim}b, \\ {\sim}^2 a \wedge b \wedge {\sim}b \leqslant a &\iff {\sim}^2 a \wedge b \wedge {\sim}b \leqslant a \wedge b \wedge {\sim}b. \end{aligned}$$

The conjunction of these gives the equality

$$a \wedge b \wedge {\sim}b = {\sim}^2 a \wedge b \wedge {\sim}b.$$

With every simple inequality $A \leqslant B$ having $2n$ atomic terms we shall associate a rectangular array having n rows and 4 columns. Reading from left to right, these columns contain

the even p's of A, the odd q's of A, the odd p's of B, the even q's of B.

Each row of the array has only two entries, those of the i-th row being p_i and q_i, and in that order except when both are odd, in which case the order is reversed. We shall call such a rectangular array a *tabulation* and denote its elements by α_{ij}.

By way of illustration, corresponding to Examples 5.2–5.6 above we have the following tabulations :

$$T_{5.2} \; : \; \left| \begin{array}{cccc} 0 & - & - & 2 \\ - & - & 1 & 2 \\ 2 & - & - & 0 \end{array} \right|$$

$$T_{5.3} \; : \; \left| \begin{array}{cccc} 2 & 1 & - & - \end{array} \right|$$

$$T_{5.4} \; : \; \left| \begin{array}{cccc} 0 & 1 & - & - \\ - & - & 1 & 0 \\ 2 & 1 & - & - \\ - & - & 1 & 2 \end{array} \right|$$

$$T_{5.5} \; : \; \left| \begin{array}{cccc} 0 & - & - & 2 \\ - & - & 1 & 2 \end{array} \right| \quad \text{and} \quad \left| \begin{array}{cccc} 2 & - & - & 0 \\ - & - & 1 & 2 \end{array} \right|$$

$$T_{5.6} \; : \; \left| \begin{array}{cccc} 0 & - & - & 2 \\ 0 & 1 & - & - \end{array} \right| \quad \text{and} \quad \left| \begin{array}{cccc} 2 & - & - & 0 \\ 0 & 1 & - & - \end{array} \right|$$

Let T be a tabulation with n rows. If we delete r of these rows (with $0 \leqslant r \leqslant n-1$) we obtain a new tabulation which we call a *subtabulation* of T. In this way we can form $2^n - 1$ subtabulations of T.

If T' is a subtabulation of T then the inequality associated with T' implies that associated with T. We shall say that two tabulations are *equivalent* if they correspond to equivalent inequalities. For example, if we permute two rows of a tabulation T then we obtain an equivalent tabulation.

A tabulation will be called *reducible* if it is equivalent to another tabulation with fewer rows. For example, the tabulation $T_{5.4}$ is reducible; in fact, because $g^0 \nparallel g$ implies $g^2 \nparallel g$, it is equivalent to the tabulation

$$\left| \begin{array}{cccc} 2 & 1 & - & - \\ - & - & 1 & 2 \end{array} \right|$$

A non-reducible tabulation will be called *irreducible*. Note that if an irreducible tabulation contains the ordered pair (m, n) as one of its rows then it cannot contain either $(m-2k, n-2k)$ for any $k \geqslant 1$ or $(n-2k-1, m-2k-1)$ for any $k \geqslant 0$. This is a direct consequence of Urquhart's theorem since each of $g^{m-2k} \geqslant g^{n-2k}$ and $g^{n-2k-1} \geqslant g^{m-2k-1}$ implies $g^m \geqslant g^n$.

A tabulation is *non-trivial* if it is such that

$$(\forall i) \qquad \alpha_{i1} \neq \alpha_{i4}, \quad \alpha_{i2} \neq \alpha_{i3}.$$

Otherwise, the corresponding inequality would be trivial, in the sense that it would contain a term of the form $\sim^n a$ on each side.

The *dual* (n_d) of an inequality (n) is obtained by replacing $\wedge, \vee, \leqslant, 0, 1$ respectively by $\vee, \wedge, \geqslant, 1, 0$. The *dual* of a tabulation T is the tabulation T' obtained from T by rewriting T from right to left; e.g. the dual of $T_{5.2}$ is

$$\begin{vmatrix} 2 & - & - & 0 \\ 2 & 1 & - & - \\ 0 & - & - & 2 \end{vmatrix}$$

and represents the inequality

$$\sim^2 a \wedge \sim b \wedge \sim^2 b \wedge c \leqslant a \vee \sim^2 c.$$

If a tabulation represents a particular inequality then it is clear that the dual tabulation represents the dual inequality. The notions of *self-dual* tabulations and inequalities are obvious. An example is provided by $T_{5.4}$ and the equality it represents.

A simple inequality has the advantage of being suitable for a direct application of Urquhart's theorem. But it has the disadvantage of involving too many atomic terms and variables. The aim of the following result is to remedy this situation.

Theorem 5.3 *Let T be the tabulation corresponding to a simple inequality I. If T has n rows and if the same entry appears r times $(r \geqslant 2)$ in the same column then I can be reduced to an equivalent inequality having $n-r+1$ variables and $2n-r+1$ atomic terms.*

Proof Suppose that the same entry appears r times in the first column of T. Without loss of generality, we may assume that this occurs in the first r rows of T, so that T has the form

$$\begin{vmatrix} p_1 & q_1 & - & - \\ p_2 & - & - & q_2 \\ \vdots & & & \\ p_r & - & - & q_r \\ \vdots & & & \end{vmatrix}$$

with $p_1 = p_2 = \cdots = p_r$. The simple inequality I is then of the form

$$(1) \quad \sim^{p_1} a_1 \wedge \sim^{q_1} a_1 \wedge \sim^{p_1} a_2 \wedge \cdots \wedge \sim^{p_1} a_r \wedge A \leqslant \sim^{q_2} a_2 \vee \cdots \vee \sim^{q_r} a_r \vee B.$$

We claim that this is equivalent to

$$(2) \qquad \sim^{p_1} a_1 \wedge \sim^{q_1} a_1 \wedge A \leqslant \sim^{q_2} a_1 \vee \cdots \vee \sim^{q_r} a_1 \vee B.$$

In fact, if in (1) we replace $a_2, \ldots, a_r$ by a_1 then we obtain (2). Conversely, if in (2) we write $a_1 \wedge a_2 \wedge \cdots \wedge a_r$ for a_1 then we obtain

$$\sim^{p_1} a_1 \wedge \sim^{p_1} a_2 \wedge \cdots \wedge \sim^{p_1} a_r \wedge (\sim^{q_1} a_1 \vee \sim^{q_1} a_2 \vee \cdots \vee \sim^{q_1} a_r) \wedge A$$
$$\leqslant (\sim^{q_2} a_1 \wedge \sim^{q_2} a_2 \wedge \cdots \wedge \sim^{q_2} a_r) \vee \cdots \vee (\sim^{q_r} a_1 \wedge \sim^{q_r} a_2 \wedge \cdots \wedge \sim^{q_r} a_r) \vee B.$$

Since the first member of the latter inequality is greater than or equal to the first member of (1) and the second member of the latter inequality is less than or equal to the second member of (1), the part of T under the r-th row remaining unaltered, we thus see that (2) implies (1).

The same procedure can be applied if an entry appears r times in the third column. As for columns 2 and 4, the only change required in the above proof is the substitution $a_1 \vee a_2 \vee \cdots \vee a_r$ for a_1. ◇

Example 5.7 Corresponding to the set

$$E = \{(0,1), (2,1), (0,2), (3,1), (1,5)\}$$

we have the tabulation

0	1	–	–
2	1	–	–
0	–	–	2
–	1	3	–
–	5	1	–

and the simple inequality

$$a \wedge \sim a \wedge \sim^2 b \wedge \sim b \wedge c \wedge \sim d \wedge \sim^5 e \leqslant \sim^2 c \vee \sim^3 d \vee \sim e.$$

Applying Theorem 5.3 to $\alpha_{11} = \alpha_{31}$ on the one hand, and to $\alpha_{22} = \alpha_{42}$ on the other, we obtain the simplified inequality

$$a \wedge \sim a \wedge \sim^2 b \wedge \sim b \wedge \sim^5 e \leqslant \sim^2 a \vee \sim^3 b \vee \sim e,$$

which has 3 variables and 8 atomic terms.

Alternatively, we could take advantage of the fact that $\alpha_{12} = \alpha_{22} = \alpha_{42}$. This yields the inequality

$$a \wedge \sim a \wedge \sim^2 a \wedge c \wedge \sim^5 e \leqslant \sim^3 a \vee \sim^2 c \vee \sim e,$$

which has the same degree of complexity as the previous simplification.

From now on we shall focus our attention on the class $\mathbf{P}_{3,1}$ and describe all of its subclasses. In $\mathbf{P}_{3,1}$ there are only 6 possible ordered pairs (p,q) involved in the set E, namely

$$(0,1),(1,0),(0,2),(2,0),(1,2),(2,1).$$

The biggest tabulation possible in this case is therefore

0	1	–	–
–	–	1	0
0	–	–	2
2	–	–	0
–	–	1	2
2	1	–	–

We observe immediately that this tabulation is reducible (since $g^0 \not\parallel g$ implies $g^2 \not\parallel g$). Its irreducible subtabulations will yield all of the desired inequalities. To obtain these, we use the following observations.

- There are 6 subtabulations consisting of a single row, but the tabulations

2	1	–	–

and

–	–	1	2

give in $\mathbf{P}_{3,1}$ the same axiom :

$$\sim^2 a \wedge \sim a = 0 \iff \sim a \vee \sim^2 a = 1.$$

- There are 15 subtabulations consisting of two rows, but we must exclude two of them, namely those that contain $(0,1)$ and $(2,1)$, or $(1,0)$ and $(1,2)$, which are clearly reducible.

- There are 20 subtabulations consisting of three rows, 8 of which are reducible.

- There are 15 subtabulations consisting of four rows, 11 of which are reducible.

- All subtabulations having more than four rows are clearly reducible.

It follows from these observations that in $\mathbf{P}_{3,1}$ we can define 34 non-equivalent axioms, and that these axioms involve at most 3 variables. A complete list of these is given below, including the corresponding tabulations and the dual space equivalents. We begin with those that are self-dual, for which we use the Greek letters $(\alpha),\dots,(\eta)$. The 14 axioms (i) that follow, together with their duals (i_d), complete the list. Note that the equality (α), which is the conjunction of (1) and (1_d), is added for sake of completeness.

tabulation	*axiom*	*dual equivalent*
	(α) $a = \sim^2 a$	$g^0 = g^2$
$\begin{vmatrix} 2 & 1 & - & - \end{vmatrix}$	(β) $\sim a \wedge \sim^2 a = 0$	$g = g^2$
$\begin{vmatrix} 0 & 1 & - & - \\ - & - & 1 & 0 \end{vmatrix}$	(γ) $a \wedge \sim a \leqslant b \vee \sim b$	$g \nparallel g^0$
$\begin{vmatrix} 2 & 1 & - & - \\ - & - & 1 & 2 \end{vmatrix}$	(δ) $\sim a \wedge \sim^2 a \leqslant \sim b \vee \sim^2 b$	$g^2 \nparallel g$
$\begin{vmatrix} 0 & - & - & 2 \\ 2 & - & - & 0 \end{vmatrix}$	(ϵ) $a \wedge \sim^2 b \leqslant \sim^2 a \vee b$	$g^2 \nparallel g^0$
$\begin{vmatrix} 0 & 1 & - & - \\ - & - & 1 & 0 \\ 0 & - & - & 2 \\ 2 & - & - & 0 \end{vmatrix}$	(ζ) $a \wedge \sim a \wedge \sim^2 b \leqslant \sim^2 a \vee b \vee \sim b$	$g \nparallel g^0 \ \bigvee \ g^2 \nparallel g^0$
$\begin{vmatrix} 0 & - & - & 2 \\ 2 & - & - & 0 \\ - & - & 1 & 2 \\ 2 & 1 & - & - \end{vmatrix}$	(η) $\sim a \wedge \sim^2 a \wedge b \leqslant a \vee \sim b \vee \sim^2 b$	$g^2 \nparallel g \ \bigvee \ g^2 \nparallel g^0$
$\begin{vmatrix} 0 & - & - & 2 \end{vmatrix}$	(1) $a \leqslant \sim^2 a$	$g^0 \geqslant g^2$
$\begin{vmatrix} - & - & 1 & 0 \end{vmatrix}$	(2) $a \vee \sim a = 1$	$g \geqslant g^0$
$\begin{vmatrix} 0 & - & - & 2 \\ - & - & 1 & 2 \end{vmatrix}$	(3) $a \leqslant \sim a \vee \sim^2 a$	$g^0 \geqslant g^2 \ \bigvee \ g \geqslant g^2$
$\begin{vmatrix} 0 & - & - & 2 \\ 0 & 1 & - & - \end{vmatrix}$	(4) $a \wedge \sim a \leqslant \sim^2 a$	$g^0 \geqslant g^2 \ \bigvee \ g^0 \geqslant g$
$\begin{vmatrix} 0 & 1 & - & - \\ - & - & 1 & 2 \end{vmatrix}$	(5) $a \wedge \sim a \leqslant \sim b \vee \sim^2 b$	$g^0 \geqslant g \ \bigvee \ g \geqslant g^2$
$\begin{vmatrix} 0 & - & - & 2 \\ 2 & 1 & - & - \end{vmatrix}$	(6) $a \wedge \sim b \wedge \sim^2 b \leqslant \sim^2 a$	$g^0 \geqslant g^2 \ \bigvee \ g^2 \geqslant g$

tabulation	axiom	dual equivalent
$\begin{vmatrix} 0 & - & - & 2 \\ - & - & 1 & 0 \end{vmatrix}$	(7) $a \leqslant {\sim^2 a} \vee b \vee {\sim b}$	$g \geqslant g^0 \bigvee g^0 \geqslant g^2$
$\begin{vmatrix} 0 & - & - & 2 \\ 2 & - & - & 0 \\ - & - & 1 & 2 \end{vmatrix}$	(8) $a \wedge {\sim^2 b} \leqslant {\sim a} \vee {\sim^2 a} \vee b$	$g^2 \not\geqslant g^0 \bigvee g \geqslant g^2$
$\begin{vmatrix} 0 & - & - & 2 \\ - & - & 1 & 0 \\ 0 & 1 & - & - \end{vmatrix}$	(9) $a \wedge {\sim a} \leqslant {\sim^2 a} \vee b \vee {\sim b}$	$g \not\geqslant g^0 \bigvee g^0 \geqslant g^2$
$\begin{vmatrix} 0 & - & - & 2 \\ 2 & - & - & 0 \\ - & - & 1 & 0 \end{vmatrix}$	(10) $a \wedge {\sim^2 b}$ $\leqslant {\sim^2 a} \vee b \vee {\sim b}$	$g^2 \not\geqslant g^0 \bigvee g \geqslant g^0$
$\begin{vmatrix} 0 & - & - & 2 \\ 2 & 1 & - & - \\ - & - & 1 & 2 \end{vmatrix}$	(11) $a \wedge {\sim b} \wedge {\sim^2 b}$ $\leqslant {\sim a} \vee {\sim^2 a}$	$g^2 \not\geqslant g \bigvee g^0 \geqslant g^2$
$\begin{vmatrix} 0 & - & - & 2 \\ - & - & 1 & 2 \\ 0 & 1 & - & - \end{vmatrix}$	(12) $a \wedge {\sim a}$ $\leqslant {\sim^2 a} \vee {\sim b} \vee {\sim^2 b}$	$g^0 \geqslant g \bigvee g^0 \geqslant g^2 \bigvee g \geqslant g^2$
$\begin{vmatrix} 0 & - & - & 2 \\ 0 & 1 & - & - \\ - & - & 1 & 2 \\ 2 & - & - & 0 \end{vmatrix}$	(13) $a \wedge {\sim a} \wedge {\sim^2 c}$ $\leqslant {\sim^2 a} \vee {\sim b} \vee {\sim^2 b} \vee c$	$g^0 \not\geqslant g^2 \bigvee g^0 \geqslant g \bigvee g \geqslant g^2$
$\begin{vmatrix} 0 & 1 & - & - \\ 2 & - & - & 0 \\ - & - & 1 & 2 \end{vmatrix}$	(14) $a \wedge {\sim a} \wedge {\sim^2 b}$ $\leqslant b \vee {\sim c} \vee {\sim^2 c}$	$g^0 \geqslant g \bigvee g^2 \geqslant g^0 \bigvee g \geqslant g^2$

These relations, together with the duals $(1_d), \ldots, (14_d)$ can be ordered by logical implication. However, to determine all the implications using only the axioms is rather difficult. For example, though it is clear that (δ) implies (η) it is not at all clear that (ζ) implies (η). Once more, resort to the dual space (either the g-inequalities or the tabulations) greatly helps.

To see for example that (ζ) implies (η), we can proceed as follows : in (ζ) write $\sim a \wedge b$ for a, and $a \vee \sim b$ for b. We obtain

$$A = \sim a \wedge b \wedge (\sim^2 a \vee \sim b) \wedge (\sim^2 a \vee \sim^3 b)$$
$$\leqslant (\sim^3 a \wedge \sim^2 b) \vee a \vee \sim b \vee (\sim a \wedge \sim^2 b) = B.$$

Since $A \geqslant \sim a \wedge \sim^2 a \wedge b$ and $B \leqslant a \vee \sim b \vee \sim^2 b$, (η) follows. In contrast, a glance at the dual equivalents of (ζ) and (η) yields a straightforward proof : the tabulation for (ζ) is a subtabulation of that for (η); equivalently, since g is order-reversing, $g \nparallel g^0$ implies $g \nparallel g^2$.

Ordered by implication, these relations and their duals give the Hasse diagram

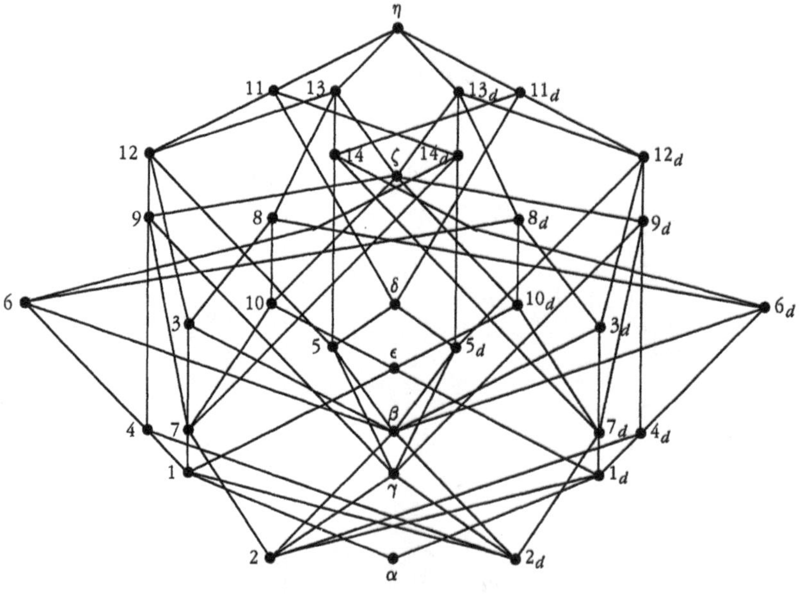

Recall that we have added to the above list the equality (α) which is the conjunction of (1) and (1_d). We write this conjunction in the form $\alpha = (1, 1_d)$. Other conjunctions worthy of note are the following :

$(1, \beta) = (2_d)$, $(1_d, \beta) = (2)$;
$(3, 4) = (1)$, $(3_d, 4_d) = (1_d)$;
$(3, 4_d) = (15)$: $a \vee \sim a = \sim^2 a \vee \sim a$;
$(3_d, 4) = (15_d)$: $a \wedge \sim a = \sim^2 a \wedge \sim a$;
$(3, 6_d) = (16)$: $a \vee \sim b \vee \sim^2 b = \sim^2 a \vee \sim b \vee \sim^2 b$;
$(3_d, 6) = (16_d)$: $a \wedge \sim b \wedge \sim^2 b = \sim^2 a \wedge \sim b \wedge \sim^2 b$.

The dual equivalents of $(15), (15_d), (16), (16_d)$ are respectively

$$g^0 = g^2 \bigvee g \geqslant g^0;$$
$$g^0 = g^2 \bigvee g^0 \geqslant g;$$
$$g^0 = g^2 \bigvee g \geqslant g^2;$$
$$g^0 = g^2 \bigvee g^2 \geqslant g.$$

Less immediate is the following equality :

$$(9, 12_d) = (17) \ : \ (a \wedge \sim a) \vee b \vee \sim b = (\sim^2 a \wedge \sim a) \vee b \vee \sim b.$$

To establish this, observe that clearly (17) implies (9), and also implies

$$a \vee b \vee \sim b \geqslant \sim^2 a \wedge \sim a,$$

which is readily seen to be equivalent to (12_d). Conversely, (9) gives

$$(a \wedge \sim a) \vee b \vee \sim b \leqslant \sim^2 a \vee b \vee \sim b,$$

and since clearly

$$(a \wedge \sim a) \vee b \vee \sim b \leqslant \sim a \vee b \vee \sim b,$$

we have, by distributivity,

$$(\star) \qquad (a \wedge \sim a) \vee b \vee \sim b \leqslant (\sim^2 a \wedge \sim a) \vee b \vee \sim b.$$

Now (12_d) yields

$$(\sim^2 a \wedge \sim a) \vee b \vee \sim b \leqslant a \vee b \vee \sim b,$$

and since clearly

$$(\sim^2 a \wedge \sim a) \vee b \vee \sim b \leqslant \sim a \vee b \vee \sim b,$$

we obtain

$$(\star\star) \quad (\sim^2 a \wedge \sim a) \vee b \vee \sim b \leqslant (a \wedge \sim a) \vee b \vee \sim b.$$

Finally, $(\star)$ and $(\star\star)$ together give (17).

It follows that the dual equivalent of (17) is

$$(g \nparallel g^0 \bigvee g^0 \geqslant g^2) \bigwedge (g \geqslant g^0 \bigvee g^2 \geqslant g^0 \bigvee g^2 \geqslant g)$$
$$= g \geqslant g^0 \bigvee [(g \leqslant g^0 \bigvee g^0 \geqslant g^2) \bigwedge (g^2 \geqslant g^0 \bigvee g^2 \geqslant g)]$$
$$= g \geqslant g^0 \bigvee g \leqslant g^0 \bigvee [g^0 \geqslant g^2 \bigwedge (g^2 \geqslant g^0 \bigvee g^2 \geqslant g)] \ \ (g \leqslant g^0 \Rightarrow g^2 \geqslant g)$$
$$= g \nparallel g^0 \bigvee g^0 = g^2 \bigvee g^0 \geqslant g^2 \geqslant g$$
$$= g \nparallel g^0 \bigvee g^0 = g^2.$$

This shows that (17) is self-dual and is implied by each of the axioms (α) and (γ). In fact, it can be shown that $(17) = (15, 15_d)$.

In an exactly similar way we can prove that

$$(12, 11_d) = (18) \; : \; (a \wedge {\sim}a) \vee {\sim}b \vee {\sim}^2 b = ({\sim}^2 a \wedge {\sim}a) \vee {\sim}b \vee {\sim}^2 b,$$

and hence also

$$(12_d, 11) = (18_d) \; : \; (a \vee {\sim}a) \wedge {\sim}b \wedge {\sim}^2 b = ({\sim}^2 a \vee {\sim}a) \wedge {\sim}b \wedge {\sim}^2 b.$$

The dual equivalents of (18) and (18_d) are respectively

$$g^2 = g^0 \bigvee g \geqslant g^2 \bigvee g^0 \geqslant g,$$
$$g^2 = g^0 \bigvee g^2 \geqslant g \bigvee g \geqslant g^0.$$

In Chapter 4 we determined the 19 non-trivial subdirectly irreducible algebras of the class $\mathbf{P}_{3,1}$, as well as their dual spaces. Using the preceding list of inequalities (and mainly their dual equivalents) we can easily give an equational basis for each of the classes generated by these 19 algebras. For notational convenience, we shall use the symbol ($\tilde{n}$) to denote the conjunction of an inequality (n) with its dual (n_d). Similarly, for every subvariety $\mathbf{A}$ we shall use the notation $\widetilde{\mathbf{A}}$ for $\mathbf{A} \vee \overline{\mathbf{A}}$.

Since $\mathbf{P}_{3,1}$ (and, more generally, $\mathbf{O}$) is congruence distributive, we can apply the following fundamental theorem of B. A. Davey **[62]** and, at least in theory, determine $\Lambda(\mathbf{P}_{3,1})$.

Theorem 5.4 *Let* $\mathbf{K} = \mathrm{Var}\, S$ *be a congruence-distributive variety generated by a finite set* S *of finite algebras, and order the set* $\mathrm{Si}(\mathbf{K})$ *of subdirectly irreducible algebras in* $\mathbf{K}$ *by*

$$A \leqslant B \iff A \textit{ is a homomorphic image of a subalgebra of } B.$$

Then $\Lambda(\mathbf{K})$ *is a finite distributive lattice and is isomorphic to* $\mathcal{O}(\mathrm{Si}(\mathbf{K}))$. *Moreover,* $\mathbf{A}$ *is a join-irreducible element of* $\Lambda(\mathbf{K})$ *if and only if* $\mathbf{A} = \mathrm{Var}\, A$ *for some* $A \in \mathrm{Si}(\mathbf{K})$. ◇

In order to apply this result to the variety $\mathbf{P}_{3,1} = \mathrm{Var}\, B_1$, we observe that if $B \in \mathrm{Si}(\mathbf{P}_{3,1})$ then every homomorphic image of every subalgebra of B is isomorphic to a subalgebra of B. Indeed, if C is a subalgebra of $B \in \mathrm{Si}(\mathbf{P}_{3,1})$ then by Corollary 1 of Theorem 3.14 we have $C \in \mathrm{Si}(\mathbf{P}_{3,1})$ and so, if A is a homomorphic image of C then by Theorem 3.14 we have $A \simeq C/\varphi$ where $\varphi \in \{\omega, \Phi, \iota\}$. Consequently, $A \simeq C$ or $A \simeq f(C)$ or A is trivial; i.e. A is isomorphic to a subalgebra of C. It follows that we can order $\mathrm{Si}(\mathbf{P}_{3,1})$ by

$$A \leqslant B \iff A \text{ is isomorphic to a subalgebra of } B,$$

and thus obtain the Hasse diagram

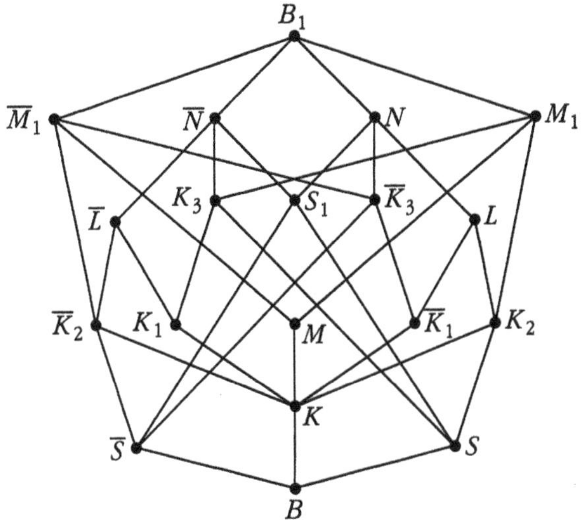

Equational bases for the corresponding varieties are the following :

$\mathbf{B}$	$(\tilde{2})$		
$\mathbf{K}$	(α, γ)		
$\mathbf{M}$	(α)		
$\mathbf{S}_1$	(β)		
$\mathbf{S}$	(2_d)	$\overline{\mathbf{S}}$	(2)
$\mathbf{K}_1$	$(1, 4_d, \gamma)$	$\overline{\mathbf{K}}_1$	$(1_d, 4, \gamma)$
$\mathbf{K}_2$	$(1, 3_d, \gamma)$	$\overline{\mathbf{K}}_2$	$(1_d, 3, \gamma)$
$\mathbf{K}_3$	$(1, 5, 6_d)$	$\overline{\mathbf{K}}_3$	$(1_d, 5_d, 6)$
$\mathbf{L}$	$(3_d, 4, \gamma) = (15_d, \gamma)$	$\overline{\mathbf{L}}$	$(3, 4_d, \gamma) = (15, \gamma)$
$\mathbf{N}$	$(3_d, 5_d, 6) = (5_d, 16_d)$	$\overline{\mathbf{N}}$	$(3, 5, 6_d) = (5, 16)$
$\mathbf{M}_1$	(1)	$\overline{\mathbf{M}}_1$	(1_d)

The task of determining $\Lambda(\mathbf{P}_{3,1})$ is rather a daunting one. Even the cardinality of this lattice is not easy to find, though in principle it can be determined by use of the following ingenious result of Berman and Köhler **[29]**.

Theorem 5.5 *Let F be a finite ordered set and let $\mathcal{O}(F)$ be the lattice of down-sets of F. Then, for every $x \in F$,*

$$|\mathcal{O}(F)| = |\mathcal{O}(F \setminus \{x\})| + |\mathcal{O}(F \setminus C_x)|. \quad \diamond$$

Although less daunting in nature, this problem requires several applications of the above result, so much so that paper and pencil calculations become unwieldy. Deference to the computer program mentioned in **[29]** would be in order. This, of course, would tell us only the size of the subvariety lattice and not what it looks like. In what follows we shall determine both by using judiciously chosen subsets of the ordered set $\mathrm{Si}(\mathbf{P}_{3,1})$.

First step We begin by determining $\Lambda(\mathbf{MS})$. Applying Davey's theorem to the down-set $M_1^{\downarrow}$ we obtain the lattice

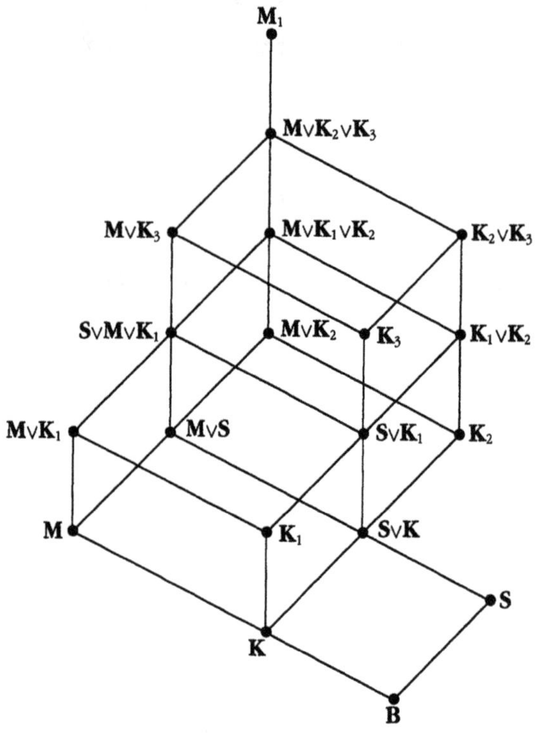

To obtain equational bases for each of the subvarieties of **MS**, we first determine those of the $\wedge$-irreducible elements of $\Lambda(\mathbf{MS})$ by applying the following obvious principle :

$(a_1, a_2, \ldots, a_r)$ is an equational basis of $\mathbf{V}_1 \vee \mathbf{V}_2 \vee \cdots \vee \mathbf{V}_s$ if and only if only those subdirectly irreducible algebras contained in $\mathbf{V}_1$ or $\mathbf{V}_2$ or ... or $\mathbf{V}_s$ satisfy $a_1, a_2, \ldots, a_r$.

We can then use these equational bases to obtain those of all the subvarieties in the lattice.

Favourably, we can exploit the following idea. Since we have the complete list of implications between all the axioms in $\mathbf{P}_{3,1}$ (up to equivalence), if we are given the equational bases of subvarieties $\mathbf{S}_1, \mathbf{S}_2, \ldots, \mathbf{S}_n$ then an equational basis of $\mathbf{S}_1 \vee \mathbf{S}_2 \vee \cdots \vee \mathbf{S}_n$ is obtained by taking the intersection of the sets of axioms satisfied by $\mathbf{S}_1, \mathbf{S}_2, \ldots, \mathbf{S}_n$ and then deleting those axioms that are consequences of others.

We illustrate this procedure with some examples.

Consider for instance the subvariety $\mathbf{M}\vee\mathbf{K}_2\vee\mathbf{K}_3$. We know that equational bases for these three subvarieties are, respectively,

$$(\alpha), \quad (\gamma, 1, 3_d), \quad (1, 5, 6_d).$$

It follows that they satisfy the following sets of axioms :

$$A = \alpha^{\uparrow}, \quad B = \gamma^{\uparrow} \cup 1^{\uparrow} \cup 3_d^{\uparrow}, \quad C = 1^{\uparrow} \cup 5^{\uparrow} \cup 6_d^{\uparrow}$$

where $n^{\uparrow}$ denotes the up-set generated by axiom (n) in the ordered set of implications. Now since

$$\begin{aligned} A \cap B &= (\alpha^{\uparrow} \cap \gamma^{\uparrow}) \cup (\alpha^{\uparrow} \cap 1^{\uparrow}) \cup (\alpha^{\uparrow} \cap 3_d^{\uparrow}) \\ &= 9^{\uparrow} \cup 9_d^{\uparrow} \cup 1^{\uparrow} \cup 3_d^{\uparrow} \end{aligned}$$

an equational basis of $\mathbf{M} \vee \mathbf{K}_2$ is $(1, 3_d, 9, 9_d)$. But, using the g-inequalities, we see that $(1, 3_d)$ implies both (9) and (9_d). So an equational basis of $\mathbf{M}\vee\mathbf{K}_2$ is $(1, 3_d)$.

Now we consider

$$\begin{aligned} (1^{\uparrow} \cup 3_d^{\uparrow}) \cap (1^{\uparrow} \cup 5^{\uparrow} \cup 6_d^{\uparrow}) &= 1^{\uparrow} \cup (3_d^{\uparrow} \cap 5^{\uparrow}) \cup (3_d^{\uparrow} \cap 6_d^{\uparrow}) \\ &= 1^{\uparrow} \cup 11_d^{\uparrow}. \end{aligned}$$

It follows that an equational basis of $\mathbf{M} \vee \mathbf{K}_2 \vee \mathbf{K}_3$ is $(1, 11_d)$.

Similarly, we can compute an equational basis for $\mathbf{M} \vee \mathbf{K}_3$. In fact, we know that equational bases for $\mathbf{M}$ and $\mathbf{K}_3$ are, respectively,

$$(\alpha), \quad (1, 5, 6_d).$$

Applying the above procedure, we consider

$$\begin{aligned} \alpha^{\uparrow} \cap (1^{\uparrow} \cup 5^{\uparrow} \cup 6_d^{\uparrow}) &= (\alpha^{\uparrow} \cap 1^{\uparrow}) \cup (\alpha^{\uparrow} \cap 5^{\uparrow}) \cup (\alpha^{\uparrow} \cap 6_d^{\uparrow}) \\ &= 1^{\uparrow} \cup 6_d^{\uparrow}, \end{aligned}$$

from which we see that an equational basis for $\mathbf{M} \vee \mathbf{K}_3$ is $(1, 6_d)$.

Using these observations, we can now establish an equational basis for

$$\mathbf{M} \vee \mathbf{S} = (\mathbf{M} \vee \mathbf{K}_2) \wedge (\mathbf{M} \vee \mathbf{K}_3).$$

Clearly, this is $(1, 3_d, 6_d)$.

In listing the equational bases for the varieties of **MS** we make a slight digression and compare our results with those obtained by M. E. Adams and H. A. Priestley in **[22]**. These authors present a 'purely algorithmic approach to the generation of equational bases' for certain subvarieties of distributive lattice ordered algebras, especially Ockham algebras. The identities they obtain by this algorithmic method are of a canonical form, being basic inequalities that involve the minimum possible number of variables and 'can be read off from the dual spaces of the free algebras'. The computation of the free algebras themselves is not necessary. Unfortunately, the theory involves not only the distributive lattice duality that we used intensively in Chapter 4, but also natural dualities based on schizophrenic objects. It is therefore impossible to give here an account of all the deep ideas contained in this long paper which we commend to the interested reader.

Adams and Priestley apply their method to the class **MS**, regarding all varieties as lying in $\mathbf{P}_{3,1}$. For this they use the 7 axioms listed on the left of the following table (the numbering is as in **[22]**). On the right of the table are our equivalents. Here the reader will observe that an inequality can appear in many forms, sometimes misleading.

1	$a \vee \sim a \vee \sim^2 a = 1$	β	$\sim a \wedge \sim^2 a = 0$
2	$\sim a \wedge \sim^2 a \leqslant a$	3_d	$\sim a \wedge \sim^2 a \leqslant a$
4	$\sim^2 a \leqslant a \vee \sim a$	4_d	$\sim^2 a \leqslant a \vee \sim a$
12	$\sim b \wedge \sim^2 b \leqslant a \vee \sim a \vee \sim^2 a \vee b$	11_d	$\sim a \wedge \sim^2 a \leqslant a \vee \sim b \vee \sim^2 b$
13	$b \wedge \sim b \wedge \sim^2 b \leqslant a \vee \sim a \vee \sim^2 a$	δ	$\sim a \wedge \sim^2 a \leqslant \sim b \vee \sim^2 b$
14	$\sim^2 b \leqslant a \vee \sim a \vee \sim^2 a \vee b \vee \sim b$	6_d	$\sim^2 a \leqslant a \vee \sim b \vee \sim^2 b$
24	$\sim a \wedge \sim^2 a \wedge \sim^2 b \leqslant a \vee b \vee \sim b$	12_d	$\sim a \wedge \sim^2 a \wedge \sim^2 b \leqslant b \vee \sim b$

The table that follows compares the equational bases obtained on the one hand by Adams and Priestley and on the other by ourselves. That these axiomatics are equivalent can be shown as follows. We know already the implications

$$(6_d) \Rightarrow (11_d), \quad (12_d) \Rightarrow (11_d), \quad (\delta) \Rightarrow (11_d), \quad (\beta) \Rightarrow (3_d).$$

Moreover, we have the following equalities :

$$(1,\delta) = (1,5), \quad (1,\beta) = (2_d), \quad (1,12_d,\delta) = (1,\gamma), \quad (1,3_d,\delta) = (1,3_d,\gamma),$$
$$(1,4_d,\delta) = (1,4_d,\gamma), \quad (\alpha,\gamma) = (\alpha,\delta), \quad (\alpha,\beta) = (\widetilde{2}).$$

By way of illustration, we show that $(1,12_d,\delta)=(1,\gamma)$. For this purpose, we observe that since (γ) implies both (12_d) and (δ) it suffices to prove that $(1,12_d,\delta)$ implies (γ). By (12_d) we have to consider three cases :

(a) $g \geqslant g^0$: clearly, (γ) is verified;

(b) $g^2 \geqslant g^0$: since by (1) we have $g^0 \geqslant g^2$ the equality $g^0 = g^2$ holds and then (δ) yields (γ);

(c) $g^2 \geqslant g$: using (1) we obtain $g^0 \geqslant g^2 \geqslant g$ and (γ) is satisfied.

Using these observations we can establish the equivalence of the sets of axioms given in the following table.

MS-*subvariety*	*Adams–Priestley*	*Blyth–Varlet*
$\mathbf{M}_1$	(1)	(1)
$\mathbf{M} \vee \mathbf{K}_2 \vee \mathbf{K}_3$	$(1,11_d)$	$(1,11_d)$
$\mathbf{M} \vee \mathbf{K}_3$	$(1,11_d,6_d)$	$(1,6_d)$
$\mathbf{M} \vee \mathbf{K}_1 \vee \mathbf{K}_2$	$(1,11_d,12_d)$	$(1,12_d)$
$\mathbf{K}_2 \vee \mathbf{K}_3$	$(1,11_d,\delta)$	$(1,5)$
$\mathbf{M} \vee \mathbf{K}_2$	$(1,3_d)$	$(1,3_d)$
$\mathbf{M} \vee \mathbf{K}_1$	$(1,4_d)$	$(1,4_d)$
$\mathbf{S}$	$(1,\beta,3_d)$	(2_d)
$\mathbf{S} \vee \mathbf{M} \vee \mathbf{K}_1$	$(1,11_d,6_d,12_d)$	$(1,6_d,12_d)$
$\mathbf{K}_3$	$(1,11_d,\delta,6_d)$	$(1,5,6_d)$
$\mathbf{K}_1 \vee \mathbf{K}_2$	$(1,11_d,\delta,12_d)$	$(1,\gamma)$
$\mathbf{M} \vee \mathbf{S}$	$(1,3_d,6_d)$	$(1,3_d,6_d)$
$\mathbf{S} \vee \mathbf{K}_1$	$(1,11_d,\delta,6_d,12_d)$	$(1,6_d,\gamma)$
$\mathbf{K}_2$	$(1,3_d,\delta)$	$(1,3_d,\gamma)$
$\mathbf{M}$	$(1,3_d,4_d)$	(α)
$\mathbf{K}_1$	$(1,4_d,\delta)$	$(1,4_d,\gamma)$
$\mathbf{S} \vee \mathbf{K}$	$(1,3_d,\delta,6_d)$	$(1,3_d,6_d,\gamma)$
$\mathbf{K}$	$(1,3_d,4_d,\delta)$	(α,γ)
$\mathbf{B}$	$(1,\beta,3_d,4_d)$	$(\widetilde{2})$

We may conclude that the two axiomatics are equivalent. Of course, the Adams–Priestley axioms are neither independent nor minimal, but their objective is different from ours in that it is more general and practical and in particular enables computer assistance.

Second step We determine $\Lambda(\widetilde{\mathbf{L}})$. The ordered set of subdirectly irreducibles in $\widetilde{\mathbf{L}}$ is

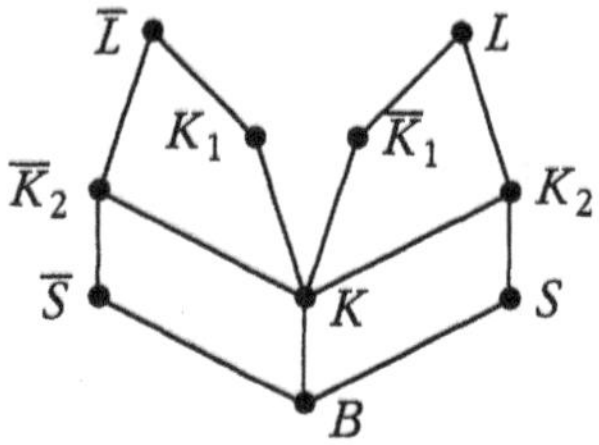

By Theorem 5.5 the number of down-sets of this ordered set is 53, which is the size of $\Lambda(\widetilde{\mathbf{L}})$. Using Davey's theorem we can construct this lattice :

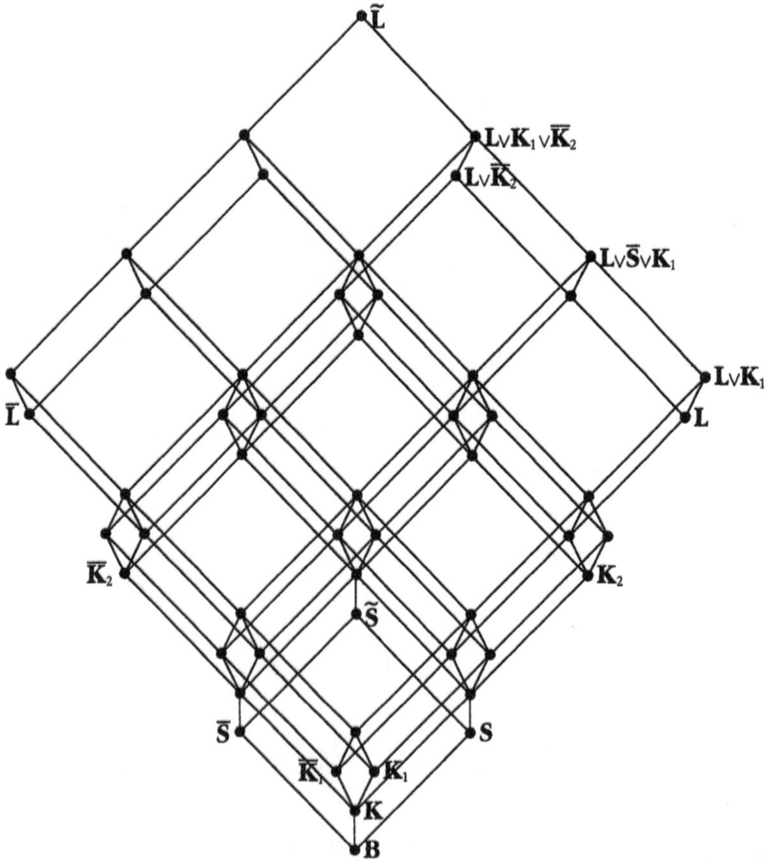

Observe that the south-east and south-west faces consist of **MS**-algebras and dual **MS**-algebras respectively.

Equational bases for the 10 $\wedge$-irreducible elements of $\Lambda(\widetilde{\mathbf{L}})$ can be found easily, and are as follows.

$\widetilde{\mathbf{L}}$	(γ)		
$\mathbf{L} \vee \mathbf{K}_1 \vee \overline{\mathbf{K}}_2$	$(\gamma, 8_d)$	$\overline{\mathbf{L}} \vee \overline{\mathbf{K}}_1 \vee \mathbf{K}_2$	$(\gamma, 8)$
$\mathbf{L} \vee \overline{\mathbf{K}}_2$	$(\gamma, 3_d)$	$\overline{\mathbf{L}} \vee \mathbf{K}_2$	$(\gamma, 3)$
$\mathbf{L} \vee \overline{\mathbf{S}} \vee \mathbf{K}_1$	$(\gamma, 6)$	$\overline{\mathbf{L}} \vee \mathbf{S} \vee \overline{\mathbf{K}}_1$	$(\gamma, 6_d)$
$\mathbf{L} \vee \mathbf{K}_1$	$(\gamma, 4)$	$\overline{\mathbf{L}} \vee \overline{\mathbf{K}}_1$	$(\gamma, 4_d)$
$\widetilde{\mathbf{S}}$	(β, γ)		

Using the fact that every element of $\Lambda(\widetilde{\mathbf{L}})$ is a meet of $\wedge$-irreducible elements, equational bases for the remaining subvarieties in this lattice can now be obtained from the above.

Third step We can use the lattice $\Lambda(\widetilde{\mathbf{L}})$ to construct the lattice $\Lambda(\widetilde{\mathbf{N}})$. In fact, to obtain the latter we have to add (as $\vee$-irreducible elements) to the former the subvarieties generated by $K_3, \overline{K}_3, S_1, N, \overline{N}$. This we do in two stages. First we add the subvarieties

$$\mathbf{K}_3,\ \overline{\mathbf{K}}_3$$

to obtain the 102-element lattice $\Lambda(\widetilde{\mathbf{L}} \vee \widetilde{\mathbf{K}}_3)$ which is shown on page 98. We then extend this lattice by adding (as $\vee$-irreducibles) the subvarieties

$$\mathbf{S}_1,\ \mathbf{N},\ \overline{\mathbf{N}}.$$

This is achieved by adding the 82-element lattice which is shown on page 99. By pasting the latter as a second layer projecting down onto the former, with $\mathbf{S}_1$ directly above $\widetilde{\mathbf{S}}$, we obtain the lattice $\Lambda(\widetilde{\mathbf{N}})$. It therefore has 184 elements.

Equational bases for the 15 $\wedge$-irreducible elements of $\Lambda(\widetilde{\mathbf{N}})$ are as follows.

$\widetilde{\mathbf{N}}$	(δ)		
$\mathbf{N} \vee \overline{\mathbf{L}} \vee \mathbf{K}_3$	$(\delta, 13_d)$	$\overline{\mathbf{N}} \vee \mathbf{L} \vee \overline{\mathbf{K}}_3$	$(\delta, 13)$
$\mathbf{N} \vee \overline{\mathbf{L}}$	(5_d)	$\overline{\mathbf{N}} \vee \mathbf{L}$	(5)
$\mathbf{N} \vee \overline{\mathbf{K}}_2 \vee \mathbf{K}_3$	$(\delta, 8_d)$	$\overline{\mathbf{N}} \vee \mathbf{K}_2 \vee \overline{\mathbf{K}}_3$	$(\delta, 8)$
$\mathbf{N} \vee \overline{\mathbf{K}}_2$	$(\delta, 3_d)$	$\overline{\mathbf{N}} \vee \mathbf{K}_2$	$(\delta, 3)$
$\mathbf{N} \vee \mathbf{K}_3$	$(\delta, 6)$	$\overline{\mathbf{N}} \vee \overline{\mathbf{K}}_3$	$(\delta, 6_d)$
$\mathbf{S}_1$	(β)		
$\widetilde{\mathbf{L}} \vee \widetilde{\mathbf{K}}_3$	(δ, ζ)		
$\mathbf{L} \vee \mathbf{K}_3$	$(\delta, 4)$	$\overline{\mathbf{L}} \vee \overline{\mathbf{K}}_3$	$(\delta, 4_d)$

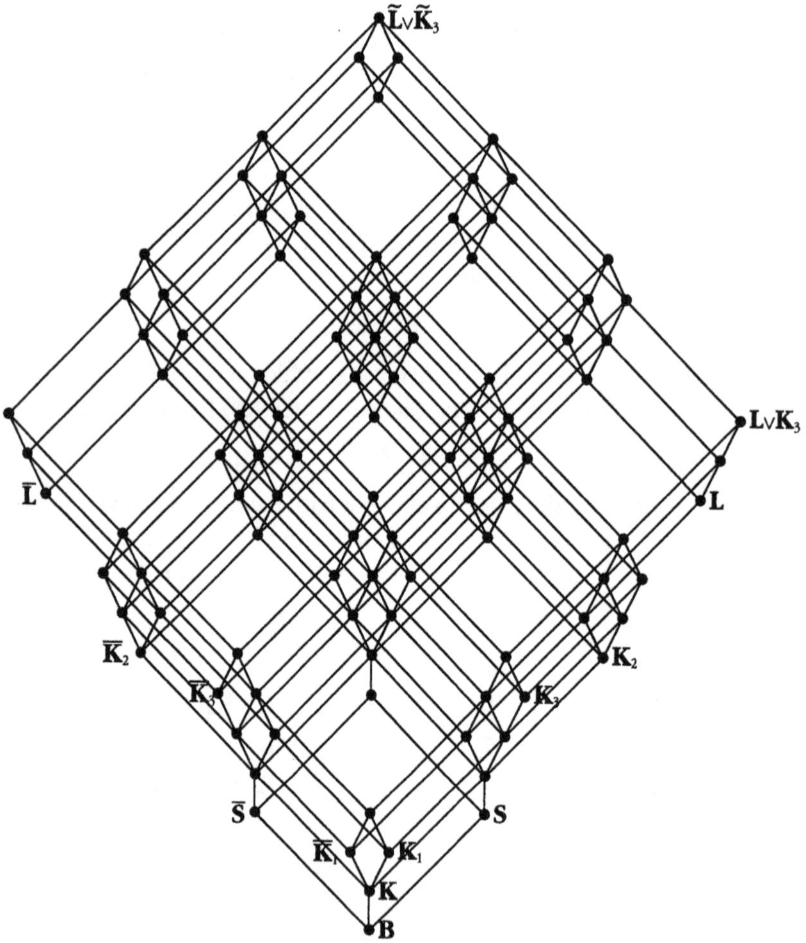
$\widetilde{L} \vee \widetilde{K}_3$
$L \vee K_3$
$\overline{L}$
L
$\overline{K}_2$
K_2
$\overline{K}_3$
K_3
$\overline{S}$
S
$\overline{K}_1$
K_1
K
B

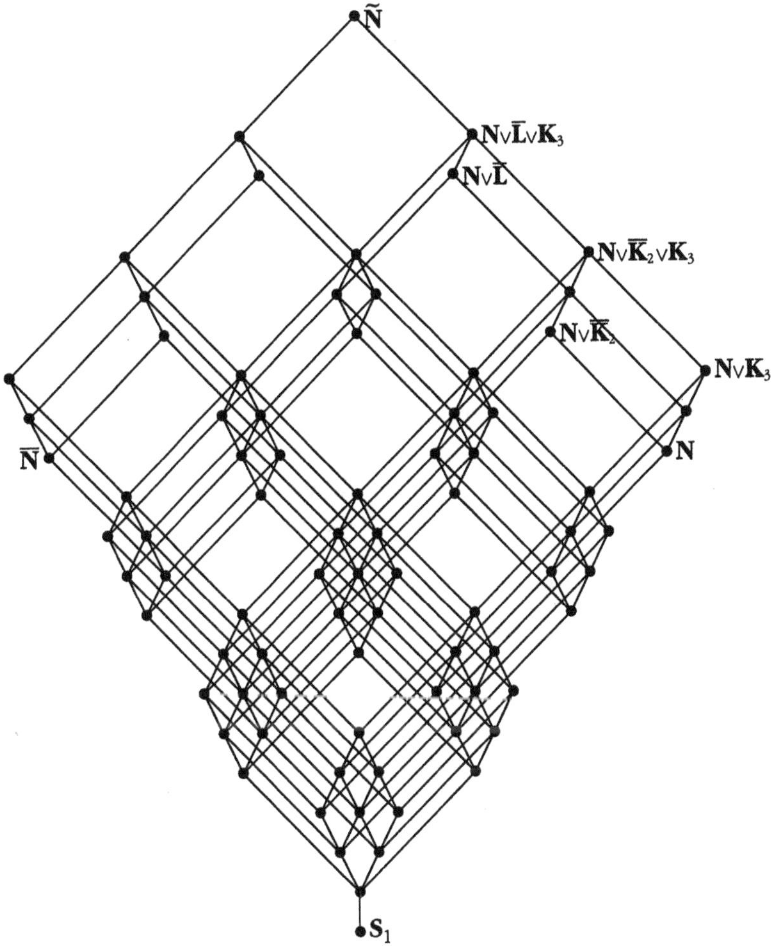
$\tilde{N}$
$N \vee \bar{L} \vee K_3$
$N \vee \bar{L}$
$N \vee \bar{\bar{K}}_2 \vee K_3$
$N \vee \bar{\bar{K}}_2$
$N \vee K_3$
$\bar{N}$
N
S_1

Fourth step Finally, we can obtain the lattice $\Lambda(\mathbf{P}_{3,1}) = \Lambda(\mathbf{B}_1)$ from the lattice $\Lambda(\widetilde{\mathbf{N}})$ by adding to the latter (as $\vee$-irreducible elements) the varieties

$$\mathbf{M},\ \mathbf{M}_1,\ \overline{\mathbf{M}}_1,\ \mathbf{B}_1.$$

We first add **M** (directly above **K**), in so doing obtaining a copy of the entire double layer lying above **K**. This adds another 180 elements.

Next, on the new top level, we add $\mathbf{M}_1$ and $\overline{\mathbf{M}}_1$ (with $\mathbf{M}_1$ directly above $\mathbf{M} \vee \mathbf{K}_2 \vee \mathbf{K}_3$). This produces another 97 elements.

Finally, we top everything off with $\mathbf{B}_1$. We conclude that, excluding as usual the trivial subvariety, $|\Lambda(\mathbf{P}_{3,1})| = 462$. (This corrects an error in **[48]**.)

Now we consider a type of subvariety that is particularly interesting in the context of Ockham algebras. A subvariety will be called *self-dual* if it contains with each of its algebras the order-duals of these algebras. Thus, for example, **B**, **K**, **M**, $\widetilde{\mathbf{S}}$ are self-dual, as is every subvariety of the form $\widetilde{\mathbf{A}} = \mathbf{A} \vee \overline{\mathbf{A}}$ for $\mathbf{A} \in \Lambda(\mathbf{O})$. The self-dual subvarieties form the spine of the lattice of subvarieties, in the sense that they constitute the axis of symmetry of this lattice. Moreover, the self-dual subvarieties form a sublattice of $\Lambda(\mathbf{O})$.

In $\mathbf{P}_{3,1}$ the self-dual subvarieties have equational bases that are conjunctions of the self-dual axioms $(\alpha), (\beta), \dots, (\eta)$ and $(\widetilde{n})$ for $n = 1, \dots, 14$. The dual equivalents of the $(\widetilde{n})$ are

$$
\begin{array}{rl}
(\alpha) = (\widetilde{1}) & g^0 = g^2 \\
(\widetilde{2}) & g = g^0 \\
(\widetilde{3}) & g^0 = g^2 \ \bigvee\ g = g^2 \ \bigvee\ g^0 \geqslant g^2 \geqslant g \ \bigvee\ g \geqslant g^2 \geqslant g^0 \\
(\widetilde{4}) & g^0 = g^2 \ \bigvee\ g^2 \geqslant g^0 \geqslant g \ \bigvee\ g \geqslant g^0 \geqslant g^2 \\
(\widetilde{5}) & g = g^2 \ \bigvee\ g \nparallel g^0 \\
(\widetilde{6}) & g^0 = g^2 \ \bigvee\ g = g^2 \ \bigvee\ \substack{g^0 \\ g} \geqslant g^2 \ \bigvee\ g^2 \geqslant \substack{g^0 \\ g} \\
(\widetilde{7}) & g^0 = g^2 \ \bigvee\ g^0 \geqslant g^2 \geqslant g \ \bigvee\ g \geqslant g^2 \geqslant g^0 \\
(\widetilde{8}) & g = g^2 \ \bigvee\ g^2 \nparallel g^0 \\
(\widetilde{9}) & g^0 = g^2 \ \bigvee\ g \nparallel g^0 \\
(\epsilon) = (\widetilde{10}) & g^2 \nparallel g^0 \\
(\widetilde{11}) & g^0 = g^2 \ \bigvee\ g^2 \nparallel g \\
(\widetilde{12}) & g^0 = g^2 \ \bigvee\ g = g^2 \ \bigvee\ g \nparallel g^0 \\
(\widetilde{13}) & g = g^2 \ \bigvee\ g^2 \nparallel g^0 \ \bigvee\ g \nparallel g^0 \\
(\widetilde{14}) & g^0 = g^2 \ \bigvee\ g = g^2 \ \bigvee\ g \nparallel g^0 \ \bigvee\ \substack{g^0 \\ g} \geqslant g^2 \ \bigvee\ g^2 \geqslant \substack{g^0 \\ g}
\end{array}
$$

If (n) implies (r) or (r_d) then $(\widetilde{n})$ implies $(\widetilde{r})$; but this condition is not necessary for $(\widetilde{n})$ to imply $(\widetilde{r})$, as is shown by the implication $(\widetilde{4}) \Rightarrow (\epsilon) = (\widetilde{10})$

which holds although neither (4) nor (4_d) implies (10). Similarly, we have $(\tilde{2}) \Rightarrow (\alpha)$, $(\tilde{6}) \Rightarrow (\widetilde{14})$, and $(\widetilde{12}) \Rightarrow (\widetilde{14})$. Moreover, it is easy to verify the following equalities and implications :

$$(\alpha, \beta) = (\tilde{2}) \qquad (\delta, \widetilde{12}) = (\tilde{5}) \qquad (\epsilon, \tilde{6}) = (\zeta, \tilde{6})$$
$$(\tilde{3}, \tilde{4}) = (\tilde{4}, \tilde{7}) = (\alpha) \qquad (\zeta, \widetilde{12}) = (\tilde{9}) \qquad (\gamma, \tilde{8}) = (\gamma, \epsilon)$$
$$(\zeta, \tilde{5}) = (\tilde{5}, \tilde{9}) = (\gamma) \qquad (\alpha, \gamma) = (\alpha, \delta) = (\alpha, \tilde{5}) \qquad (\epsilon, \tilde{9}) = (\epsilon, \widetilde{12})$$
$$(\zeta, \tilde{8}) = (\epsilon) \qquad (\gamma, \tilde{4}) = (\delta, \tilde{4}) = (\tilde{4}, \tilde{5}) \qquad (\beta, \gamma) = (\beta, \epsilon)$$
$$(\delta, \tilde{9}) = (\gamma) \qquad (\epsilon, \tilde{5}) = (\gamma, \epsilon) \qquad (\tilde{3}, \tilde{9}) = (\epsilon, \tilde{3}) = (\tilde{7})$$
$$(\widetilde{11}, \widetilde{13}) = (\widetilde{14})$$
$$(\gamma, \tilde{3}) \Rightarrow (\epsilon) \qquad (\delta, \tilde{3}) \Rightarrow (\tilde{5}) \qquad (\gamma, \tilde{6}) \Rightarrow (\epsilon)$$

The lattice of (non-trivial) self-dual subvarieties of $\Lambda(\mathbf{P}_{3,1})$ is of cardinality 47 and has the following diagram.

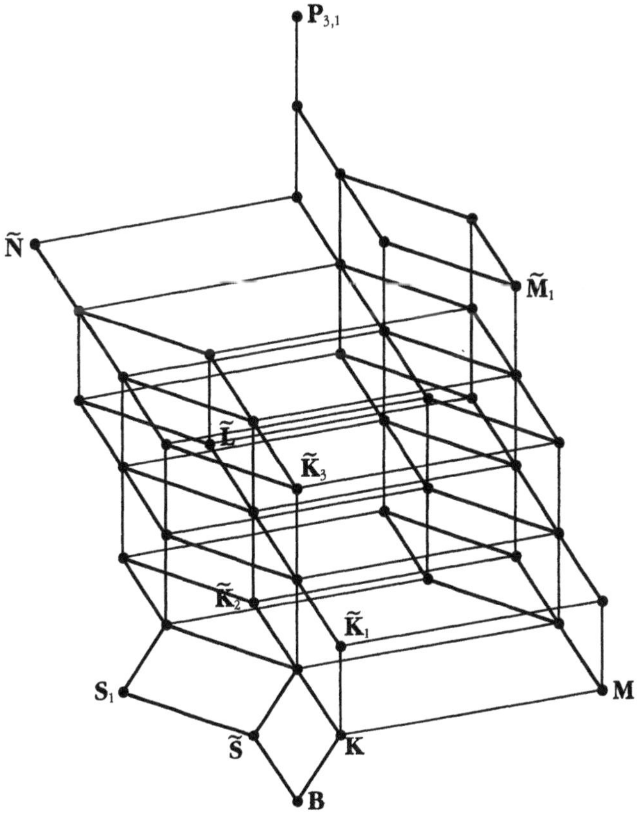

Equational bases for the $\wedge$-irreducible elements of this lattice are as follows.

$\widetilde{\mathbf{M}}_1 \vee \widetilde{\mathbf{N}}$	(η)	$\widetilde{\mathbf{K}}_3 \vee \mathbf{M} \vee \mathbf{S}_1$	$(\widetilde{6})$
$\mathbf{M} \vee \widetilde{\mathbf{N}}$	$(\widetilde{11})$	$\widetilde{\mathbf{K}}_2 \vee \mathbf{M} \vee \mathbf{S}_1$	$(\widetilde{3})$
$\widetilde{\mathbf{L}} \vee \widetilde{\mathbf{M}}_1 \vee \mathbf{S}_1$	$(\widetilde{13})$	$\widetilde{\mathbf{K}}_1 \vee \mathbf{M}$	$(\widetilde{4})$
$\widetilde{\mathbf{L}} \vee \widetilde{\mathbf{M}}_1$	(ζ)	$\widetilde{\mathbf{N}}$	(δ)
$\widetilde{\mathbf{M}}_1 \vee \mathbf{S}_1$	$(\widetilde{8})$	$\mathbf{S}_1$	(β)
$\widetilde{\mathbf{L}} \vee \mathbf{M} \vee \mathbf{S}_1$	$(\widetilde{12})$		

Those that involve only one axiom and do not already appear above are

$\mathbf{M}$	(α)	$\widetilde{\mathbf{L}} \vee \mathbf{S}_1$	$(\widetilde{5})$
$\widetilde{\mathbf{L}}$	(γ)	$\widetilde{\mathbf{K}}_2 \vee \mathbf{M}$	$(\widetilde{7})$
$\widetilde{\mathbf{M}}_1$	(ϵ)	$\widetilde{\mathbf{L}} \vee \mathbf{M}$	$(\widetilde{9})$
$\mathbf{B}$	$(\widetilde{2})$	$\widetilde{\mathbf{K}}_3 \vee \widetilde{\mathbf{L}} \vee \mathbf{M} \vee \mathbf{S}_1$	$(\widetilde{14})$

In the same spirit as in Chapter 2, we shall use the notation $\mathbf{V}(L)$ to denote the smallest subvariety of $\mathbf{O}$ to which L belongs.

We close this chapter with some illustrative examples.

Example 5.8 On the 4-element chain $0 < a < b < 1$ we can define 10 non-isomorphic Ockham algebras (see Example 1.8). The various operations $\sim$ and the corresponding classes are summarised as follows :

0	a	b	1	
1	0	0	0	$\mathbf{S}$
1	1	1	0	$\overline{\mathbf{S}}$
1	b	b	0	$\mathbf{K}_1$
1	a	a	0	$\overline{\mathbf{K}}_1$
1	a	0	0	$\mathbf{K}_2$
1	1	b	0	$\overline{\mathbf{K}}_2$
1	b	a	0	$\mathbf{K}$
1	1	0	0	$\widetilde{\mathbf{S}}$
1	b	0	0	$\mathbf{P}_{3,2}$
1	1	a	0	$\mathbf{P}_{3,2}$

The first eight subvarieties above represent $\mathbf{V}(L)$.

Example 5.9 In Example 1.9 we showed that the unit cube C can be made into an algebra in $\mathbf{P}_{3,1}$ in two ways, namely

$$(C;\sim_1) \;:\; \sim_1(x,y,z) = (1-z, 1-y, 1-z);$$
$$(C;\sim_2) \;:\; \sim_2(x,y,z) = (1-y, 1-z, 1-y).$$

As far as $\sim_1$ is concerned, we have $\sim_1^2(x,y,z) = (z,y,z)$ and so

$$\sim_1^2(x,y,z) \wedge \sim_1(x,y,z) \leqslant (\tfrac{1}{2},\tfrac{1}{2},\tfrac{1}{2}) \leqslant \sim_1^2(a,b,c) \vee \sim_1(a,b,c).$$

Thus $(C;\sim_1)$ satisfies (δ) and so belongs to $\widetilde{\mathbf{N}}$.

Now (13) fails in $(C;\sim_1)$, as can be seen by taking $a = (\frac{3}{4},\frac{1}{2},\frac{1}{4})$, $b = (\frac{1}{2},\frac{1}{2},\frac{1}{2})$, $c = (\frac{1}{4},\frac{1}{2},\frac{3}{4})$. Hence $(C;\sim_1) \notin \overline{\mathbf{N}} \vee \mathbf{L} \vee \overline{\mathbf{K}}_3$. Likewise, (13_d) fails, as can be seen by taking $a = (\frac{1}{4},\frac{1}{2},\frac{3}{4})$, $b = (\frac{1}{2},\frac{1}{2},\frac{1}{2})$, $c = (\frac{3}{4},\frac{1}{2},\frac{1}{4})$. Thus $(C;\sim_1) \notin \mathbf{N} \vee \overline{\mathbf{L}} \vee \mathbf{K}_3$. It follows that $\mathbf{V}(C;\sim_1) = \widetilde{\mathbf{N}}$.

As for $(C;\sim_2)$, we have that $\mathbf{V}(C;\sim_2) = \mathbf{B}_1$. To see this, it suffices to show that the axiom (η), which is an equational basis for $\widetilde{\mathbf{M}}_1 \vee \widetilde{\mathbf{N}}$, fails. For this purpose, take $a = (\frac{1}{2},\frac{1}{4},\frac{3}{4})$ and $b = (\frac{3}{4},\frac{1}{2},\frac{1}{2})$.

We refer the interested reader to **[46]** for further properties.

Example 5.10 The algebra of Example 1.10 satisfies the axiom (γ); for

$$\begin{cases} x \wedge \sim x = x \wedge (b \vee x') \wedge a = x \wedge b \wedge a = x \wedge b; \\ y \vee \sim y = y \vee b \vee (y' \wedge a) = y \vee b \vee a = y \vee a. \end{cases}$$

It also satisfies (4) and (4_d); for

$$\sim^2 x = (b \vee x) \wedge a \begin{cases} \geqslant x \wedge a \geqslant x \wedge b = x \wedge \sim x; \\ \leqslant b \vee x \vee a = x \vee a = x \vee \sim x. \end{cases}$$

Hence we have that $L \in (\mathbf{L} \vee \mathbf{K}_1) \wedge (\overline{\mathbf{L}} \vee \overline{\mathbf{K}}_1) = \mathbf{K}_1 \vee \overline{\mathbf{K}}_1 = \widetilde{\mathbf{K}}_1$. Since neither (1) nor (1_d) is satisfied, $L \notin \mathbf{K}_1$ and $L \notin \overline{\mathbf{K}}_1$. Therefore $\mathbf{V}(L) = \widetilde{\mathbf{K}}_1$.

Example 5.11 Referring to Example 1.11, consider first the Ockham algebra L_1 that corresponds to the polarity p. This does not belong to $\widetilde{\mathbf{M}}_1$, an equational basis for which is (ϵ), since

$$\begin{cases} a_{2n} \wedge \sim^2 b_{2n} = a_{2n} \wedge a_{2n+1} = a_{2n}; \\ \sim^2 a_{2n} \vee b_{2n} = a_{2n-1} \vee b_{2n} = b_{2n}. \end{cases}$$

Likewise, it does not belong to $\widetilde{\mathbf{N}}$, an equational basis for which is (δ); for

$$\sim^2 a_0 \wedge \sim a_0 = a_0 \parallel b_0 = \sim^2 b_0 \vee \sim b_0.$$

However, the relation (η) holds. This follows from the observation that if $a \notin \{a_0, b_0\}$ then $\sim a \wedge \sim^2 a$ is of the form a_k with $k < 0$, and if $b \notin \{a_0, b_0\}$ then $\sim b \vee \sim^2 b$ is of the form a_k with $k > 0$. Hence L_1 belongs to $\widetilde{\mathbf{N}} \vee \widetilde{\mathbf{M}}_1$.

As for the Ockham algebra L_2 that corresponds to the polarity p', it can be seen that this belongs to $\widetilde{\mathbf{L}}$.

Example 5.12 In Example 1.12 we have

$$\sim^2 X = \begin{cases} X \cup \{a\} & \text{if } b \in X; \\ X \cap \{a\}' & \text{if } b \notin X, \end{cases}$$

hence $\sim^2 X \cap \sim X = \emptyset$ for every X. Thus (β) is satisfied and the algebra belongs to $\mathbf{S}_1$. Now $\mathbf{S}_1$ covers only $\widetilde{\mathbf{S}}$ which is characterised by (β) and (γ). Since $\{a\} \wedge \sim\{a\} = \{a\}$ and $\{b\} \vee \sim\{b\} = \{a\}'$ we see that (γ) fails. Hence $\mathbf{V}(L) = \mathbf{S}_1$.

Example 5.13 In Example 1.13 we have

$$\begin{cases} \sim(x_1, x_2, \ldots, x_n) = (x'_n, x'_1, \ldots, x'_1); \\ \sim^2(x_1, x_2, \ldots, x_n) = (x_1, x_n, \ldots, x_n). \end{cases}$$

Taking into account that $x_1 \leqslant x_n$ implies $x_1 \wedge x'_n = 0$ and $x'_1 \vee x_n = 1$ we readily see that axioms (3), (5) and (6_d) are satisfied, so that $L \in \overline{\mathbf{N}}$. For instance, taking $a = (x_1, x_2, \ldots, x_n)$ and $b = (y_1, y_2, \ldots, y_n)$ we have $\sim^2 a \leqslant a \vee \sim b \vee \sim^2 b$, i.e. axiom ($6_d$), since

$$\begin{aligned}(x_1, x_n, \ldots, x_n) &\leqslant (x_1 \vee y'_n \vee y_1,\ x_2 \vee y'_1 \vee y_n, \ldots,\ x_n \vee y'_1 \vee y_n) \\ &= (x_1 \vee y'_n \vee y_1,\ 1,\ \ldots,\ 1).\end{aligned}$$

We even have $\mathbf{V}(L) = \overline{\mathbf{N}}$ since $\overline{\mathbf{N}}$ covers only $\mathbf{K}_3 \vee \widetilde{\mathbf{L}} \vee \mathbf{S}_1$ which satisfies (13_d), an axiom that generally does not hold in $(L; \sim)$.

Example 5.14 In Example 1.14 let $n = \prod_{i \in I} p_i^{\alpha_i}$, $a = \prod_{i \in J} p_i^{\beta_i}$, $b = \prod_{i \in K} p_i^{\gamma_i}$. Then

$$a \wedge a^+ = \prod_{i \in I \cup J} p_i^{\min\{\beta_i, \alpha_i - \min\{\alpha_i, \beta_i\}\}}, \quad b \vee b^+ = \prod_{i \in I \cup K} p_i^{\max\{\gamma_i, \alpha_i - \min\{\alpha_i, \gamma_i\}\}}.$$

Let p_i be any prime factor of a, b, or n and, deleting subscripts, consider

$$A = \min\{\beta, \alpha - \min\{\alpha, \beta\}\}, \quad B = \max\{\gamma, \alpha - \min\{\alpha, \gamma\}\}.$$

Considering separately the cases (1) $\alpha \leqslant \beta$, (2) $\beta \leqslant \alpha \leqslant \gamma$, (3) $\beta \leqslant \alpha$ and $\gamma \leqslant \alpha$, we see easily that $A \leqslant B$. Consequently $a \wedge a^+ \leqslant b \vee b^+$ and axiom (γ) holds. Moreover,

$$a \wedge a^+ = \gcd\left\{a, \frac{n}{\gcd\{n, a\}}\right\} \leqslant \gcd\{a, n\} = a^{++}$$

whence axiom (4) holds. We conclude from this that $\mathbf{V}(L; {}^+) = \overline{\mathbf{K}_1}$.

6 Fixed points

In an Ockham algebra $(L;f)$ the notion of a *fixed point* of f was introduced in Chapter 2 as an element $a \in L$ such that $f(a) = a$. Dually, we say that an element x of an Ockham space $(X;g)$ is a *fixed point* of g if $g(x) = x$. We shall denote by Fix L [resp. Fix X] the subset of L [resp. X] formed by the fixed points of f [resp. g].

For every $L \in \mathbf{O}$, Fix L is an antichain (possibly empty). In fact, if a, b are fixed points of f with $a \leqslant b$ then we also have $b = f(b) \leqslant f(a) = a$.

We shall now show that a fixed point of f corresponds to a special down-set of the dual space X, and that a fixed point of g corresponds to a particular prime ideal of L that has been already considered in the framework of de Morgan algebras [103].

Theorem 6.1 *Let $(X;g)$ be the dual space of $(L;f) \in \mathbf{O}$. Then $a \in L$ is a fixed point of f if and only if the graph of g is bipartite with one of the blocks in the partition the down-set A corresponding to a.*

Proof Given $a \in L$ let $A \in \mathcal{O}(X)$ be the down-set that corresponds to a. Then we have

$$\begin{aligned} a = f(a) &\iff A = f(A) = X \setminus g^{-1}(A) \\ &\iff \{A, g^{-1}(A)\} \text{ is a bipartition of } X \\ &\iff g(X \setminus A) \subseteq A \text{ and } g(A) \subseteq X \setminus A \\ &\iff \text{the graph of } g \text{ is bipartite and } A \text{ is a block.} \quad \Diamond \end{aligned}$$

Definition A down-set A of X that satisfies the conditions

$$g(A) \subseteq X \setminus A, \quad g(X \setminus A) \subseteq A$$

will be called a *distinguished down-set* of X.

Observe that if g is surjective (which is the case in particular for algebras in $\mathbf{P}_{n,0}$), then a distinguished down-set A of X satisfies the equalities

$$g(A) = X \setminus A, \quad g(X \setminus A) = A.$$

If, moreover, X is finite then the existence of a distinguished down-set requires $|X|$ to be even.

Corollary *If $(X;g)$ has a fixed point then $(L;f)$ is fixed point free.*

Proof This follows from the fact that a distinguished down-set cannot contain an element of X together with its image under g. $\Diamond$

The converse of the above corollary is in general false, but as we shall see below it is true for $L \in \mathbf{P}_{3,1}$. For this purpose, we extend to the class $\mathbf{O}$ the following notion, which was considered in de Morgan algebras by Belnap and Spencer **[27]**.

Definition A filter F of $L \in \mathbf{O}$ is a *truth filter* if, for every $a \in L$, F contains exactly one of the elements a and $f(a)$.

In **[27]** it is shown that a de Morgan algebra is fixed point free if and only if it has a truth filter.

Clearly, a truth filter is a prime filter, and so its complement is a prime ideal which enjoys the same defining property. We shall therefore call such an ideal a *falsity ideal*. Every falsity ideal is of course closed under the operation $x \mapsto f^2(x)$.

Lemma 6.1 *If $L \in \mathbf{O}$ then L and $f(L)$ have the same fixed points.* ◇

Lemma 6.2 *If $L \in \mathbf{O}$ and if I is a falsity ideal of L then $I \cap f(L)$ is a falsity ideal of $f(L)$. Conversely, if I is a falsity ideal of $f(L)$ then*

$$I^{\downarrow} = \{x \in L \mid (\exists y \in I)\ x \leqslant y\}$$

is a falsity ideal of L. ◇

Theorem 6.2 *Let $(X; g)$ be the dual space of $(L; f) \in \mathbf{O}$. Then $x \in X$ is a fixed point of g if and only if the prime ideal P that corresponds to x is a falsity ideal.*

Proof We have

$$\begin{aligned} (X;g) \text{ has a fixed point} &\iff (\exists P \in I_p(L))\ g(P) = P \\ &\iff (\exists P \in I_p(L))\ a \in P \Leftrightarrow f(a) \notin P \\ &\iff (\exists P \in I_p(L))\ P \text{ is a falsity ideal of } L. \quad \diamondsuit \end{aligned}$$

Theorem 6.3 *Let $L \in \mathbf{P}_{3,1}$. If $(L; f)$ is fixed point free then $(X; g)$ has a fixed point.*

Proof If $(L; f)$ is fixed point free then, by Lemma 6.1, so is $(f(L); f)$. Since $(L; f) \in \mathbf{P}_{3,1}$ we have that $f(L)$ is a de Morgan algebra. It follows that $f(L)$ has a falsity ideal Q. By Lemma 6.2, L also has a falsity ideal P to which there corresponds in $(X; g)$ a fixed point. ◇

We now apply these results to the finite subdirectly irreducible Ockham algebras.

Theorem 6.4 *Let $(L; f)$ be a subdirectly irreducible Ockham algebra belonging to $\mathbf{P}_{m,n}$ and let $(X; g)$ be its dual space. If the cardinality $m - n$ of the* *loop of X is even then L has at most two fixed points; and if it is odd then L is fixed point free.*

Proof Suppose that X contains a distinguished down-set A. This requires that if A contains $g^i(x)$ then A contains neither $g^{i-1}(x)$ nor $g^{i+1}(x)$, but must contain $g^{i+2}(x)$. In other words, the bipartition of the graph of g is

$$\{\{x, g^2(x), g^4(x), \ldots\}, \{g(x), g^3(x), g^5(x), \ldots\}\}.$$

This bipartition is possible (and then is unique) if and only if the cardinality of the loop of X is even. In this case, the complement of A in X can also be a down-set, which means that L has at most two fixed points. ◇

The possible cardinalities of Fix L for an algebra $(L; f) \in \mathbf{O}$ depend heavily on the subvariety of $\mathbf{O}$ to which L belongs. Clearly, every totally ordered Ockham algebra has at most one fixed point. Some classes of $\mathbf{O}$ enjoy the same property. In what follows the algebras that are considered will be assumed to be non-trivial (i.e. not to reduce to a single element).

Theorem 6.5 *All algebras in subvarieties of $\widetilde{\mathbf{N}}$ have at most one fixed point, and all algebras in subvarieties of $\mathbf{S}_1$ are fixed point free.*

Proof If $L \in \mathbf{O}$ has at least two fixed points then the axiom

$$(\delta) \qquad {\sim}a \wedge {\sim}^2 a \leqslant {\sim}b \vee {\sim}^2 b,$$

which characterises the subvariety $\widetilde{\mathbf{N}}$, cannot be satisfied.

If now L has at least one fixed point then the axiom

$$(\beta) \qquad {\sim}a \wedge {\sim}^2 a = 0,$$

which characterises the subvariety $\mathbf{S}_1$, cannot be satisfied. ◇

For the moment we shall focus our attention on the class $\mathbf{M}$ of de Morgan algebras. It has the (non-trivial) proper subclasses $\mathbf{B}$ of boolean algebras, and $\mathbf{K}$ of Kleene algebras. Obviously, if $L \in \mathbf{B}$ then Fix $L = \emptyset$. The cases where $L \in \mathbf{K} \setminus \mathbf{B}$ and $L \in \mathbf{M} \setminus \mathbf{K}$ have to be treated separately, the latter being much more complicated.

Theorem 6.6 *If $L \in \mathbf{K} \setminus \mathbf{B}$ then* $|\text{Fix } L| \in \{0, 1\}$.

Proof By Theorem 6.5, L has at most one fixed point. Both the possibilities of being fixed point free, and of having precisely one fixed point, occur. For example, the subdirectly irreducible algebra $K \in \mathbf{K}$ has a single fixed point; and the 4-element chain $0 < a < b < 1$ can be made into a fixed point free Kleene algebra by defining $\overline{0} = 1$, $\overline{a} = b$, $\overline{b} = a$, $\overline{1} = 0$. ◇

We recall that $(L;f) \in \mathbf{M}$ if and only if f is a polarity, in which case g is also a polarity in the dual space $(X;g)$. Defining $\mathcal{P}(L)$ to be the set of all polarities on L, let

$$\min|\mathrm{Fix}\ L| = \min\{|\mathrm{Fix}\,(L;f)| \ :\ f \in \mathcal{P}(L)\},$$

and let $\max|\mathrm{Fix}\ L|$ be defined similarly.

In what follows we shall employ by abuse of language the expression 'odd chain' (resp. 'even chain') to mean a chain with an odd (resp. even) number of elements. Also, for every positive real number x the notation $\lfloor x \rfloor$ will denote the greatest integer not exceeding x.

Theorem 6.7 [103] *Let* $L = \mathbf{n}_1^{r_1} \times \cdots \times \mathbf{n}_k^{r_k}$ *with* $(L;f) \in \mathbf{M}$. *Then*

$$\min|\mathrm{Fix}\ L| = \begin{cases} 1 & \textit{if all } n_i \textit{ are odd,} \\ 0 & \textit{otherwise;} \end{cases}$$

$$\max|\mathrm{Fix}\ L| = \begin{cases} \prod_{i=1}^{k} n_i^{\lfloor \frac{1}{2} r_i \rfloor} & \textit{if } r_i \textit{ is even whenever } n_i \textit{ is even,} \\ 0 & \textit{otherwise.} \end{cases}$$

Proof The dual space X of L consists of the disjoint union of r_1 chains of $n_1 - 1$ elements, r_2 chains of $n_2 - 1$ elements, ..., r_k chains of $n_k - 1$ elements. Every polarity g of X maps each of these chains either onto itself or onto another chain of the same cardinality.

If now all the n_i are odd then each component of X is an even chain and so, whatever the mapping g may be, X has a distinguished down-set. It follows that $(L;f)$ has exactly one fixed point. If, on the contrary, some n_i is even then X has an odd chain C as a component, and g can be defined in such a way that $g(C) = C$. Then g has a fixed point and, by the Corollary to Theorem 6.1, $(L;f)$ is fixed point free.

Suppose now that, for some i, n_i is even and r_i is odd. Then every polarity g on X maps at least one odd chain of X onto itself; so X has a fixed point, and L is fixed point free. If however n_i even always implies r_i even then, for $|\mathrm{Fix}\ L|$ to be maximum, g has to link the odd chains of X in pairs. As to the even chains, g has to link these in pairs also, with the possible exception of one 'isolated' chain. Since a pair of g-linked chains $\mathbf{n}$ contributes a factor $n + 1$ to the number of distinguished down-sets of X it follows that the contribution from the r_i chains $\mathbf{n}_i - \mathbf{1}$ to the number of distinguished down-sets is $n_i^{\lfloor \frac{1}{2} r_i \rfloor}$, and the result follows. ◇

Corollary *If* $L = \mathbf{n}^2$ *and* $(L;f) \in \mathbf{M}$ *then* $\max|\mathrm{Fix}\ L| = n$. ◇

Example 6.1 The following table illustrates Theorem 6.7 :

L	min $\lvert\text{Fix } L\rvert$	max $\lvert\text{Fix } L\rvert$
$\mathbf{3}^3$	1	3
$\mathbf{3}^4$	1	9
$\mathbf{3}^2 \times \mathbf{5}$	1	3
$\mathbf{4}^2 \times \mathbf{5}^2$	0	20
$\mathbf{2}^5 \times \mathbf{4}^3$	0	0

To complete the information about $|\text{Fix } L|$ when $L \in \mathbf{M} \setminus \mathbf{K}$, we require some results concerning the subsets $L^\wedge$ and $L^\vee$ defined as follows for every $L \in \mathbf{O}$:

$$L^\wedge = \{a \in L \mid a \leqslant {\sim}a\} = \{a \wedge {\sim}a \mid a \in L\};$$
$$L^\vee = \{a \in L \mid a \geqslant {\sim}a\} = \{a \vee {\sim}a \mid a \in L\}.$$

Clearly, $L^\wedge$ is a down-set and $L^\vee$ is an up-set of L.

Lemma 6.3 *Let $(L;f) \in \mathbf{O}$. Then the fixed points of f are minimal elements of $L^\vee$ and maximal elements of $L^\wedge$.*

If $L \in \mathbf{M}$ and Fix $L \neq \emptyset$ *then $L^\vee$ and $L^\wedge$ are respectively the up-set and the down-set of L generated by* Fix L.

Proof If $a \in \text{Fix } L$ then $a \in L^\vee \cap L^\wedge$. Moreover, if $b < a < c$ then ${\sim}b \geqslant {\sim}a = a > b$ gives $b \in L^\wedge \setminus L^\vee$, and ${\sim}c \leqslant {\sim}a = a < c$ gives $c \in L^\vee \setminus L^\wedge$.

If now $L \in \mathbf{M}$ and $a \in \text{Fix } L$ consider an element $b \in L^\vee$ such that $b \parallel a$. The element $c = \overline{b} \vee (b \wedge a)$ belongs to Fix L since

$$c = (\overline{b} \vee b) \wedge (\overline{b} \vee a) = b \wedge (\overline{b} \vee a) = \overline{c}.$$

It now suffices to observe that $b \geqslant c$. ◇

Note that the second part of Lemma 6.3 is not true for arbitrary Ockham algebras, as is shown for instance by the subdirectly irreducible algebra K_3.

Lemma 6.4 *Let $L \in \mathbf{P}_{3,1}$. Then $L^\wedge$ is an ideal of L if and only if $L \in \overline{\mathbf{N}} \vee \mathbf{L}$.*

Proof $L^\wedge$ is an ideal if and only if, for all $a, b \in L$,

$$\begin{aligned}(a \wedge {\sim}a) \vee (b \wedge {\sim}b) &\leqslant {\sim}[(a \wedge {\sim}a) \vee (b \wedge {\sim}b)] \\ &= ({\sim}a \vee {\sim}^2 a) \wedge ({\sim}b \vee {\sim}^2 b).\end{aligned}$$

This holds if and only if

$$(\forall a, b \in L) \qquad a \wedge {\sim}a \leqslant {\sim}b \vee {\sim}^2 b$$

which is the basic inequality (5) that characterises $\overline{\mathbf{N}} \vee \mathbf{L}$. ◇

Corollary *If $L \in \mathbf{M}_1$ (resp. $L \in \mathbf{M}$) then $L^\wedge$ is an ideal of L if and only if $L \in \mathbf{K}_2 \vee \mathbf{K}_3$ (resp. $L \in \mathbf{K}$).*

Proof It suffices to note that $\mathbf{M}_1 \wedge (\overline{\mathbf{N}} \vee \mathbf{L}) = \mathbf{K}_2 \vee \mathbf{K}_3$ and $\mathbf{M} \wedge (\overline{\mathbf{N}} \vee \mathbf{L}) = \mathbf{K}$. ◇

Dually, we have the following results.

Lemma 6.5 *If $L \in \mathbf{P}_{3,1}$ then $L^\vee$ is a filter of L if and only if $L \in \mathbf{N} \vee \overline{\mathbf{L}}$.* ◇

By considering $(\overline{\mathbf{N}} \vee \mathbf{L}) \wedge (\mathbf{N} \vee \overline{\mathbf{L}})$ we can deduce from Lemmas 6.4 and 6.5 the following :

Lemma 6.6 *If $L \in \mathbf{P}_{3,1}$ then $L^\wedge$ is an ideal of L and $L^\vee$ is a filter of L if and only if $L \in \mathbf{S}_1 \vee \overline{\mathbf{L}}$.*

Proof We have

$$\begin{aligned}(\overline{\mathbf{N}} \vee \mathbf{L}) \wedge (\mathbf{N} \vee \overline{\mathbf{L}}) &= (\overline{\mathbf{N}} \wedge \mathbf{N}) \vee (\mathbf{L} \wedge \mathbf{N}) \vee (\overline{\mathbf{N}} \wedge \overline{\mathbf{L}}) \vee (\mathbf{L} \wedge \overline{\mathbf{L}}) \\ &= \mathbf{S}_1 \vee \mathbf{K} \vee \mathbf{L} \vee \overline{\mathbf{L}} \\ &= \mathbf{S}_1 \vee \mathbf{L} \vee \overline{\mathbf{L}}. \quad \diamond\end{aligned}$$

Theorem 6.8 *For every $n \in \mathbb{N} \setminus \{1\}$ there exists $L \in \mathbf{M} \setminus \mathbf{K}$ with $|\mathrm{Fix}\, L| = n$.*

Proof First, let $n = 0$. We know that there are fixed point free algebras that belong to $\mathbf{M} \setminus \mathbf{K}$ (for example, the subdirectly irreducible algebra M).

Now let $n = 1$. In this case, if $L \in \mathbf{M}$ and $\mathrm{Fix}\, L = \{a\}$ then by Lemma 6.3, $L^\wedge = a^\downarrow$ and so, by the Corollary to Lemma 6.4, we must have $L \in \mathbf{K}$.

Finally, let $n \geqslant 2$. By the Corollary to Theorem 6.7 we can obtain a de Morgan algebra with n fixed points, and this algebra does not belong to $\mathbf{K}$ because of Theorem 6.6. ◇

We shall now extend the results obtained for $\mathbf{M}$ to a larger class, namely $\mathbf{P}_{3,1}$. It is almost obvious (and will be shown in Theorem 8.16) that for every $L \in \mathbf{P}_{3,1}$ the smallest congruence for which the quotient algebra belongs to $\mathbf{M}$ is Φ_1, i.e. the congruence given by

$$(x, y) \in \Phi_1 \iff {\sim}x = {\sim}y.$$

Moreover, L/Φ_1 is dually isomorphic to ${\sim}L$.

Lemma 6.7 *If $L \in \mathbf{P}_{3,1}$ then*

(1) $\mathbf{V}(L) \in [\mathbf{B}, \mathbf{S}_1]$ *if and only if* $\mathbf{V}({\sim}L) = \mathbf{B}$;
(2) $\mathbf{V}(L) \in \widetilde{\mathbf{N}} \setminus \mathbf{S}_1$ *if and only if* $\mathbf{V}({\sim}L) = \mathbf{K}$;
(3) $\mathbf{V}(L) \in \mathbf{P}_{3,1} \setminus \widetilde{\mathbf{N}}$ *if and only if* $\mathbf{V}({\sim}L) = \mathbf{M}$.

Proof (1) : $L \in \mathbf{S}_1$ if and only if the axiom (β) is satisfied; i.e. L has a boolean skeleton.

(2) : $L \in \widetilde{\mathbf{N}}$ if and only if the axiom (δ) is satisfied; i.e. L has a Kleene skeleton.

(3) : $L \in \mathbf{P}_{3,1}$ if and only if $\sim^3 a = a$ for every $a \in L$; i.e. L has a de Morgan skeleton. ◇

Since L and $\sim L$ have the same fixed points, and since we know the possible cardinalities of Fix L when $L \in \mathbf{M}$, the following result is straightforward.

Theorem 6.9 *Let* $L \in \mathbf{P}_{3,1}$. *Then*

(1) *if* $\mathbf{V}(L) \in [\mathbf{B}, \mathbf{S}_1]$ *then* Fix $L = \emptyset$;
(2) *if* $\mathbf{V}(L) \in \widetilde{\mathbf{N}} \setminus \mathbf{S}_1$ *then* $|\text{Fix } L| \in \{0, 1\}$;
(3) *if* L *is countable and* $\mathbf{V}(L) \in \mathbf{P}_{3,1} \setminus \widetilde{\mathbf{N}}$ *then* $|\text{Fix } L| \in \mathbb{N} \setminus \{1\}$. ◇

Corollary *Let* $L \in \mathbf{M}_1$. *Then*

(1) *if* $\mathbf{V}(L) \in [\mathbf{B}, \mathbf{S}]$ *then* Fix $L = \emptyset$;
(2) *if* $\mathbf{V}(L) \in [\mathbf{K}, \mathbf{K}_2 \vee \mathbf{K}_3]$ *then* $|\text{Fix } L| \in \{0, 1\}$;
(3) *if* L *is countable and* $\mathbf{V}(L) \in [\mathbf{M}, \mathbf{M}_1]$ *then* $|\text{Fix } L| \in \mathbb{N} \setminus \{1\}$.

Proof We have $\mathbf{S}_1 \wedge \mathbf{M}_1 = \mathbf{S}$ and

$$\widetilde{\mathbf{N}} \wedge \mathbf{M}_1 = (\mathbf{N} \vee \overline{\mathbf{N}}) \wedge \mathbf{M}_1 = (\mathbf{N} \wedge \mathbf{M}_1) \vee (\overline{\mathbf{N}} \wedge \mathbf{M}_1) = \mathbf{K}_2 \vee \mathbf{K}_3. \quad ◇$$

Our purpose now is to sharpen the results in Theorem 6.9(2),(3) for some subvarieties of $\mathbf{P}_{3,1}$.

Theorem 6.10 *Let* $L \in \mathbf{P}_{3,1}$. *If* $\mathbf{V}(L)$ *satisfies axioms* $(5, 3_d, 6)$ *but not axiom* (4) *then* L *is fixed point free.*

Proof Observe that $(5, 3_d, 6) = (5, 16_d)$. Since L satisfies (5), $L^\wedge$ is an ideal by Lemma 6.4. Assume, by way of obtaining a contradiction, that L has a fixed point e. The axiom (5) characterises $\mathbf{L} \vee \overline{\mathbf{N}} \subset \widetilde{\mathbf{N}}$ and therefore, by Theorem 6.5, this fixed point is unique. By Lemma 6.3, e is a maximal element of the ideal $L^\wedge$, hence is the generating element of $L^\wedge$; i.e. $L^\wedge = e^\downarrow$. Since L satisfies (16_d), $a \leqslant e$ implies that $a = \sim^2 a$. It follows that for every $b \in L$ we have

$$b \wedge \sim b = \sim^2(b \wedge \sim b) = \sim^2 b \wedge \sim b,$$

which means that $(15_d) = (3_d, 4)$ is satisfied, contrary to the hypothesis. ◇

Corollary *If* $\mathbf{V}(L)$ *is any of the subvarieties*

$$\mathbf{L} \vee \overline{\mathbf{S}},\ \overline{\mathbf{S}} \vee \overline{\mathbf{K}}_1 \vee \mathbf{K}_2,\ \widetilde{\mathbf{S}} \vee \overline{\mathbf{K}}_1,\ \overline{\mathbf{S}} \vee \overline{\mathbf{K}}_1,\ \overline{\mathbf{S}} \vee \mathbf{K}_2,\ \overline{\mathbf{S}} \vee \mathbf{K},\ \widetilde{\mathbf{S}} \vee \mathbf{K},$$
$$\overline{\mathbf{L}} \vee \mathbf{S},\ \mathbf{S} \vee \mathbf{K}_1 \vee \overline{\mathbf{K}}_2,\ \widetilde{\mathbf{S}} \vee \mathbf{K}_1,\ \mathbf{S} \vee \mathbf{K}_1,\ \mathbf{S} \vee \overline{\mathbf{K}}_2,\ \mathbf{S} \vee \overline{\mathbf{K}},$$

then L *is fixed point free.*

Proof Consider the first seven subvarieties listed. Each of these satisfies the axioms $(5, 3_d, 6)$. To show that axiom (4) is not satisfied, it suffices to replace axiom (6) by axiom (4) in the equational basis of the subvariety and observe that in so doing we obtain a subvariety that is smaller than (in fact, covered by) the original. This procedure is summarised in the following table.

subvariety	equational basis	changed to	giving
$\mathbf{L} \vee \overline{\mathbf{S}}$	$(\gamma, 3_d, 6)$	$(\gamma, 3_d, 4)$	$\mathbf{L}$
$\overline{\mathbf{S}} \vee \overline{\mathbf{K}}_1 \vee \mathbf{K}_2$	$(\gamma, 3_d, 6, 8)$	$(\gamma, 3_d, 4, 8)$	$\overline{\mathbf{K}}_1 \vee \mathbf{K}_2$
$\widetilde{\mathbf{S}} \vee \overline{\mathbf{K}}_1$	$(\gamma, 3_d, 6, 6_d)$	$(\gamma, 3_d, 4, 6_d)$	$\mathbf{S} \vee \overline{\mathbf{K}}_1$
$\overline{\mathbf{S}} \vee \overline{\mathbf{K}}_1$	$(\gamma, 1_d, 6)$	$(\gamma, 1_d, 4)$	$\overline{\mathbf{K}}_1$
$\overline{\mathbf{S}} \vee \mathbf{K}_2$	$(\gamma, 3, 3_d, 6)$	$(\gamma, 3, 3_d, 4) = (\gamma, 1, 3_d)$	$\mathbf{K}_2$
$\overline{\mathbf{S}} \vee \mathbf{K}$	$(\gamma, 1_d, 3, 6)$	$(\gamma, 1_d, 3, 4) = (\gamma, 1, 1_d) = (\alpha, \gamma)$	$\mathbf{K}$
$\widetilde{\mathbf{S}} \vee \mathbf{K}$	$(\gamma, 3, 3_d, 6, 6_d)$	$(\gamma, 3, 3_d, 4, 6_d) = (\gamma, 1, 3_d, 6_d)$	$\mathbf{S} \vee \mathbf{K}$

By duality, the same conclusion holds for the other subvarieties. ◇

For $L \in \mathbf{M}_1$ a more sophisticated procedure can be used to decide when L is fixed point free.

Theorem 6.11 [35] *If* $\mathbf{V}(L) = \mathbf{S} \vee \mathbf{M}$ *then* L *is fixed point free.*

Proof L satisfies the axioms $(1, 3_d, 6_d)$. Suppose, by way of obtaining a contradiction, that L has a fixed point e. By Lemma 6.3, we have $e^{\downarrow} \subseteq L^{\wedge}$. Since (1) implies (4), L satisfies $(15_d) = (3_d, 4)$, which says that

$$a \wedge {\sim}a = {\sim}^2 a \wedge {\sim}a$$

and means that $L^{\wedge} \subseteq {\sim}L$. It follows that every $a \leqslant e$ is such that $a = {\sim}^2 a$.

Moreover, (1) implies (3) and L satisfies $(16) = (3, 6_d)$, that is

$$a \vee {\sim}b \vee {\sim}^2 b = {\sim}^2 a \vee {\sim}b \vee {\sim}^2 b.$$

Hence every $a \geqslant e$ is such that $a = {\sim}^2 a$.

Finally, if $a \parallel e$ then $a \wedge e$ and $a \vee e$ both belong to ${\sim}L$ and, owing to the semiconvexity of ${\sim}L$ in any MS-algebra, we have $a \in {\sim}L$.

Thus ${\sim}L = L$ and so $L \in \mathbf{M}$, a contradiction. ◇

Theorem 6.12 [35] *If* $\mathbf{V}(L) = \mathbf{S} \vee \mathbf{M} \vee \mathbf{K}_1$ *then* L *is fixed point free.*

Proof L satisfies $(1, 6_d, 12_d)$. Suppose, by way of obtaining a contradiction, that L has a fixed point e. Then the axiom $(16) = (3, 6_d)$, namely

$$a \vee {\sim}b \vee {\sim}^2 b = {\sim}^2 a \vee {\sim}b \vee {\sim}^2 b,$$

is satisfied and, for every $a \in L$, we have $a \vee e \in {\sim}L$. Hence we have also

$$(\star) \qquad (\forall a \in L) \quad a \vee {\sim}a \vee e \in {\sim}L.$$

Now L also satisfies the axiom $(18_d) = (11, 12_d)$, namely

$$(a \vee {\sim}a) \wedge {\sim}b \wedge {\sim}^2 b = ({\sim}^2 a \vee {\sim}a) \wedge {\sim}b \wedge {\sim}^2 b,$$

which gives

$$(\forall a, b \in L) \qquad (a \vee {\sim}a) \wedge {\sim}b \wedge {\sim}^2 b \in {\sim}L.$$

In particular, we have

$$(\star\star) \qquad (\forall a \in L) \quad (a \vee {\sim}a) \wedge e \in {\sim}L.$$

By $(\star)$, $(\star\star)$, and the convexity of ${\sim}L$, it follows that $a \vee {\sim}a \in {\sim}L$ for every $a \in L$, which is $(15) = (3, 4_d)$, namely

$$a \vee {\sim}a = {\sim}^2 a \vee {\sim}a,$$

and gives the contradiction $L \in \mathbf{M} \vee \mathbf{K}_1 \subset \mathbf{S} \vee \mathbf{M} \vee \mathbf{K}_1$. ◇

Example 6.2 Here we shall illustrate the fact that if L is countable and such that $\mathbf{V}(L) = \mathbf{M}_1$ then $|\text{Fix } L| \in \mathbb{N} \setminus \{1\}$ (refer to the Corollary to Theorem 6.9). First, an example of $L \in \mathbf{M}_1$ that is fixed point free. Consider the lattice $L = \mathbf{2}^2 \times \mathbf{3}$ made into an MS-algebra as follows :

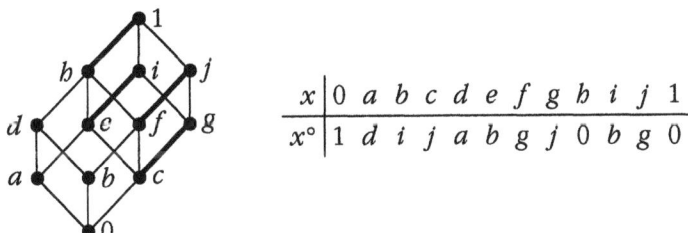

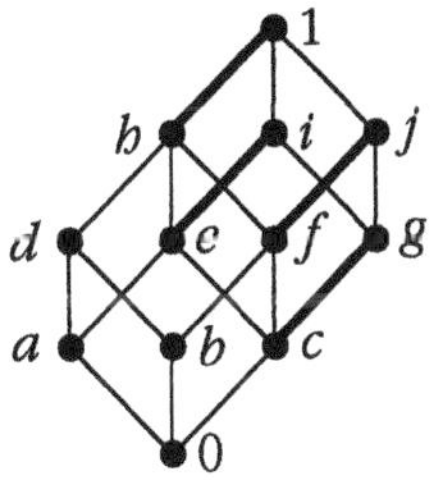

x	0	a	b	c	d	e	f	g	h	i	j	1
x°	1	d	i	j	a	h	g	j	0	h	g	0

The thick lines indicate the Φ-classes. To see that $\mathbf{V}(L) = \mathbf{M}_1$ we observe that the axiom (11_d), namely

$${\sim}a \wedge {\sim}^2 a \leqslant a \vee {\sim}b \vee {\sim}^2 b,$$

which appears in the equational basis for $\mathbf{M} \vee \mathbf{K}_2 \vee \mathbf{K}_3$, fails to hold. In fact we have

$${\sim}c \wedge {\sim}^2 c = j \wedge g = g,$$

whereas

$$c \vee {\sim}a \vee {\sim}^2 a = c \vee d \vee a = h.$$

Next we observe that the subdirectly irreducible MS-algebra M_1 itself has two fixed points. For examples of algebras L such that $\mathbf{V}(L) = \mathbf{M}_1$ and having r fixed points for $r \geqslant 3$, we consider the lattice $L = \mathbf{n} \times \mathbf{n}$ with $n \geqslant 4$, made into an MS-algebra as indicated in the following diagram, the thick lines indicating the Φ-classes.

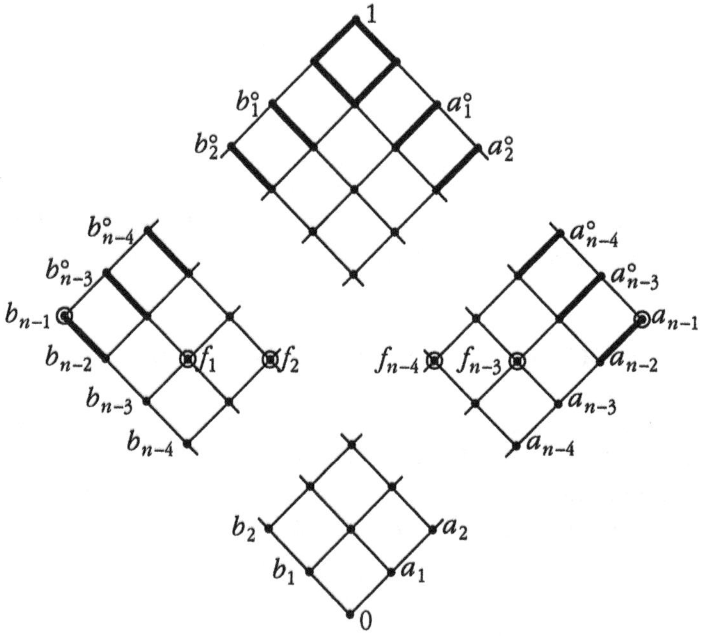

There are $n-1$ fixed points, marked with circles, namely

$$b_{n-1}, f_1, f_2, \ldots, f_{n-4}, f_{n-3}, a_{n-1}.$$

That $\mathbf{V}(L) = \mathbf{M}_1$ follows from the observation that it contains copies of the subdirectly irreducible algebra M_1, namely the intervals $[b_{n-3}, b_{n-3}^\circ]$ and $[a_{n-3}, a_{n-3}^\circ]$.

7 Fixed point separating congruences

If $(L;f)$ is an Ockham algebra and x,y are distinct elements of L then we shall say that a congruence ϑ on L *separates* x and y if $(x,y) \notin \vartheta$. By a *fixed point separating congruence* on L we shall mean a congruence that separates every pair of fixed points of L. We shall denote by $\mathcal{F}(L)$ the set of fixed point separating congruences on L. We let $\operatorname{Fix} L = \{\alpha_i \,;\, i \in I\}$ be the set of fixed points of $(L;f)$, and we shall assume throughout this chapter that $|\operatorname{Fix} L| \geqslant 2$. All of the results that follow appear in **[56]**.

Theorem 7.1 *Let* $(L;\,^{-}) \in \mathbf{M}$. *If* α, β *are distinct fixed points of* L *and* ϑ *is a fixed point separating congruence on* L *then* $\vartheta|_{[\alpha\wedge\beta,\alpha\vee\beta]} = \omega$.

Proof We prove first that $\vartheta|_{[\alpha,\alpha\vee\beta]} = \omega$. For this purpose, suppose that

$$\alpha \leqslant x \leqslant y \leqslant \alpha \vee \beta$$

with $(x,y) \in \vartheta$. Define

$$x_\beta = x \wedge (\overline{x} \vee \beta).$$

Then we have

$$\overline{x_\beta} = \overline{x} \vee (x \wedge \beta) = (\overline{x} \vee x) \wedge (\overline{x} \vee \beta) = x \wedge (\overline{x} \vee \beta) = x_\beta$$

and so x_β is a fixed point. Likewise, so is

$$y_\beta = y \wedge (\overline{y} \vee \beta).$$

Since $(x,y) \in \vartheta$ we have $(x_\beta, y_\beta) \in \vartheta$ whence $x_\beta = y_\beta$. Then $x_\beta \wedge \beta = y_\beta \wedge \beta$ gives $x \wedge \beta = y \wedge \beta$, whence

$$x = x \wedge (\alpha \vee \beta) = \alpha \vee (x \wedge \beta) = \alpha \vee (y \wedge \beta) = y \wedge (\alpha \vee \beta) = y.$$

Thus $\vartheta|_{[\alpha,\alpha\vee\beta]} = \omega$. Similarly, we can show that $\vartheta|_{[\alpha\wedge\beta,\alpha]} = \omega$.

Suppose now that $\alpha \wedge \beta \leqslant x \leqslant y \leqslant \alpha \vee \beta$ with $(x,y) \in \vartheta$. Then

$$\alpha \leqslant \alpha \vee x \leqslant \alpha \vee y \leqslant \alpha \vee \beta \quad \text{with} \quad (\alpha \vee x, \alpha \vee y) \in \vartheta;$$
$$\alpha \wedge \beta \leqslant \alpha \wedge x \leqslant \alpha \wedge y \leqslant \alpha \quad \text{with} \quad (\alpha \wedge x, \alpha \wedge y) \in \vartheta.$$

It follows that $\alpha \vee x = \alpha \vee y$ and $\alpha \wedge x = \alpha \wedge y$, whence $x = y$. ◊

The above result can be extended to $\mathbf{K}_\omega$ as follows.

Theorem 7.2 *Let $(L;f) \in \mathbf{K}_\omega$ and let α, β be distinct fixed points of L. If $\vartheta \in \mathcal{F}(L)$ then every $x \in (\alpha, \alpha \vee \beta]$ that is separated from α by Φ_ω is also separated from α by ϑ.*

Proof Suppose that $x \in (\alpha, \alpha \vee \beta]$ with $(x, \alpha) \notin \Phi_\omega$, i.e. $f^n(x) \neq \alpha$ for every n. Since $L \in \mathbf{K}_\omega$ there exist positive integers m, n such that $f^{2m+2n}(x) = f^{2n}(x)$, and clearly

$$f^{2n+1}(x) < \alpha < f^{2n}(x).$$

Consider the elements

$$s = \bigwedge_{i=0}^{m-1} f^{2n+2i+1}(x), \quad t = \bigvee_{i=0}^{m-1} f^{2n+2i}(x).$$

We have $\alpha \wedge \beta \leqslant s < \alpha < t \leqslant \alpha \vee \beta$ with $f(s) = t$ and $f(t) = s$. Consequently, the subalgebra $M = \langle \alpha, \beta, s, t \rangle$ is de Morgan. Since $\vartheta|_M$ separates fixed points, it follows by Theorem 7.1 that $(s, t) \notin \vartheta$.

We now show that ϑ separates α and x. In fact, if we had $(\alpha, x) \in \vartheta$ then on the one hand $(\alpha, f^{2k}(x)) \in \vartheta$ would give $(\alpha, t) \in \vartheta$, and on the other hand $(\alpha, f^{2k+1}(x)) \in \vartheta$ would give $(\alpha, s) \in \vartheta$, whence we would have the contradiction $(s, t) \in \vartheta$. ◇

Theorem 7.3 *If $(L;f) \in \mathbf{K}_\omega$ then $\mathcal{F}(L)$ forms a complete ideal of* Con L.

Proof It is clear that $\mathcal{F}(L)$ is a down-set of Con L, so it suffices to prove that if $A = (\vartheta_i)_{i \in I}$ is a family of fixed point separating congruences then $\bigvee_{i \in I} \vartheta_i$ is fixed point separating. Suppose, by way of obtaining a contradiction, that α, β are distinct fixed points such that $(\alpha, \beta) \in \bigvee_{i \in I} \vartheta_i$. Then $(\alpha, \alpha \vee \beta) \in \bigvee_{i \in I} \vartheta_i$ and there exist $z_0, \dots, z_n \in L$ and $\vartheta_1, \dots, \vartheta_n \in A$ such that

$$\alpha = z_0 \overset{\vartheta_1}{<} z_1 \overset{\vartheta_2}{<} z_2 \overset{\vartheta_3}{<} \cdots \overset{\vartheta_n}{<} z_n = \alpha \vee \beta.$$

By Theorem 7.2 there exists k_1 such that $f^{k_1}(z_1) = \alpha$. Applying f^{2k_1} to the above chain we obtain, for some p and q,

$$\alpha = f^{2k_1}(z_1) = \cdots = f^{2k_1}(z_{p-1}) \overset{\vartheta_p}{<} f^{2k_1}(z_p) = \cdots = f^{2k_1}(z_{q-1}) \overset{\vartheta_q}{<} \cdots \overset{\vartheta_n}{<} \alpha \vee \beta.$$

Applying Theorem 7.2 again, there exists k_2 such that

$$f^{2k_2}(f^{2k_1}(z_p)) = \alpha.$$

Continuing this argument, we arrive at the existence of t such that

$$f^{2t}(\alpha \vee \beta) = \alpha,$$

i.e. $\alpha \vee \beta = \alpha$. This provides the required contradiction. ◇

In general, for $L \notin \mathbf{K}_\omega$ the down-set $\mathcal{F}(L)$ of fixed point separating congruences on L need not be an ideal of Con L. This can be illustrated in the following way.

Example 7.1 Consider the lattice

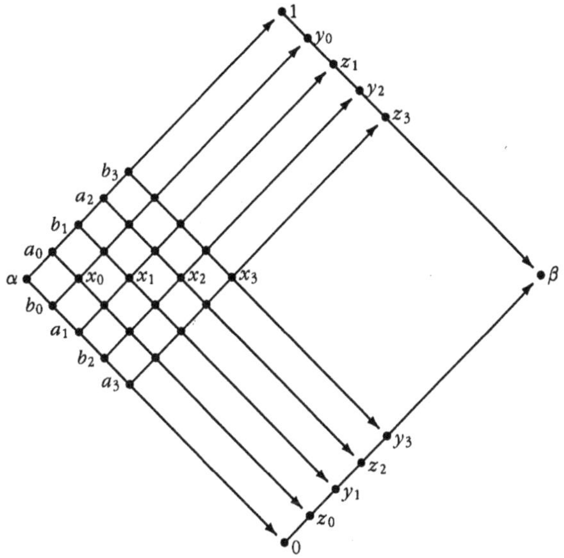

made into an Ockham algebra by defining

$$f(0) = 1, \quad f(1) = 0, \quad f(\alpha) = \alpha, \quad f(\beta) = \beta,$$
$$(\forall i \geqslant 0) \ f(x_i) = x_{i+1},$$

and extending to the whole of L. Then we have

$$(\forall i \geqslant 0) \quad f(a_i) = a_{i+1}, \quad f(b_i) = b_{i+1}, \quad f(y_i) = y_{i+1}, \quad f(z_i) = z_{i+1}.$$

Clearly, $(L; f) \notin \mathbf{K}_\omega$ and $\varphi_1 = \vartheta(x_0, \beta)$ is fixed point separating with classes

$$\{\alpha\}, \quad [a_0, 1], \quad [0, b_0], \quad [z_0, y_0].$$

Now the relation φ_2 whose classes are

$$A = [0, \beta), \quad B = \{\beta\}, \quad C = (\beta, 1], \quad L\backslash\{A \cup B \cup C\}$$

is also a fixed point separating congruence. Since clearly $(\alpha, x_0) \in \varphi_2$ and $(x_0, \beta) \in \varphi_1$ we have $(\alpha, \beta) \in \varphi_1 \vee \varphi_2$ and therefore

$$\varphi_1 \vee \varphi_2 = \iota \notin \mathcal{F}(L).$$

Definition We shall say that an Ockham algebra $(L;f)$ is *fixed point complete* if $\alpha^\star = \bigvee_{i\in I} \alpha_i$ and $\alpha_\star = \bigwedge_{i\in I} \alpha_i$ exist with $f(\alpha^\star) = \alpha_\star$ and $f(\alpha_\star) = \alpha^\star$.

It is clear that every finite Ockham algebra is fixed point complete.

Example 7.2 For an example of an Ockham algebra that is not fixed point complete, consider the lattice

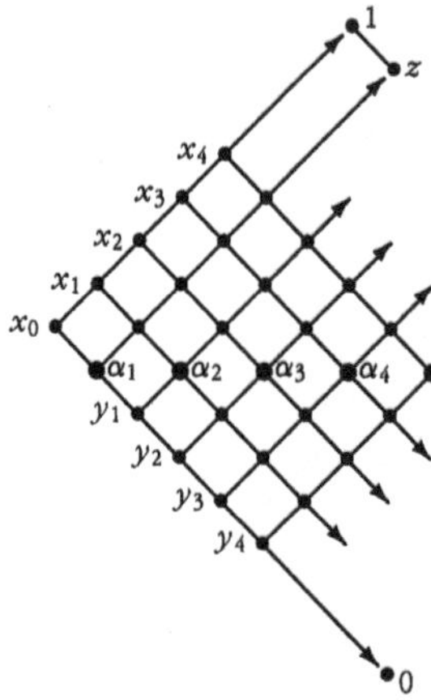

made into an Ockham algebra by defining

$$f(x_0) = \alpha_1, \quad (\forall i)\ f(\alpha_i) = \alpha_i, \quad f(z) = 0,$$

and extending to the whole of L. Here we have

$$f(\bigwedge_{i\geqslant 1} \alpha_i) = f(0) = 1 \neq z = \bigvee_{i\geqslant 1} \alpha_i.$$

It is immediate from Theorem 7.3 that for $L \in \mathbf{K}_\omega$ there exists a maximum fixed point separating congruence on L. We shall denote this by Ψ. Since Φ_ω is clearly fixed point separating, we have that $\Phi_\omega \leqslant \Psi$.

Theorem 7.4 *Let* $L \in \mathbf{K}_\omega$ *be fixed point complete. If* $\Psi = \Phi_\omega$ *then* $\bigvee_{i\in I} \alpha_i = 1$.

Proof Suppose, by way of obtaining a contradiction, that $\alpha^\star = \bigvee_{i\in I} \alpha_i < 1$. Then we have $\alpha_\star = \bigwedge_{i\in I} \alpha_i > 0$. Denote by A the subalgebra generated by $\{\alpha_\star, \alpha^\star\} \cup \operatorname{Fix} L$. The congruence φ on A whose classes are

$$\{0, \alpha_\star\}, \ \{\alpha^\star, 1\}, \text{ and singletons otherwise}$$

is fixed point separating with $\varphi > \omega$, and extends to a fixed point separating congruence $\widehat{\varphi}$ on L. Then $\Psi \geqslant \widehat{\varphi}$ and therefore

$$\Psi|_A \geqslant \widehat{\varphi}|_A = \varphi > \omega.$$

But we have $\Phi_\omega|_A = \omega$ and so, by the hypothesis, $\Psi|_A = \omega$. We deduce from this contradiction that $\alpha^\star = 1$. ◊

The converse of Theorem 7.4 does not hold in general, as is exhibited by the following example.

Example 7.3 *The angel fish.* Consider the lattice

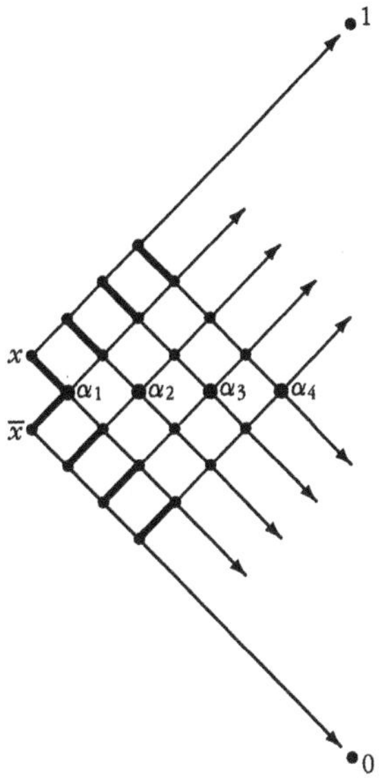

made into a fixed point complete de Morgan algebra in the obvious way with fixed points α_i. Here we have $\bigvee_{i\geqslant 1} \alpha_i = 1$ and $\Phi_\omega = \omega$. But $\Psi \neq \omega$. The Ψ-classes are shown by thick lines; for example, $(\overline{x}, x) \in \Psi$.

Definition We shall say that L is *fixed point distributive* if, for every $x \in L$, whenever $\bigvee_{i\in I} \alpha_i$, $\bigwedge_{i\in I} \alpha_i$, $\bigvee_{i\in I}(x \wedge \alpha_i)$ and $\bigwedge_{i\in I}(x \vee \alpha_i)$ exist we have

$$x \wedge \bigvee_{i\in I} \alpha_i = \bigvee_{i\in I}(x \wedge \alpha_i), \quad x \vee \bigwedge_{i\in I} \alpha_i = \bigwedge_{i\in I}(x \vee \alpha_i).$$

When L is fixed point complete and fixed point distributive we have the following analogue of Theorem 7.1.

Theorem 7.5 *Let* $(L; ^-) \in \mathbf{M}$. *Suppose that L is fixed point complete and fixed point distributive. Then every* $\vartheta \in \mathcal{F}(L)$ *is such that* $\vartheta|_{[\alpha_\star, \alpha^\star]} = \omega$.

Proof Suppose that

$$\alpha_\star = \bigwedge_{i\in I} \alpha_i \leqslant x \leqslant y \leqslant \bigvee_{i\in I} \alpha_i = \alpha^\star$$

with $(x, y) \in \vartheta$. Then for all $i, j \in I$ we have

$$\alpha_i \wedge \alpha_j \leqslant (x \vee \alpha_i) \wedge \alpha_j \leqslant (y \vee \alpha_i) \wedge \alpha_j \leqslant \alpha_j$$

with

$$((x \vee \alpha_i) \wedge \alpha_j, (y \vee \alpha_i) \wedge \alpha_j) \in \vartheta.$$

It follows by Theorem 7.1 that

$$(x \vee \alpha_i) \wedge \alpha_j = (y \vee \alpha_i) \wedge \alpha_j$$

whence we have

$$\begin{aligned} x \vee \alpha_i &= (x \vee \alpha_i) \wedge \bigvee_{j \in I} \alpha_j \\ &= \bigvee_{j \in I} [(x \vee \alpha_i) \wedge \alpha_j] \\ &= \bigvee_{j \in I} [(y \vee \alpha_i) \wedge \alpha_j] \\ &= (y \vee \alpha_i) \wedge \bigvee_{j \in I} \alpha_j \\ &= y \vee \alpha_i \end{aligned}$$

and consequently

$$x = x \vee \bigwedge_{i \in I} \alpha_i = \bigwedge_{i \in I} (x \vee \alpha_i) = \bigwedge_{i \in I} (y \vee \alpha_i) = y \vee \bigwedge_{i \in I} \alpha_i = y. \quad \diamond$$

Suppose now that L is fixed point complete and consider the relation Θ defined on L by

$$(x, y) \in \Theta \iff (x \vee \alpha_\star) \wedge \alpha^\star = (y \vee \alpha_\star) \wedge \alpha^\star.$$

It is clear that Θ is a fixed point separating congruence on L.

When L is also fixed point distributive, we have the following situation.

Theorem 7.6 *Let $(L; f)$ be an Ockham algebra that is fixed point complete and fixed point distributive. For $i, j \in I$ define the relation ϑ_{ij} by*

$$(x, y) \in \vartheta_{ij} \iff (x \vee \alpha_i) \wedge \alpha_j = (y \vee \alpha_i) \wedge \alpha_j.$$

Then $\Theta = \bigwedge_{i,j} \vartheta_{ij}$.

Proof That $\Theta \leqslant \bigwedge_{i,j} \vartheta_{ij}$ follows from the observation that

$$([(x \vee \alpha_\star) \wedge \alpha^\star] \vee \alpha_i) \wedge \alpha_j = (x \vee \alpha_i) \wedge \alpha_j,$$

and that $\bigwedge_{i,j} \vartheta_{ij} \leqslant \Theta$ follows from the observation that

$$\begin{aligned}(x \vee \alpha_\star) \wedge \alpha^\star &= \big(x \vee \bigwedge_i \alpha_i\big) \wedge \alpha^\star \\ &= \bigwedge_i (x \vee \alpha_i) \wedge \alpha^\star \\ &= \bigwedge_i [(x \vee \alpha_i) \wedge \alpha^\star] \\ &= \bigwedge_i [(x \vee \alpha_i) \wedge \bigvee_j \alpha_j] \\ &= \bigwedge_i \bigvee_j [(x \vee \alpha_i) \wedge \alpha_j]. \quad \diamond\end{aligned}$$

When $L \in \mathbf{K}_\omega$ is fixed point complete and fixed point distributive we have the following description of the congruence Ψ.

Theorem 7.7 *If* $(L;f) \in \mathbf{K}_\omega$ *is fixed point complete and fixed point distributive then*

$$\Psi = \Phi_\omega \vee \Theta.$$

Proof Suppose first that $L \in \mathbf{K}_{p,q}$. Then we have $f^q(L) \in \mathbf{K}_{p,0}$. Recalling the notation used in Chapter 3, let $M = T_2\big(f^q(L)\big)$ be the biggest de Morgan subalgebra of $f^q(L)$.

Suppose that $x, y \in M$ are such that $(x,y) \in \Psi|_M$. Then, for all $i, j \in I$,

$$\big((x \vee \alpha_i) \wedge \alpha_j, (y \vee \alpha_i) \wedge \alpha_j\big) \in \Psi|_M.$$

It follows by Theorem 7.5 that

$$(\forall i, j \in I) \qquad (x \vee \alpha_i) \wedge \alpha_j = (y \vee \alpha_i) \wedge \alpha_j,$$

and therefore, by Theorem 7.6, $(x,y) \in \Theta|_M$. Thus $\Psi|_M \leqslant \Theta|_M$ and consequently, since $\Theta \leqslant \Psi$, we have $\Psi|_M = \Theta|_M$. Since, by Theorem 3.11, $f^q(L)$ is a strong extension of M it follows that $\Psi|_{f^q(L)} = \Theta|_{f^q(L)}$ and hence

$$\Psi|_{f^q(L)} = \Phi_q|_{f^q(L)} \vee \Psi|_{f^q(L)} = \Phi_q|_{f^q(L)} \vee \Theta|_{f^q(L)} \leqslant (\Phi_q \vee \Theta)|_{f^q(L)}.$$

The reverse inequality being trivial, we have that $\Psi|_{f^q(L)} = (\Phi_q \vee \Theta)|_{f^q(L)}$. Since $\operatorname{Con} f^q(L) \simeq [\Phi_q, \iota]$, we deduce that $\Psi = \Phi_q \vee \Theta$.

Suppose now that $L \in \mathbf{K}_\omega$. Let $(x,y) \in \Psi$ and let A be the subalgebra generated by $\{x, y, \alpha_\star, \alpha^\star\} \cup \operatorname{Fix} L$. Then by Theorem 3.6 applied first to the elements x, y we see that $A \in \mathbf{K}_{p,q}$ for some p, q. It follows by the above that

$$(x,y) \in \Psi|_A = \Phi_q|_A \vee \Theta|_A,$$

whence $(x,y) \in \Phi_\omega \vee \Theta$. Thus $\Psi \leqslant \Phi_\omega \vee \Theta$ whence we have equality. $\diamond$

Note, in fact, that in Theorems 7.5, 7.6, 7.7 only the existence of $\alpha_\star$ and $\alpha^\star$ is used; we do not require the properties $f(\alpha_\star) = \alpha^\star$ and $f(\alpha^\star) = \alpha_\star$.

Theorem 7.8 *If $(L; f) \in \mathbf{K}_\omega$ is fixed point complete and fixed point distributive then the following statements are equivalent:*

(1) $\Psi = \Phi_\omega$;

(2) $\bigvee_{i \in I} \alpha_i = 1$;

(3) $\Theta = \omega$.

Proof (1) $\Rightarrow$ (2) follows from Theorem 7.4.

(2) $\Rightarrow$ (3) : If (2) holds then $\bigwedge_{i \in I} \alpha_i = 0$ and clearly $\Theta = \omega$.

(3) $\Rightarrow$ (1) follows from Theorem 7.7. ◇

It is of course possible to have $\Psi = \Theta$. By Theorem 7.7, this occurs precisely when $\Phi_\omega \leqslant \Theta$. An example of this situation is the following.

Example 7.4 Consider the lattice

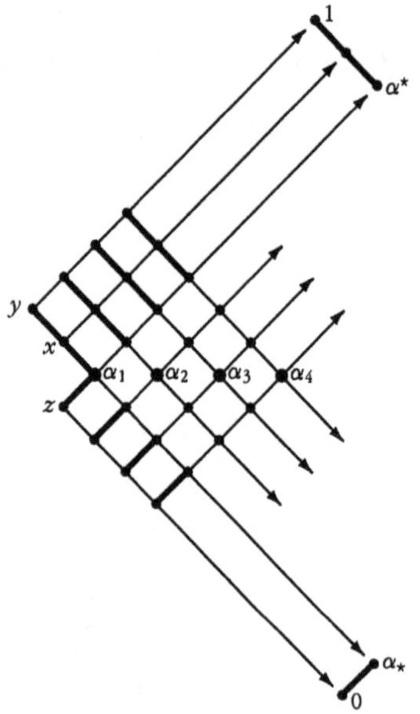

made into an Ockham algebra by defining

$$f(x) = f(y) = z, \quad f(z) = y, \quad (\forall i)\ f(\alpha_i) = \alpha_i, \quad f(\alpha_\star) = \alpha^\star, \quad f(\alpha^\star) = \alpha_\star$$

and extending to the whole of L. Then L is fixed point complete and fixed point distributive. Here we have $\Psi = \Theta$, the classes being indicated by thick lines. The Φ_ω-classes are the interval $[x,y]$, the intervals 'parallel' to it, and singletons otherwise.

Consider now the congruence

$$\Gamma = \vartheta(\alpha_\star, \alpha^\star).$$

If L is fixed point complete then by Theorem 2.1 we have

$$\Gamma = \vartheta_{\text{lat}}(\alpha_\star, \alpha^\star).$$

Theorem 7.9 *If $L \in \mathbf{K}_\omega$ is fixed point complete then Γ is the complement of Θ in* Con L.

Proof Since $(0,\alpha_\star) \in \Theta$, $(\alpha_\star, \alpha^\star) \in \Gamma$, and $(\alpha^\star, 1) \in \Theta$, we have

$$\Theta \vee \Gamma = \iota.$$

Now for all $x, y \in L$ we have

$$\vartheta_{\text{lat}}(x,y) \wedge \Gamma = \vartheta_{\text{lat}}(x,y) \wedge \vartheta_{\text{lat}}(\alpha_\star, \alpha^\star) = \vartheta_{\text{lat}}\big((x \vee \alpha_\star) \wedge y \wedge \alpha^\star,\ y \wedge \alpha^\star\big).$$

It follows immediately that

$$(x,y) \in \Theta \ \Rightarrow\ \vartheta_{\text{lat}}(x,y) \wedge \Gamma = \omega.$$

Since Θ is a congruence it follows by Theorem 2.1 that

$$(x,y) \in \Theta \ \Rightarrow\ \vartheta(x,y) \wedge \Gamma = \omega$$

and therefore that

$$\Theta \wedge \Gamma = \bigvee_{(x,y)\in\Theta} \vartheta(x,y) \wedge \Gamma = \bigvee_{(x,y)\in\Theta} \big(\vartheta(x,y) \wedge \Gamma\big) = \omega.$$

Consequently, Γ is the complement of Θ. ◇

Corollary $\Theta = \vartheta(0,\alpha_\star) = \vartheta(\alpha^\star, 1)$.

Proof The principal congruence $\Gamma = \vartheta_{\text{lat}}(\alpha_\star, \alpha^\star)$ is complemented. Its complement is $\vartheta_{\text{lat}}(0,\alpha_\star) \vee \vartheta_{\text{lat}}(\alpha^\star, 1) = \vartheta(0,\alpha_\star) = \vartheta(\alpha^\star, 1)$. ◇

Theorem 7.10 *If $L \in \mathbf{K}_\omega$ is fixed point complete and fixed point distributive then* Con L *contains the sublattice*

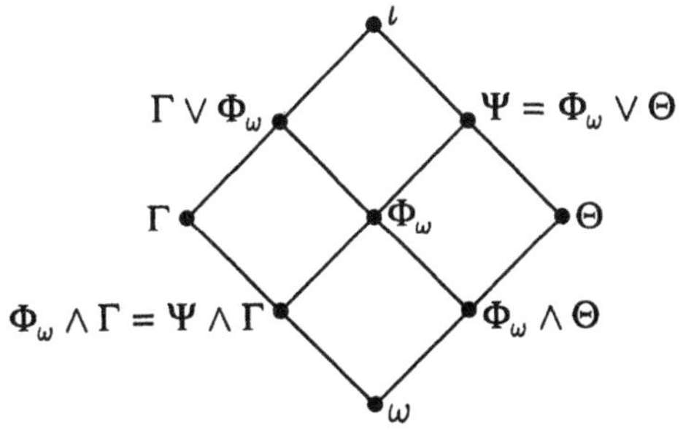

Proof By Theorems 7.7 and 7.9 we have that

$$\Psi \wedge \Gamma = (\Phi_\omega \vee \Theta) \wedge \Gamma = \Phi_\omega \wedge \Gamma,$$

whence it follows that

$$\begin{aligned}(\Gamma \vee \Phi_\omega) \wedge \Psi &= (\Gamma \wedge \Psi) \vee (\Phi_\omega \wedge \Psi)\\ &= (\Gamma \wedge \Phi_\omega) \vee (\Phi_\omega \wedge \Psi)\\ &= \Phi_\omega \wedge (\Gamma \vee \Psi)\\ &= \Phi_\omega \wedge \iota\\ &= \Phi_\omega,\end{aligned}$$

and that

$$(\Phi_\omega \wedge \Gamma) \vee (\Phi_\omega \wedge \Theta) = \Phi_\omega \wedge (\Gamma \vee \Theta) = \Phi_\omega \wedge \iota = \Phi_\omega.$$

We then have the sublattice illustrated. ◇

Corollary 1 $[\Phi_\omega, \iota] \simeq [\Phi_\omega, \Psi] \times [\Psi, \iota]$.

Proof Ψ is a complemented element of $[\Phi_\omega, \iota]$. ◇

Corollary 2 *The interval* $[\Phi_\omega, \iota]$ *is boolean if and only if both the intervals* $[\Phi_\omega, \Psi]$ *and* $[\Psi, \iota]$ *are boolean.* ◇

Definition We shall say that L is *fixed point compact* if it is fixed point complete and there is a finite subset $I^\star$ of I such that $\alpha_\star = \bigwedge_{i \in I^\star} \alpha_i$ (equivalently, $\alpha^\star = \bigvee_{i \in I^\star} \alpha_i$).

If L is fixed point compact then necessarily L is fixed point distributive. In fact,

$$x \wedge \bigvee_{i \in I} \alpha_i = x \wedge \bigvee_{i \in I^\star} \alpha_i = \bigvee_{i \in I^\star} (x \wedge \alpha_i) \leqslant \bigvee_{i \in I} (x \wedge \alpha_i),$$

and the reverse inequality is trivial.

Theorem 7.11 *If* $(L; f) \in \mathbf{K}_\omega$ *is fixed point compact then every congruence* $\varphi \in [\Psi, \iota]$ *is such that*

$$\varphi = \Psi \vee \bigvee_{(\alpha_i, \alpha_j) \in \varphi} \vartheta(\alpha_i, \alpha_j).$$

Proof Suppose first that $L \in \mathbf{K}_{p,q}$, so that $f^q(L) \in \mathbf{K}_{p,0}$. Consider the de Morgan algebra $M = T_2(f^q(L))$. For $x \in M$ and $i, j \in I$ define

$$x_{ij} = (x \vee \alpha_i) \wedge (f(x) \vee \alpha_j) \wedge (\alpha_i \vee \alpha_j).$$

Then a simple calculation reveals that x_{ij} is a fixed point of M. Given $\varphi \in [\Psi, \iota]$, let

$$\varphi^\star = \bigvee_{(\alpha_i,\alpha_j)\in\varphi} \vartheta(\alpha_i, \alpha_j).$$

If $x, y \in M$ are such that $(x, y) \in \varphi$ then we have $(x_{ij}, y_{ij}) \in \varphi$. It follows from the equality

$$x_{ij} \wedge \alpha_j = (x \vee \alpha_i) \wedge \alpha_j$$

that we therefore have

$$\big((x \vee \alpha_i) \wedge \alpha_j, (y \vee \alpha_i) \wedge \alpha_j\big) \in \vartheta(x_{ij}, y_{ij}) \leqslant \varphi^\star.$$

Since L is fixed point compact we have

$$(x \vee \alpha_\star) \wedge \alpha^\star = \bigvee_{j\in I^\star} \bigwedge_{i\in I^\star} [(x \vee \alpha_i) \wedge \alpha_j]$$

and so it follows from the above that

$$\big((x \vee \alpha_\star) \wedge \alpha^\star, (y \vee \alpha_\star) \wedge \alpha^\star\big) \in \varphi^\star.$$

Since $\big(x, (x \vee \alpha_\star) \wedge \alpha^\star\big) \in \Theta$ we deduce from this that

$$(x, y) \in \Theta \vee \varphi^\star \leqslant \Psi \vee \varphi^\star$$

and hence that

$$\varphi|_M = \bigvee_{(x,y)\in\varphi|_M} \vartheta(x, y)|_M \leqslant (\Psi \vee \varphi^\star)|_M.$$

The reverse inequality being trivial we therefore have, using Theorem 7.7,

$$\varphi|_M = (\Psi \vee \varphi^\star)|_M = (\Phi_q \vee \Theta \vee \varphi^\star)|_M.$$

Arguing as in the proof of Theorem 7.7, we deduce that

$$\varphi = \Phi_q \vee \Theta \vee \varphi^\star \leqslant \Psi \vee \varphi^\star,$$

whence we have equality and so the result holds for $L \in \mathbf{K}_{p,q}$.

Suppose now that $L \in \mathbf{K}_\omega$. Let $(x, y) \in \varphi \geqslant \Psi$ and consider the subalgebra B generated by $\{x, y, \alpha_\star, \alpha^\star\} \cup \operatorname{Fix} L$. We have $B \in \mathbf{K}_{p,q}$ for some p, q so, by the above,

$$(x, y) \in \varphi|_B = \Psi|_B \vee \varphi^\star|_B \leqslant (\Psi \vee \varphi^\star)|_B$$

whence $(x, y) \in \Psi \vee \varphi^\star$ and hence $\varphi \leqslant \Psi \vee \varphi^\star$. The reverse inequality is trivial. ◇

As the following example shows, Theorem 7.11 does not hold in general if L is not fixed point compact.

Example 7.5 Consider the Ockham algebra obtained by adding to the angel fish of Example 7.3 a fixed point α_0 as a complement of α_1 in the interval

$[\overline{x}, x]$. We obtain an Ockham algebra that is fixed point complete and fixed point distributive, but not fixed point compact. In this we have $\Psi = \omega$ and the congruence $\bigvee_{i,j} \vartheta(\alpha_i, \alpha_j)$ has three classes, namely $\{0\}$, $\{1\}$, and $L\backslash\{0,1\}$. The equality of Theorem 7.11 therefore fails for $\varphi = \iota$.

We now give an example of a fixed point compact de Morgan algebra in which the interval $[\Phi_\omega, \Psi] = [\omega, \Psi]$ of Con L is boolean but the interval $[\Psi, \iota]$ is not boolean.

Example 7.6 Consider the lattice

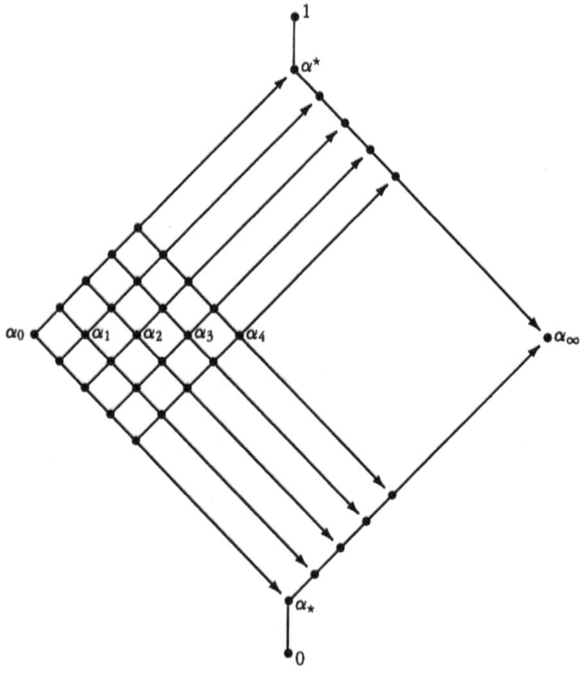

made into a fixed point compact de Morgan algebra in the obvious way with fixed points $\alpha_0, \ldots, \alpha_\infty$. In this, the congruence $\varphi > \Psi$ whose classes are

$$A = [0, \alpha_\infty), \quad B = \{\alpha_\infty\}, \quad C = (\alpha_\infty, 1], \quad D = L\backslash\{A \cup B \cup C\}$$

has no complement in $[\Psi, \iota]$. In contrast, the interval $[\omega = \Phi_\omega, \Psi] \simeq \mathbf{2}$.

Theorem 7.12 *Let $L \in \mathbf{K}_\omega$ be fixed point complete and fixed point distributive. Then*

$$[\Psi, \iota] \simeq \text{Con } T_2(A)/\Theta$$

where A is the subalgebra $\{0\} \oplus [\alpha_*, \alpha^*] \oplus \{1\}$.

Proof Suppose first that $L \in \mathbf{M}$ so that $\Psi = \Theta$ and $T_2(A) = A$. Let

$$A^\star = ([\alpha_\star, \alpha^\star]; f)$$

and note that, by the Corollary to Theorem 7.9, we have $A/\Theta \simeq A^\star$. Define

$$\mu : [\Psi, \iota] \to \mathrm{Con}\, A^\star$$

by setting (with a slight abuse of notation) $\mu(\varphi) = \varphi|_{A^\star}$ where $\varphi|_{A^\star}$ is the congruence induced on $A^\star$ by $\varphi|_A$.

It is clear that μ is a morphism for $\wedge$. To see that it is also a morphism for $\vee$, let $a, b \in A^\star$ be such that $a \leqslant b$ and $(a, b) \in (\varphi_1 \vee \varphi_2)|_{A^\star}$. Then $(a, b) \in \varphi_1 \vee \varphi_2$ and so there exist $x_0, \ldots, x_n \in L$ such that

$$a = x_0 \equiv x_1 \equiv \cdots \equiv x_n = b$$

where each $\equiv$ is either φ_1 or φ_2. Defining $y_i = (x_i \vee \alpha_\star) \wedge \alpha^\star$ we have

$$a = y_0 \equiv y_1 \equiv \cdots \equiv y_n = b$$

with each $y_i \in A^\star$. Consequently, $(a, b) \in \varphi_1|_{A^\star} \vee \varphi_2|_{A^\star}$. Thus

$$(\varphi_1 \vee \varphi_2)|_{A^\star} \leqslant \varphi_1|_{A^\star} \vee \varphi_2|_{A^\star}.$$

Since the reverse inequality is trivial, we have that μ is a $\vee$-morphism.

To see that μ is injective, let $\varphi_1, \varphi_2 \in [\Psi, \iota]$ be such that $\varphi_1|_{A^\star} = \varphi_2|_{A^\star}$. If $(x, y) \in \varphi_1$ then we have

(1) $((x \vee \alpha_\star) \wedge \alpha^\star, (y \vee \alpha_\star) \wedge \alpha^\star) \in \varphi_1|_{A^\star} = \varphi_2|_{A^\star}$;

and, by the Corollary to Theorem 7.9,

(2) $(x \vee \alpha^\star, y \vee \alpha^\star) \in \vartheta(\alpha^\star, 1) = \Theta \leqslant \Psi \leqslant \varphi_2$;

(3) $(x \wedge \alpha^\star, y \wedge \alpha^\star) \in \vartheta(0, \alpha_\star) = \Theta \leqslant \Psi \leqslant \varphi_2$.

By (1) and (2) we have $(x \vee \alpha_\star, y \vee \alpha_\star) \in \varphi_2$ which, together with (3) gives $(x, y) \in \varphi_2$. Hence $\varphi_1 \leqslant \varphi_2$. Similarly, $\varphi_2 \leqslant \varphi_1$ and so $\varphi_1 = \varphi_2$.

To see that μ is surjective, let $\vartheta \in \mathrm{Con}\, A^\star$. Let $\overline{\vartheta} \in \mathrm{Con}\, A$ be such that $\overline{\vartheta}|_{A^\star} = \vartheta$; i.e. let

$$[0]\overline{\vartheta} = [\alpha_\star]\vartheta \cup \{0\} \quad \text{and} \quad [1]\overline{\vartheta} = [\alpha^\star]\vartheta \cup \{1\}.$$

By the congruence extension property there exists $\varphi \in \mathrm{Con}\, L$ such that $\varphi|_A = \overline{\vartheta}$. Then

$$\mu(\varphi \vee \Psi) = \varphi|_{A^\star} \vee \Psi|_{A^\star} = \vartheta$$

since, by Theorem 7.5, $\Psi|_{A^\star} = \omega$.

Thus μ is an isomorphism and consequently the result holds for $L \in \mathbf{M}$.

Suppose now that $L \in \mathbf{K}_\omega$. We have

$$[\Phi_\omega, \iota] \simeq \mathrm{Con}\, L/\Phi_\omega \simeq \mathrm{Con}\, T(L) \simeq \mathrm{Con}\, T_2(L) = [\Phi_\omega|_{T_2(L)}, \iota|_{T_2(L)}].$$

For $\varphi \geqslant \Phi_\omega$ the correspondence $\varphi \leftrightarrow \varphi|_{T_2(L)}$ is therefore a bijection. Thus we see that

$$[\Psi, \iota] \simeq [\Psi|_{T_2(L)}, \iota|_{T_2(L)}].$$

Now since the result holds for $T_2(L) \in \mathbf{M}$ we have

$$[\Psi|_{T_2(L)}, \iota|_{T_2(L)}] \simeq \mathrm{Con}\ T_2(C)/\Theta$$

where $C = T_2(L) \cap A$. Since $T_2(C) = T_2(A)$ the result for $L \in \mathbf{K}_\omega$ follows. ◇

Corollary *With L as above, the following statements are equivalent:*

(1) $[\Psi, \iota]$ *is boolean;*
(2) $T_2(A)$ *is finite;*
(3) L *has finitely many fixed points.*

Proof (1) ⇒ (2) : If (1) holds then $T_2(A)$ has finitely many Θ-classes and it follows by the definition of Θ that $T_2(A)$ must be finite.

(2) ⇒ (3) : This is clear.

(3) ⇒ (1) : Observe that for fixed points α_i, α_j we have

$$\vartheta(\alpha_i, \alpha_j) = \vartheta(\alpha_i \wedge \alpha_j, \alpha_i \vee \alpha_j) = \vartheta_{\mathrm{lat}}(\alpha_i \wedge \alpha_j, \alpha_i \vee \alpha_j).$$

Since then

$$\begin{aligned}\vartheta(\alpha_i \wedge \alpha_j,\ &\alpha_i \vee \alpha_j) \wedge \vartheta(\alpha_i \vee \alpha_j, 1)\\ &= \vartheta_{\mathrm{lat}}(\alpha_i \wedge \alpha_j, \alpha_i \vee \alpha_j) \wedge [\vartheta_{\mathrm{lat}}(\alpha_i \vee \alpha_j, 1) \vee \vartheta_{\mathrm{lat}}(0, \alpha_i \wedge \alpha_j)]\\ &= \omega,\end{aligned}$$

$$\begin{aligned}\vartheta(\alpha_i \wedge \alpha_j,\ &\alpha_i \vee \alpha_j) \vee \vartheta(\alpha_i \vee \alpha_j, 1)\\ &= \vartheta_{\mathrm{lat}}(\alpha_i \wedge \alpha_j, \alpha_i \vee \alpha_j) \vee \vartheta_{\mathrm{lat}}(\alpha_i \vee \alpha_j, 1) \vee \vartheta_{\mathrm{lat}}(0, \alpha_i \wedge \alpha_j)\\ &= \iota\end{aligned}$$

we have that $\vartheta(\alpha_i, \alpha_j)$ is complemented in Con L with complement

$$\vartheta(\alpha_i, \alpha_j)' = \vartheta(\alpha_i \vee \alpha_j, 1) = \vartheta(0, \alpha_i \wedge \alpha_j).$$

If L has finitely many fixed points it therefore follows by Theorem 7.11 that every congruence $\varphi \geqslant \Psi$ has a complement in $[\Psi, \iota]$. Hence $[\Psi, \iota]$ is boolean. ◇

Example 7.7 Consider the lattice

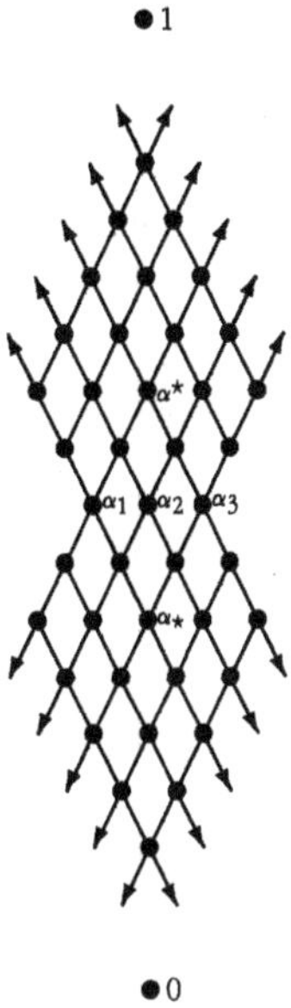

made into a de Morgan algebra with fixed points $\alpha_1, \alpha_2, \alpha_3$. Here $\Phi_\omega = \omega$ and so, by Theorem 7.7 and the Corollary to Theorem 7.9,

$$\Psi = \Theta = \vartheta(0, \alpha_\star) = \vartheta(\alpha^\star, 1).$$

By Theorem 7.12, $[\Psi, \iota]$ is boolean. In contrast, $[\omega = \Phi_\omega, \Psi]$ is not boolean. This follows from Corollary 2 to Theorem 7.10 and the fact that, since L is infinite, Con $L = [\omega = \Phi_\omega, \iota]$ is not boolean.

Theorem 7.13 *If L is fixed point compact then* $\Gamma = \bigvee_{i,j} \vartheta(\alpha_i, \alpha_j)$.

Proof By the hypothesis there exists a finite subset $I^\star$ of I such that

$$\Gamma = \vartheta(\alpha_\star, \alpha^\star) = \vartheta\Big(\bigwedge_{i \in I^\star} \alpha_i, \bigvee_{i \in I^\star} \alpha_i\Big).$$

Now, given any $j \in I^\star$, we have

$$(\forall i \in I^\star) \qquad (\alpha_i, \alpha_j) \in \bigvee_{i \in I^\star} \vartheta(\alpha_i, \alpha_j)$$

and therefore

$$\Big(\bigwedge_{i \in I^\star} \alpha_i, \alpha_j\Big) \in \bigvee_{i \in I^\star} \vartheta(\alpha_i, \alpha_j) \leqslant \bigvee_{i,j} \vartheta(\alpha_i, \alpha_j).$$

It follows that

$$\Big(\bigwedge_{i \in I^\star} \alpha_i, \bigvee_{j \in I^\star} \alpha_j\Big) \in \bigvee_{i,j} \vartheta(\alpha_i, \alpha_j)$$

and hence that

$$\Gamma = \vartheta\big(\bigwedge_{i\in I^*} \alpha_i, \bigvee_{j\in I^*} \alpha_j\big) \leqslant \bigvee_{i,j} \vartheta(\alpha_i,\alpha_j).$$

The reverse inequality being trivial, we deduce that

$$\Gamma = \bigvee_{i,j} \vartheta(\alpha_i,\alpha_j). \quad \diamond$$

The following example shows that Theorem 7.13 fails when L is not fixed point compact.

Example 7.8 Let L be the lattice obtained by adding to the cartesian ordered set $\mathbb{Z}\times\mathbb{Z}$ a biggest element 1 and a smallest element 0, and make L into a de Morgan algebra in the obvious way with fixed points $\alpha_n = (n, -n)$ for every $n \in \mathbb{Z}$. Then L is fixed point complete and fixed point distributive, but not fixed point compact. Since $\alpha^\star = 1$, we have $\Theta = \omega$ and $\Gamma = \iota$. In contrast, the congruence $\bigvee_{i,j} \vartheta(\alpha_i,\alpha_j)$ has three classes, namely $\{0\}$, $\mathbb{Z}\times\mathbb{Z}$, and $\{1\}$.

We now consider the question of precisely when the interval $[\Phi_\omega, \Psi]$ is boolean. For this purpose, we establish the following analogue of Theorem 7.12.

Theorem 7.14 *Let $L \in \mathbf{K}_\omega$ be fixed point complete and fixed point distributive. Then*

$$[\Phi_\omega, \Psi] \simeq \text{Con } T_2(B)/\Gamma$$

where B is the subalgebra $[0,\alpha_\star] \oplus [\alpha^\star, 1]$.

Proof Suppose first that $L \in \mathbf{M}$, so that $\Phi_\omega = \omega$, $\Psi = \Theta$, and $B = T_2(B)$. Consider the mapping $\lambda : [\omega, \Psi] \to \text{Con } B$ given by $\lambda(\varphi) = \varphi|_B$. Clearly, λ is a morphism for $\wedge$. To see that it is also a morphism for $\vee$, let $(a,b) \in B$ be such that $a \leqslant b$ and $(a,b) \in (\varphi_1 \vee \varphi_2)|_B$. Since $\varphi_1, \varphi_2 \leqslant \Psi$ we have that $\varphi_1 \vee \varphi_2 \leqslant \Psi$ so $\varphi_1 \vee \varphi_2$ is fixed point separating. It follows by Theorem 7.5 that $(\alpha_\star, \alpha^\star) \notin \varphi_1 \vee \varphi_2$. Consequently, either $a, b \in [0, \alpha_\star]$ or $a, b \in [\alpha^\star, 1]$. Now there exist $x_0, \dots, x_n \in L$ such that

$$a = x_0 \equiv x_1 \equiv \cdots \equiv x_n = b$$

where each $\equiv$ is either φ_1 or φ_2. Defining $y_i = (x_i \vee a) \wedge b$ we have

$$a = y_0 \equiv y_1 \equiv \cdots \equiv y_n = b$$

and, from the above observation, every $y_i \in [0, \alpha_\star]$ or every $y_i \in [\alpha^\star, 1]$, i.e. every $y_i \in B$. Consequently, $(a,b) \in \varphi_1|_B \vee \varphi_2|_B$ and so

$$(\varphi_1 \vee \varphi_2)|_B \leqslant \varphi_1|_B \vee \varphi_2|_B.$$

Since the reverse inequality is trivial, we have that λ is a $\vee$-morphism.

To see that λ is injective, suppose that $\varphi_1, \varphi_2 \in [\omega, \Psi]$ are such that $\varphi_1|_B = \varphi_2|_B$. If $(x,y) \in \varphi_1$ then since $\varphi_1 \leqslant \Psi = \Theta$ we have

(1) $(x \vee \alpha_\star) \wedge \alpha^\star = (y \vee \alpha_\star) \wedge \alpha^\star$;

(2) $(x \wedge \alpha_\star, y \wedge \alpha_\star) \in \varphi_1|_B = \varphi_2|_B$;

(3) $(x \vee \alpha^\star, y \vee \alpha^\star) \in \varphi_1|_B = \varphi_2|_B$.

By (1) and (3) we have $(x \vee \alpha_\star, y \vee \alpha_\star) \in \varphi_2$ which, together with (2) gives $(x,y) \in \varphi_2$. Hence $\varphi_1 \leqslant \varphi_2$. Similarly, $\varphi_2 \leqslant \varphi_1$ and so $\varphi_1 = \varphi_2$.

Observing that $\operatorname{Im} \lambda \subseteq [\omega|_B, \Theta|_B]$, suppose now that $\vartheta \in [\omega|_B, \Theta|_B]$. Then there exists $\varphi \in \operatorname{Con} L$ with $\varphi|_B = \vartheta$. Let $\overline{\vartheta} = \varphi \wedge \Psi$. Then we have

$$\lambda(\overline{\vartheta}) = \overline{\vartheta}|_B = \varphi|_B \wedge \Psi|_B = \vartheta \wedge \Theta|_B = \vartheta.$$

We thus see that λ induces an isomorphism $[\Phi_\omega, \Psi] \simeq [\omega|_B, \Theta|_B]$. But Theorem 7.9 gives

$$[\omega|_B, \Theta|_B] \simeq [\Gamma|_B, \iota|_B] \simeq \operatorname{Con} B/\Gamma.$$

Hence the result holds for $L \in \mathbf{M}$.

Suppose now that $L \in \mathbf{K}_\omega$. We have

$$[\Phi_\omega, \iota] \simeq \operatorname{Con} L/\Phi_\omega \simeq \operatorname{Con} T(L) \simeq \operatorname{Con} T_2(L) = [\Phi_\omega|_{T_2(L)}, \iota|_{T_2(L)}].$$

For $\varphi \geqslant \Phi_\omega$ the correspondence $\varphi \leftrightarrow \varphi|_{T_2(L)}$ is therefore a bijection. Hence

$$[\Phi_\omega, \Psi] \simeq [\Phi_\omega|_{T_2(L)}, \Psi|_{T_2(L)}].$$

Now since the result holds for $T_2(L) \in \mathbf{M}$ we have

$$[\Phi_\omega|_{T_2(L)}, \Psi|_{T_2(L)}] \simeq \operatorname{Con} T_2(C)/\Gamma$$

where $C = T_2(L) \cap B$. Since $T_2(C) = T_2(B)$, the result for $L \in \mathbf{K}_\omega$ now follows. ◇

Corollary $[\Phi_\omega, \Psi]$ *is boolean if and only if* $T_2(B)$ *is finite.*

Proof From the above, $[\Phi_\omega, \Psi]$ is boolean if and only if $T_2(B)$ has finitely many Γ-classes. It is clear that this is so if and only if $T_2(B)$ is finite. ◇

Finally, we give an example of a $\mathbf{K}_\omega$-algebra that is fixed point compact, with the congruences Φ_ω, Ψ, Θ, Γ distinct and the interval $[\Phi_\omega, \iota]$ boolean.

Example 7.9 Consider the lattice

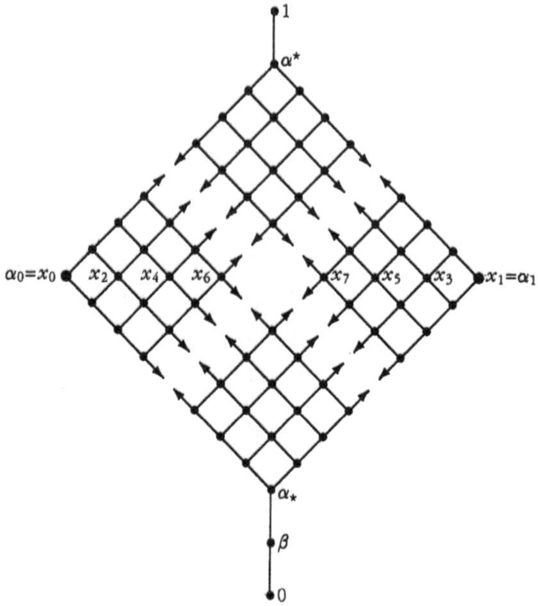

made into a $\mathbf{K}_\omega$-algebra by defining

$$f(0) = f(\beta) = 1, \quad f(1) = 0, \quad f(x_0) = x_0, \quad f(x_1) = x_1,$$
$$(\forall i \geqslant 1) \; f(x_{2i}) = x_{2i-2}, \; f(x_{2i+1}) = x_{2i-1},$$

and extending to the whole of L. Then L, though not a complete lattice, is fixed point compact. It can readily be seen that here the congruences Φ_ω, Ψ, Θ, Γ are all distinct. In fact, if we denote the four parts of $[\alpha_*, \alpha^*]$ by the cardinal points W, E, N, S then

the Φ_ω-classes are $W, E, N, S, \{0, \beta\}, \{1\}$;
the Ψ-classes are $W, E, N \cup \{1\}, S \cup \{0, \beta\}$;
the Θ-classes are $\{\alpha^*, 1\}, \{0, \beta, \alpha_*\}$, singletons otherwise;
the Γ-classes are $[\alpha_*, \alpha^*]$, singletons otherwise.

Here $[\Phi_\omega, \iota]$ is the four-element boolean lattice.

8 Congruences on $\mathbf{K}_{1,1}$-algebras

We now take a close look at congruences on a $\mathbf{K}_{1,1}$-algebra L. We begin by characterising the principal congruences. Observe that in any lattice every congruence φ satisfies

$$(\vartheta_{\text{lat}}(a,b) \vee \varphi)/\varphi = \vartheta_{\text{lat}}([a]\varphi, [b]\varphi)$$

(see, for example, [**12**, page 137]) and consequently

$$(x,y) \in \vartheta_{\text{lat}}(a,b) \vee \varphi \iff ([x],[y]) \in \vartheta_{\text{lat}}([a],[b]) \quad \text{in } L/\varphi.$$

Theorem 8.1 *If* $(L;{}^{\sim}) \in \mathbf{K}_{1,1}$ *and* $a, b \in L$ *with* $a \leqslant b$ *then* $(x,y) \in \vartheta(a,b)$ *if and only if*

(1) $x \wedge a \wedge {\sim}^2 a \wedge {\sim} b = y \wedge a \wedge {\sim}^2 a \wedge {\sim} b$;
(2) $(x \wedge a \wedge {\sim} b) \vee {\sim}^2 b = (y \wedge a \wedge {\sim} b) \vee {\sim}^2 b$;
(3) $[(x \wedge a) \vee {\sim} a] \wedge {\sim}^2 a = [(y \wedge a) \vee {\sim} a] \wedge {\sim}^2 a$;
(4) $(x \wedge a) \vee {\sim} a \vee {\sim}^2 b = (y \wedge a) \vee {\sim} a \vee {\sim}^2 b$;
(5) $(x \vee b) \wedge {\sim} b \wedge {\sim}^2 a = (y \vee b) \wedge {\sim} b \wedge {\sim}^2 a$;
(6) $[(x \vee b) \wedge {\sim} b] \vee {\sim}^2 b = [(y \vee b) \wedge {\sim} b] \vee {\sim}^2 b$;
(7) $(x \vee b \vee {\sim} a) \wedge {\sim}^2 a = (y \vee b \vee {\sim} a) \wedge {\sim}^2 a$;
(8) $x \vee {\sim} a \vee b \vee {\sim}^2 b = y \vee {\sim} a \vee b \vee {\sim}^2 b$.

Proof By Theorem 2.1 we have

$$\vartheta(a,b) = \vartheta_{\text{lat}}(a,b) \vee \vartheta_{\text{lat}}({\sim} b, {\sim} a) \vee \vartheta_{\text{lat}}({\sim}^2 a, {\sim}^2 b).$$

Denote by ψ the lattice congruence $\vartheta_{\text{lat}}({\sim} b, {\sim} a) \vee \vartheta_{\text{lat}}({\sim}^2 a, {\sim}^2 b)$. Then

$$\begin{aligned}
(x,y) \in \vartheta(a,b) &\iff ([x],[y]) \in \vartheta_{\text{lat}}([a],[b]) \quad \text{in } L/\psi \\
&\iff [x] \wedge [a] = [y] \wedge [a], \quad [x] \vee [b] = [y] \vee [b] \quad \text{in } L/\psi \\
&\iff (x \wedge a, y \wedge a) \in \psi, \quad (x \vee b, y \vee b) \in \psi.
\end{aligned}$$

Now we have

$$\begin{aligned}
(x \wedge a, y \wedge a) \in \psi &\iff ([x \wedge a], [y \wedge a]) \in \vartheta_{\text{lat}}({\sim} b, {\sim} a) \quad \text{in } L/\vartheta_{\text{lat}}({\sim}^2 a, {\sim}^2 b) \\
&\iff \begin{cases} [x \wedge a] \wedge [{\sim} b] = [y \wedge a] \wedge [{\sim} b], \\ [x \wedge a] \vee [{\sim} a] = [y \wedge a] \vee [{\sim} a] \end{cases} \\
&\iff \begin{cases} (x \wedge a \wedge {\sim} b, y \wedge a \wedge {\sim} b) \in \vartheta_{\text{lat}}({\sim}^2 a, {\sim}^2 b), \\ ((x \wedge a) \vee {\sim} a, (y \wedge a) \vee {\sim} a) \in \vartheta_{\text{lat}}({\sim}^2 a, {\sim}^2 b) \end{cases} \\
&\iff (1) \to (4) \text{ hold in } L.
\end{aligned}$$

Similarly, we have

$$(x \vee b, y \vee b) \in \psi \iff \begin{cases} ((x \vee b) \wedge {\sim} b, (y \vee b) \wedge {\sim} b) \in \vartheta_{\text{lat}}({\sim}^2 a, {\sim}^2 b), \\ (x \vee b \vee {\sim} a, y \vee b \vee {\sim} a) \in \vartheta_{\text{lat}}({\sim}^2 a, {\sim}^2 b) \end{cases}$$
$$\iff (5) \to (8) \text{ hold in } L.$$

Thus $(x, y) \in \vartheta(a, b)$ if and only (1) → (8) hold. ◇

Since (2) and (4) hold trivially if $(L; {}^\circ) \in \mathbf{MS}$, we have :

Corollary 1 [36] *If* $(L; {}^\circ) \in \mathbf{MS}$ *and* $a, b \in L$ *with* $a \leqslant b$ *then* $(x, y) \in \vartheta(a, b)$ *if and only if*

(1′) $x \wedge a \wedge b^\circ = y \wedge a \wedge b^\circ$;
(3′) $(x \wedge a) \vee (a^\circ \wedge a^{\circ\circ}) = (y \wedge a) \vee (a^\circ \wedge a^{\circ\circ})$;
(5′) $(x \vee b) \wedge b^\circ \wedge a^{\circ\circ} = (y \vee b) \wedge b^\circ \wedge a^{\circ\circ}$;
(6′) $(x \wedge b^\circ) \vee b^{\circ\circ} = (y \wedge b^\circ) \vee b^{\circ\circ}$;
(7′) $(x \vee b \vee a^\circ) \wedge a^{\circ\circ} = (y \vee b \vee a^\circ) \wedge a^{\circ\circ}$;
(8′) $x \vee a^\circ \vee b^{\circ\circ} = y \vee a^\circ \vee b^{\circ\circ}$. ◇

Since (5′) and (7′) hold trivially if $(L; {}^-) \in \mathbf{M}$, we have :

Corollary 2 [84] *If* $(L; {}^-) \in \mathbf{M}$ *and* $a, b \in L$ *with* $a \leqslant b$ *then* $(x, y) \in \vartheta(a, b)$ *if and only if*

(1″) $x \wedge a \wedge \overline{b} = y \wedge a \wedge \overline{b}$;
(3″) $(x \vee \overline{a}) \wedge a = (y \vee \overline{a}) \wedge a$;
(6″) $(x \wedge \overline{b}) \vee b = (y \wedge \overline{b}) \vee b$;
(8″) $x \vee \overline{a} \vee b = y \vee \overline{a} \vee b$. ◇

Corollary 3 [74] *If* $(L; {}^\star) \in \mathbf{S}$ *and* $a, b \in L$ *with* $a \leqslant b$ *then* $(x, y) \in \vartheta(a, b)$ *if and only if*

(*i*) $x \wedge a = y \wedge a$;
(*ii*) $(x \vee b) \wedge (a^{\star\star} \vee b^\star) = (y \vee b) \wedge (a^{\star\star} \vee b^\star)$.

Proof Writing ⋆ for ∘ in the equalities of Corollary 1, we see that (1′) and (5′) hold trivially, (3′) gives (*i*), and (6′) implies (8′). As (6′) implies that $x \wedge b^\star = y \wedge b^\star$, (*ii*) follows from (6′) and (7′). Conversely, taking the supremum with $b^{\star\star}$ and the infimum with $a^{\star\star}$ of both sides of (*ii*), we obtain (6′) and (7′) respectively. ◇

Corollary 4 *If* $(L; {}') \in \mathbf{B}$ *and* $a, b \in L$ *with* $a \leqslant b$ *then* $\vartheta(a, b) = \vartheta_{\text{lat}}(a, b)$.

Proof Writing ′ for ⋆ in (*ii*) of Corollary 3 and using distributivity, we obtain $x \vee b = y \vee b$. ◇

Theorem 8.2 *If* $(L;\sim) \in \mathbf{K}_{1,1}$ *and* $a, b \in L$ *with* $a \leqslant b$ *and* $a \wedge {\sim}a = b \wedge {\sim}b$ *then*

$$\vartheta(a,b) = \vartheta_{\text{lat}}(a,b) \vee \vartheta_{\text{lat}}({\sim}b, {\sim}a) = \vartheta_{\text{lat}}(a \vee {\sim}b, b \vee {\sim}a).$$

Proof Note that $a \wedge {\sim}a = b \wedge {\sim}b$ gives ${\sim}^2 a \wedge {\sim}a = {\sim}^2 b \wedge {\sim}b$ and ${\sim}a \vee {\sim}^2 a = {\sim}b \vee {\sim}^2 b$. Hence $\vartheta_{\text{lat}}({\sim}b, {\sim}a) = \vartheta_{\text{lat}}({\sim}^2 a, {\sim}^2 b)$ and the first equality follows.

Now $(a,b) \in \vartheta(a,b)$ and $({\sim}b, {\sim}a) \in \vartheta(a,b)$, so $(a \vee {\sim}b, b \vee {\sim}a) \in \vartheta(a,b)$ and consequently

$$\vartheta_{\text{lat}}(a \vee {\sim}b, b \vee {\sim}a) \leqslant \vartheta(a,b).$$

But since $b \wedge {\sim}b \leqslant a \leqslant b$ we have

$$a \wedge (a \vee {\sim}b) = a = (b \wedge a) \vee (b \wedge {\sim}b) = b \wedge (a \vee {\sim}b);$$
$$a \vee b \vee {\sim}a = b \vee {\sim}a = b \vee b \vee {\sim}a,$$

which shows that $(a,b) \in \vartheta_{\text{lat}}(a \vee {\sim}b, b \vee {\sim}a)$. Since $a \wedge {\sim}a \leqslant {\sim}b \leqslant {\sim}a$ we have likewise that $({\sim}b, {\sim}a) \in \vartheta_{\text{lat}}(a \vee {\sim}b, b \vee {\sim}a)$. Hence

$$\vartheta(a,b) = \vartheta_{\text{lat}}(a,b) \vee \vartheta_{\text{lat}}({\sim}b, {\sim}a) \leqslant \vartheta_{\text{lat}}(a \vee {\sim}b, b \vee {\sim}a). \quad \diamondsuit$$

Theorem 8.3 *The class* **S** *is the largest subvariety of* **MS** *in which every principal congruence is a principal lattice congruence.*

Proof It follows immediately from Theorem 8.2 that if $(L;\,^\star) \in \mathbf{S}$ then

$$\vartheta(a,b) = \vartheta_{\text{lat}}(a \vee b^\star, b \vee a^\star),$$

a fact that was first observed in **[74]**. To complete the proof, we need therefore only exhibit an algebra in **K** in which not every principal congruence is a principal lattice congruence. For this purpose, consider the four-element chain $0 < a < b < 1$ with ${\sim}0 = 1$, ${\sim}a = b$, ${\sim}b = a$, ${\sim}1 = 0$. Here

$$\vartheta(0,a) \equiv \{\{0,a\},\{b,1\}\}$$

is not a principal lattice congruence. $\diamondsuit$

We shall now consider the question of when a principal congruence $\vartheta(a,b)$ is complemented in Con L. For this purpose, we first concentrate on the case where $L \in \mathbf{M}$. Here the situation is described by the following results of Sankappanavar **[84]**.

Theorem 8.4 *Let* $(L, \,^-) \in \mathbf{M}$ *and let* $a,b,c,d \in L$ *be such that* $a \leqslant b$ *and* $c \leqslant d$. *Then*

$$\vartheta(a,b) \wedge \vartheta(c,d) = \vartheta(a \vee c, a \vee c \vee (b \wedge d)) \vee \vartheta(a \vee \overline{d}, a \vee \overline{d} \vee (b \wedge \overline{c})).$$

Proof Using the formula

$$\vartheta_{\text{lat}}(a,b) \wedge \vartheta_{\text{lat}}(c,d) = \vartheta_{\text{lat}}((a \vee c) \wedge b \wedge d, b \wedge d)$$

and the fact that $\vartheta_{\mathrm{lat}}(x \wedge y, x) = \vartheta_{\mathrm{lat}}(y, x \vee y)$ we have, by Theorem 2.1,

$$\begin{aligned}
&\vartheta(a,b) \wedge \vartheta(c,d)\\
&\quad= [\vartheta_{\mathrm{lat}}(a,b) \vee \vartheta_{\mathrm{lat}}(\overline{b},\overline{a})] \wedge [\vartheta_{\mathrm{lat}}(c,d) \vee \vartheta_{\mathrm{lat}}(\overline{d},\overline{c})]\\
&\quad= [\vartheta_{\mathrm{lat}}(a,b) \wedge \vartheta_{\mathrm{lat}}(c,d)] \vee [\vartheta_{\mathrm{lat}}(a,b) \wedge \vartheta_{\mathrm{lat}}(\overline{d},\overline{c})]\\
&\qquad\quad \vee [\vartheta_{\mathrm{lat}}(\overline{b},\overline{a}) \wedge \vartheta_{\mathrm{lat}}(c,d)] \vee [\vartheta_{\mathrm{lat}}(\overline{b},\overline{a}) \wedge \vartheta_{\mathrm{lat}}(\overline{d},\overline{c})]\\
&\quad= \vartheta_{\mathrm{lat}}\big((a \vee c) \wedge b \wedge d, b \wedge d\big) \vee \vartheta_{\mathrm{lat}}\big((a \vee \overline{d}) \wedge b \wedge \overline{c}, b \wedge \overline{c}\big)\\
&\qquad\quad \vee \vartheta_{\mathrm{lat}}\big((\overline{b} \vee c) \wedge \overline{a} \wedge d, \overline{a} \wedge d\big) \vee \vartheta_{\mathrm{lat}}\big((\overline{b} \vee \overline{d}) \wedge \overline{a} \wedge \overline{c}, \overline{a} \wedge \overline{c}\big)\\
&\quad= \vartheta_{\mathrm{lat}}\big(a \vee c, a \vee c \vee (b \wedge d)\big) \vee \vartheta_{\mathrm{lat}}\big(a \vee \overline{d}, a \vee \overline{d} \vee (b \wedge \overline{c})\big)\\
&\qquad\quad \vee \vartheta_{\mathrm{lat}}\big((\overline{b} \vee c) \wedge \overline{a} \wedge d, \overline{a} \wedge d\big) \vee \vartheta_{\mathrm{lat}}\big((\overline{b} \vee \overline{d}) \wedge \overline{a} \wedge \overline{c}, \overline{a} \wedge \overline{c}\big)\\
&\quad= \vartheta\big(a \vee c, a \vee c \vee (b \wedge d)\big) \vee \vartheta\big(a \vee \overline{d}, a \vee \overline{d} \vee (b \wedge \overline{c})\big). \quad \diamond
\end{aligned}$$

Theorem 8.5 *Every principal congruence on $(L, {}^{-}) \in \mathbf{M}$ is complemented. For $a, b \in L$ with $a \leqslant b$ we have*

$$\vartheta(a,b)' = \vartheta(\overline{a} \vee b, 1) \vee \vartheta(b \wedge \overline{b}, \overline{b}) \vee \vartheta(\overline{a}, \overline{a} \vee a).$$

Proof Consider the congruence

$$\varphi = \vartheta(\overline{a} \vee b, 1) \vee \vartheta(b \wedge \overline{b}, \overline{b}) \vee \vartheta(\overline{a}, \overline{a} \vee a).$$

That $\vartheta(a,b) \vee \varphi = \iota$ follows from the observations

$$\begin{aligned}
&(0, \overline{b} \wedge a) \in \vartheta(\overline{a} \vee b, 1), \quad (\overline{b} \wedge a, \overline{b} \wedge b) \in \vartheta(a,b),\\
&(\overline{b} \wedge b, \overline{b}) \in \vartheta(b \wedge \overline{b}, \overline{b}), \quad (\overline{b}, \overline{a}) \in \vartheta(a,b),\\
&(\overline{a}, \overline{a} \vee a) \in \vartheta(\overline{a}, \overline{a} \vee a), \quad (\overline{a} \vee a, \overline{a} \vee b) \in \vartheta(a,b),\\
&(\overline{a} \vee b, 1) \in \vartheta(\overline{a} \vee b, 1).
\end{aligned}$$

That $\vartheta(a,b) \wedge \varphi = \omega$ follows from a routine application of Theorem 8.4 which we leave to the reader. Hence we have $\vartheta(a,b)' = \varphi$. ◇

The above results provide the following characterisation of the class **M** of de Morgan algebras.

Theorem 8.6 *The class* **M** *is the largest subvariety of* $\mathbf{K}_{1,1}$ *in which every principal congruence is complemented.*

Proof By Theorem 8.5, every principal congruence on a de Morgan algebra is complemented. To complete the proof it therefore suffices to exhibit algebras in $\mathbf{S}$, $\overline{\mathbf{S}}$, $\mathbf{K}_1$, $\overline{\mathbf{K}}_1$ in which the property fails. For this purpose, consider the subdirectly irreducible algebras S, $\overline{S}$, K_1, $\overline{K}_1$. We have Con $S \simeq$ Con $\overline{S} \simeq$ Con $K_1 \simeq$ Con $\overline{K}_1 \simeq \mathbf{3}$, the non-complemented element in each case being the principal congruence Φ_1. ◇

Our objective now is to use the description of $\vartheta(a,b)'$ in a de Morgan algebra to determine precisely when a principal congruence on a $\mathbf{K}_{1,1}$-algebra is complemented. For this purpose, we require the following two results.

Theorem 8.7 *If* $L \in \mathbf{O}$ *and* $\vartheta \in Z(\text{Con } L)$ *then* $\vartheta|_{\sim^2 L} \in Z(\text{Con } {\sim^2} L)$.

Proof Let ϑ' be the complement of ϑ in Con L. From $\vartheta \wedge \vartheta' = \omega$ it follows that $\vartheta|_{\sim^2 L} \wedge \vartheta'|_{\sim^2 L} = \omega_{\sim^2 L}$. Since $\vartheta \vee \vartheta' = \iota$ there exist $x_0, \dots, x_n \in L$ such that

$$0 = x_0 \ \vartheta \ x_1 \ \vartheta' \ x_2 \ \vartheta \ \cdots \ \vartheta' \ x_n = 1.$$

This implies that

$$0 = x_0 \ \vartheta \ {\sim^2} x_1 \ \vartheta' \ {\sim^2} x_2 \ \vartheta \ \cdots \ \vartheta' \ {\sim^2} x_n = 1,$$

whence $\vartheta|_{\sim^2 L} \vee \vartheta'|_{\sim^2 L} = \iota_{\sim^2 L}$. ◇

Theorem 8.8 *Let* $(L; \sim) \in \mathbf{K}_{1,1}$. *Then we have* $({\sim^2}x, {\sim^2}y) \in \vartheta(a,b)$ *if and only if* $({\sim^2}x, {\sim^2}y) \in \vartheta({\sim^2}a, {\sim^2}b)$.

Proof If $({\sim^2}x, {\sim^2}y) \in \vartheta(a,b)$ then ${\sim^2}x$ and ${\sim^2}y$ satisfy the eight equations of Theorem 8.1. Applying ${\sim^2}$ to each, we obtain $({\sim^2}x, {\sim^2}y) \in \vartheta({\sim^2}a, {\sim^2}b)$.

The converse follows from the fact that $({\sim^2}a, {\sim^2}b) \in \vartheta(a,b)$ and therefore $\vartheta({\sim^2}a, {\sim^2}b) \leqslant \vartheta(a,b)$. ◇

In what follows we shall use the fact that if $(L, \sim) \in \mathbf{K}_{1,1}$ then ${\sim^2}L = {\sim}L$.

Theorem 8.9 *Let* $(L, \sim) \in \mathbf{K}_{1,1}$ *and let* $a, b \in L$ *be such that* $a \leqslant b$. *Define*

$$\varphi(a,b) = \vartheta(0, {\sim}b \wedge {\sim^2}a) \vee \vartheta({\sim}b \wedge {\sim^2}b, {\sim}b) \vee \vartheta({\sim}a, {\sim}a \vee {\sim^2}a).$$

Then we have $\vartheta(a,b) \vee \varphi(a,b) = \iota$; *and if* $\vartheta(a,b)$ *is complemented then necessarily* $\vartheta(a,b)' = \varphi(a,b)$.

Proof Consider the following chain of elements in L :

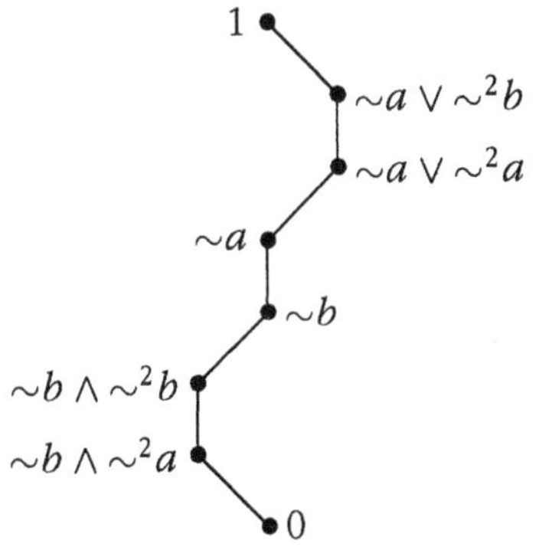

Each pair of elements that are joined by a vertical line segment are congruent modulo $\vartheta(a,b)$ and, by Theorem 2.1, $\vartheta(0,\sim b\wedge\sim^2 a)=\vartheta(\sim a\vee\sim^2 b,1)$. The above chain therefore shows that $(0,1)\in\vartheta(a,b)\vee\varphi(a,b)$ and so

$$\vartheta(a,b)\vee\varphi(a,b)=\iota.$$

Suppose now that $\vartheta(a,b)$ is complemented. Then from what we have just proved it follows that

$$\vartheta(a,b)'\leqslant\varphi(a,b).$$

To obtain the reverse inequality, we observe by Theorem 8.7 that $\vartheta(a,b)|_{\sim L}$ is $\vartheta(\sim^2 a,\sim^2 b)$ on $\sim L$, and therefore by Theorem 8.5 that the complement of $\vartheta(a,b)|_{\sim L}$ is $\varphi(a,b)|_{\sim L}$. Thus we have that

$$\vartheta(a,b)'|_{\sim L}=\big(\vartheta(a,b)|_{\sim L}\big)'=\varphi(a,b)|_{\sim L},$$

and so $\vartheta(a,b)'$ is an extension to L of $\varphi(a,b)|_{\sim L}$. Now it is clear that $\varphi(a,b)|_{\sim L}$ identifies each of the pairs

$$(0,\sim b\wedge\sim^2 a),\ (\sim b\wedge\sim^2 b,\sim b),\ (\sim a,\sim a\vee\sim^2 a)$$

of elements of $\sim L$ and that so does any extension to L of $\varphi(a,b)|_{\sim L}$. By its definition, therefore, $\varphi(a,b)$ is the smallest extension to L of $\varphi(a,b)|_{\sim L}$. Consequently, we have that

$$\varphi(a,b)\leqslant\vartheta(a,b)'$$

as required. ◇

The above knowledge of what $\vartheta(a,b)'$ is (whenever it exists) allows us to determine precisely under what conditions it does exist.

Theorem 8.10 *If* $(L;\sim)\in\mathbf{K}_{1,1}$ *then* $\vartheta(a,b)$ *is complemented if and only if*

(1) $b\wedge\sim b\wedge\sim^2 a\leqslant a$;
(2) $b\leqslant a\vee\sim a\vee\sim^2 b$;
(3) $b\wedge\sim b\leqslant a\vee\sim^2 b$;
(4) $b\wedge(\sim a\vee\sim^2 a)\leqslant a\vee\sim a$.

Proof By Theorem 8.9 we have that $\vartheta(a,b)$ is complemented if and only if $\vartheta(a,b)\wedge\varphi(a,b)=\omega$. Now, by Theorem 2.1 and the fact that $\vartheta_{\text{lat}}(x\wedge y,x)=\vartheta_{\text{lat}}(y,x\vee y)$, we have

$$\begin{aligned}\vartheta(0,\sim b\wedge\sim^2 a)&=\vartheta_{\text{lat}}(0,\sim b\wedge\sim^2 a)\vee\vartheta_{\text{lat}}(\sim^2 b\vee\sim a,1);\\ \vartheta(\sim b\wedge\sim^2 b,\sim b)&=\vartheta_{\text{lat}}(\sim b\wedge\sim^2 b,\sim b);\\ \vartheta(\sim a,\sim a\vee\sim^2 a)&=\vartheta_{\text{lat}}(\sim a,\sim a\vee\sim^2 a).\end{aligned}$$

It is now a routine matter to show that

$$\vartheta(a,b)\wedge\vartheta(0,\sim b\wedge\sim^2 a)=\omega \iff \text{(1) and (2) are satisfied;}$$
$$\vartheta(a,b)\wedge\vartheta(\sim b\wedge\sim^2 b,\sim b)=\omega \iff \text{(3) is satisfied;}$$
$$\vartheta(a,b)\wedge\vartheta(\sim a,\sim a\vee\sim^2 a)=\omega \iff \text{(4) is satisfied,}$$

from which the result follows. ◇

Corollary 1 *If $(L;\sim)\in\mathbf{MS}$ then $\vartheta(a,b)$ is complemented if and only if*

(1′) $a\wedge\sim b=\sim^2 a\wedge b\wedge\sim b$;
(2′) $b\wedge(a\vee\sim a)=b\wedge(\sim^2 a\vee\sim a)$.

Proof If $(L;\sim)\in\mathbf{MS}$ then conditions (2) and (3) of Theorem 8.10 hold, whereas conditions (1) and (4) are equivalent to (1′) and (2′) respectively. ◇

Recalling from Chapter 2 the principal congruence ϑ_a and the fact that

$$\vartheta_a=\vartheta_{\mathrm{lat}}(a\wedge\sim a,\sim a),$$

we deduce the following

Corollary 2 *If $(L;\sim)\in\mathbf{MS}$ then for every $b\in L$ the principal congruence $\vartheta(0,b)$ is complemented with $\vartheta(0,b)'=\vartheta_{\sim^2 b}$.*

Proof The four conditions of Theorem 8.10 are satisfied, so $\vartheta(0,b)'$ exists. Taking $a=0$ in Theorem 8.9 we have

$$\vartheta(0,b)'=\varphi(0,b)=\vartheta_{\mathrm{lat}}(\sim^2 b\wedge\sim b,\sim b)=\vartheta_{\sim^2 b}.\quad ◇$$

We can now apply the above to determine precisely the complemented principal congruences in a Stone algebra.

Theorem 8.11 *If $(L;\,{}^\star)\in\mathbf{S}$ then $\vartheta(a,b)$ is complemented if and only if a is complemented in $[0,b]$. Moreover, when it exists,*

$$\vartheta(a,b)'=\vartheta_{\mathrm{lat}}(a^\star\wedge b^{\star\star},1).$$

Proof By Corollary 1 of Theorem 8.10, $\vartheta(a,b)$ is complemented if and only if

(2″) $b\leqslant a\vee a^\star$.

Now (2″) gives $b\wedge a^{\star\star}\leqslant(a\vee a^\star)\wedge a^{\star\star}=a$ whence $b\wedge a^{\star\star}=a$; and this property implies (2″). Hence $\vartheta(a,b)'$ exists if and only if $b\wedge a^{\star\star}=a$.

Now if this holds then $a^\star\vee a=a^\star\vee b$ and consequently

$$b=b\wedge(a^\star\vee b)=b\wedge(a^\star\vee a)=(b\wedge a^\star)\vee a.$$

Since $b\wedge a^\star\wedge a=0$, it follows that $b\wedge a^\star$ is the complement of a in $[0,b]$. Conversely, if x is such that $a\wedge x=0$ and $a\vee x=b$ then $x\leqslant a^\star$ and $b\leqslant a\vee a^\star$ so that

$$b=b\wedge(a\vee a^\star)=a\vee(b\wedge a^\star)$$

and consequently

$$b \wedge a^{\star\star} = (a \wedge a^{\star\star}) \vee (b \wedge a^{\star} \wedge a^{\star\star}) = a \vee 0 = a.$$

As for the final statement, we first observe that $a \wedge b^{\star} = 0$ and hence $a^{\star\star} \wedge b^{\star} = 0$. We can now compute $\vartheta(a,b)'$ using Theorem 8.9. Denoting complements in $\mathrm{Con}_{\mathrm{lat}} L$ by ^{c} and recalling that

$$\vartheta_{\mathrm{lat}}(x,y)^{c} = \vartheta_{\mathrm{lat}}(0,x) \vee \vartheta_{\mathrm{lat}}(y,1),$$

we have

$$\begin{aligned}
\vartheta(a,b)' &= \vartheta(0,b^{\star}) \vee \vartheta(a^{\star},1) \\
&= \vartheta_{\mathrm{lat}}(0,b^{\star}) \vee \vartheta_{\mathrm{lat}}(b^{\star\star},1) \vee \vartheta_{\mathrm{lat}}(a^{\star},1) \vee \vartheta_{\mathrm{lat}}(0,a^{\star\star}) \\
&= \vartheta_{\mathrm{lat}}(b^{\star},a^{\star})^{c} \vee \vartheta_{\mathrm{lat}}(a^{\star\star},b^{\star\star})^{c} \\
&= [\vartheta_{\mathrm{lat}}(b^{\star},a^{\star}) \wedge \vartheta_{\mathrm{lat}}(a^{\star\star},b^{\star\star})]^{c} \\
&= [\vartheta_{\mathrm{lat}}(a^{\star} \wedge b^{\star\star} \wedge (b^{\star} \vee a^{\star\star}), a^{\star} \wedge b^{\star\star})]^{c} \\
&= [\vartheta_{\mathrm{lat}}(0, a^{\star} \wedge b^{\star\star})]^{c} \\
&= \vartheta_{\mathrm{lat}}(a^{\star} \wedge b^{\star\star}, 1). \quad \diamond
\end{aligned}$$

Corollary 1 $(\forall a \in (L,{}^{\star}))\quad \vartheta(a, a \vee a^{\star}) = \vartheta(a^{\star\star}, 1)$.

Proof From the above, $\vartheta(a, a \vee a^{\star})' = \vartheta_{\mathrm{lat}}(a^{\star},1) = \vartheta(a^{\star\star},1)'$. ◇

Corollary 2 *If $L \in \mathbf{B}$ then all principal congruences are complemented.* ◇

We shall now consider more closely, for $L \in \mathbf{K}_{1,1}$, the structure of Con L and its relationship to Con $\sim\!L$. We know that the lattice Con L is algebraic and distributive. As for Con $\sim\!L$, it follows from Theorem 8.6 that if $\sim\!L$ is finite then, every congruence being a finite join of principal congruences, Con $\sim\!L$ is boolean. It is remarkable that the converse of this is also true. As mentioned in Chapter 3, this was shown by Sankappanavar **[84]** using techniques from universal algebra. We thus have :

Theorem 8.12 *If $L \in \mathbf{M}$ then* Con L *is boolean if and only if L is finite.* ◇

Theorem 8.13 *If $L \in \mathbf{M}$ then* Con L *is a distributive algebraic lattice whose compact elements form a boolean sublattice.*

Proof By Theorem 8.4 and the fact that Con L is distributive, it follows that the set of compact elements of Con L is closed under intersection whence it forms a sublattice of Con L. That it is boolean follows from Theorem 8.5. ◇

In a $\mathbf{K}_{1,1}$-algebra L we clearly have $\Phi_1 = \cdots = \Phi_\omega$ and we shall denote this congruence simply by Φ. Every lattice congruence that is contained in

Φ is a congruence; so, as observed in Chapter 2, when L is finite the interval $[\omega, \Phi]$ of Con L is boolean. Furthermore,

$$[\Phi, \iota] \simeq \text{Con } L/\Phi \simeq \text{Con } {\sim}L$$

so, by Theorem 8.12, $[\Phi, \iota]$ is boolean if and only if ${\sim}L$ is finite. A further important property of the congruence Φ is the following.

Theorem 8.14 *Let* $L \in \mathbf{K}_{1,1}$. *Then* Φ *is the greatest dually dense element of* Con L.

Proof For every $\vartheta \in$ Con L we have that

$$(0,1) \in \vartheta \vee \Phi \iff (0,1) \in \vartheta.$$

It follows that $\vartheta \vee \Phi = \iota$ implies $\vartheta = \iota$, and so Φ is dually dense. Since the dually dense elements of a bounded lattice form an ideal, it remains to prove that if $\vartheta > \Phi$ then ϑ is not dually dense. Since the interval $[\Phi, \iota] \simeq$ Con ${\sim}L$ is an algebraic lattice, ϑ is a join of compact elements. One of these, say φ, has to be distinct from Φ. By Theorem 8.13, φ is complemented in $[\Phi, \iota]$, with complement φ' say. Then we have $\vartheta \vee \varphi' = \iota$ with $\varphi' \neq \iota$, whence ϑ is not dually dense. ◇

We can now extend Theorem 8.12 to the whole of $\mathbf{K}_{1,1}$.

Theorem 8.15 *If* $L \in \mathbf{K}_{1,1}$ *then* Con L *is boolean if and only if* L *is a finite de Morgan algebra.*

Proof It suffices to prove that if $L \in \mathbf{K}_{1,1}$ is such that Con L is boolean then necessarily $L \in \mathbf{M}$. But if $L \notin \mathbf{M}$ then $\Phi > \omega$ and, by Theorem 8.14, Φ is dually dense. Hence Φ has no complement, a contradiction. ◇

We shall now proceed to consider the following question. Given a $\mathbf{K}_{1,1}$-algebra L and a subvariety $\mathbf{R}$ of $\mathbf{K}_{1,1}$, what is the least congruence ϑ for which $L/\vartheta \in \mathbf{R}$? Put another way, what is the greatest homomorphic image of L that belongs to $\mathbf{R}$? Here we consider this question for $\mathbf{R} \in \{\mathbf{B}, \mathbf{K}, \mathbf{M}, \mathbf{S}\}$, the least such congruence being denoted by $\vartheta_{\mathbf{R}}$.

Theorem 8.16 $\vartheta_{\mathbf{M}} = \Phi$.

Proof Clearly, $L/\Phi \in \mathbf{M}$. Conversely, if $L/\vartheta \in \mathbf{M}$ then

$$(a,b) \in \Phi \Rightarrow [a]\vartheta = {\sim}^2([a]\vartheta) = [{\sim}^2 a]\vartheta = [{\sim}^2 b]\vartheta = {\sim}^2([b]\vartheta) = [b]\vartheta$$

whence $\Phi \leqslant \vartheta$. ◇

If we denote by $\vartheta_{\mathbf{R}}^{\text{si}}$ any congruence on L such that $L/\vartheta_{\mathbf{R}}^{\text{si}} \in \mathbf{R}$ with $L/\vartheta_{\mathbf{R}}^{\text{si}}$ subdirectly irreducible then, as proved in [99], we have

$$\vartheta_{\mathbf{R}} = \bigwedge\{\vartheta_{\mathbf{R}}^{\text{si}}\}.$$

Thus, if **R** has finitely many subdirectly irreducible algebras, say $R_1, \ldots, R_n$, and if ϑ_{R_i} is any congruence such that $L/\vartheta_{R_i} \simeq R_i$ then

$$\vartheta_{\mathbf{R}} = \bigwedge_{i=1}^{n} \vartheta_{R_i}.$$

Denoting by B, K, S, M the subdirectly irreducible algebras in **B**, **K**, **S**, **M** respectively, we therefore deduce the following result.

Theorem 8.17 *For every* $L \in \mathbf{MS}$,

(1) $\vartheta_{\mathbf{B}} = \bigwedge\{\vartheta_B\}$;
(2) $\vartheta_{\mathbf{K}} = \bigwedge\{\vartheta_B, \vartheta_K\}$;
(3) $\vartheta_{\mathbf{S}} = \bigwedge\{\vartheta_B, \vartheta_S\}$;
(4) $\vartheta_{\mathbf{M}} = \bigwedge\{\vartheta_B, \vartheta_K, \vartheta_M\}$. ◇

Corollary *The coatoms of* Con L *are of the form* ϑ_B, ϑ_K, ϑ_M, *and their intersection is* Φ. ◇

We shall now use the above results in considering the possible existence of a non-trivial node in Con L, i.e. a congruence $\vartheta \notin \{\omega, \iota\}$ that is comparable with all elements of Con L. For this purpose, we shall say that a lattice is *local* if it contains a unique coatom. This terminology is borrowed from ring theory : a local ring is a commutative ring having a unique maximal ideal. We shall say that a congruence has a *trivial kernel* if its kernel reduces to $\{0\}$.

Theorem 8.18 *For a* $\mathbf{K}_{1,1}$*-algebra* $(L, \sim)$ *the following statements are equivalent* :

(1) Con L *is a local lattice*;
(2) Φ *is maximal in* Con L;
(3) Con L *has comonolith* Φ;
(4) *the de Morgan algebra* L/Φ *is simple.*

If, moreover, $(L, \sim) \in \mathbf{MS}$ *then the conditions*

(5) *every congruence on* L, *other than* ι, *has a trivial kernel*;
(6) $(\forall x \in L \setminus \{0\}) \quad {\sim}x \leqslant {\sim}^2 x$,

are equivalent to the above.

Proof (1) ⇒ (2) : By the Corollary to Theorem 8.17, if there is a unique coatom then this is necessarily Φ.

(2) ⇒ (3) : If Φ is a coatom then it is the only one.

(3) ⇒ (1) : This is clear.

(2) ⇔ (4) : This is clear.

(3) ⇒ (5) : If (3) holds then (5) follows from the fact that, in **MS**, Φ has trivial kernel.

(5) $\Rightarrow$ (1) : Observe that if F is a family of congruences on L each of which has a trivial kernel then so does sup F in the complete lattice Con L. In fact, if $(0,x) \in \sup F$ then there exist $a_0, \ldots, a_n \in L$ and $\vartheta_1, \ldots, \vartheta_{n+1} \in F$ such that

$$0 \; \vartheta_1 \; a_1 \; \vartheta_2 \; a_2 \; \vartheta_3 \; \cdots \; \vartheta_{n-1} \; a_{n-1} \; \vartheta_n \; a_n \; \vartheta_{n+1} \; x.$$

Since ϑ_1 has a trivial kernel, $a_1 = 0$ and so, since ϑ_2 has trivial kernel, $a_2 = 0$ and so on. We deduce in this way that $x = 0$ and hence that sup F has trivial kernel.

If now (5) holds, choose $F = \mathrm{Con}\, L \setminus \{\iota\}$. Then by the above we have that $\sup F \neq \iota$ and hence, by its definition, sup F is the unique coatom of Con L.

(5) $\Leftrightarrow$ (6) : Observe first that (5) is equivalent to the condition

$$x \neq 0 \;\Rightarrow\; \vartheta(0,x) = \iota.$$

Now for $x \neq 0$ we have $\vartheta(0,x) = \iota$ if and only if $(0,1) \in \vartheta(0,x)$ and by Theorem 8.1 this is the case if and only if

$$(x \wedge {\sim}x) \vee {\sim}^2 x = {\sim}x \vee {\sim}^2 x,$$

i.e. if and only if

$${\sim}x \vee {\sim}^2 x \leqslant x \vee {\sim}^2 x.$$

Applying $\sim$ to this, we see that it is equivalent to ${\sim}x \leqslant {\sim}^2 x$. Consequently, (5) and (6) are equivalent. ◇

Corollary *If* $L \in \mathbf{MS}$ *then* Con L *has at most one non-trivial node. When such exists, it is necessarily* $\boldsymbol{\Phi}$ *and is covered by* ι.

Proof If Con L has a non-trivial node φ then the centre of Con L reduces to $\{\omega, \iota\}$. Since, by Corollary 2 of Theorem 8.10, every principal congruence $\vartheta(0,x)$ is complemented, we must have $\vartheta(0,x) = \iota$ for all $x \neq 0$. It follows that every congruence $\varphi \neq \iota$ has a trivial kernel; for otherwise we have $(0,x) \in \varphi$ for some $x \neq 0$, whence the contradiction $\iota = \vartheta(0,x) \leqslant \varphi$. Thus condition (5) above holds. Consequently, $\boldsymbol{\Phi}$ is a non-trivial node of Con L which is covered by ι.

Suppose that $\varphi < \boldsymbol{\Phi}$. Every lattice congruence contained in $\boldsymbol{\Phi}$ is a congruence, so $\boldsymbol{\Phi}^{\downarrow}$ is a principal ideal of $\mathrm{Con}_{\mathrm{lat}} L$. Since this lattice is algebraic, we can find compact congruences ϑ_1, ϑ_2 such that

$$\omega < \vartheta_1 \leqslant \varphi \leqslant \vartheta_2 \leqslant \boldsymbol{\Phi}.$$

But for any distributive lattice it is known **[15]** that the compact elements form a relatively complemented sublattice. Since clearly ϑ_1 cannot have a complement in $[\omega, \vartheta_2]$, it follows that we must have $\varphi = \boldsymbol{\Phi}$. ◇

It goes without saying that $\sim L$ is the most significant subalgebra of any $\mathbf{K}_{1,1}$-algebra. Since the class $\mathbf{K}_{1,1}$ enjoys the congruence extension property, it is quite natural to consider on the one hand the restriction $\vartheta|_{\sim L}$ to $\sim L$ of any congruence $\vartheta \in \operatorname{Con} L$, and on the other the smallest extension $\overline{\varphi}$ to $\operatorname{Con} L$ of $\varphi \in \operatorname{Con} \sim L$.

Theorem 8.19 *Let $L \in \mathbf{K}_{1,1}$ and $\vartheta \in \operatorname{Con} L$. Then*

$$(\sim^2 x, \sim^2 y) \in \vartheta \iff (x, y) \in \vartheta \vee \Phi.$$

Proof If $(\sim^2 x, \sim^2 y) \in \vartheta$ then $(\sim^2 x, \sim^2 y) \in \vartheta \vee \Phi$. As $(x, \sim^2 x) \in \Phi \leqslant \vartheta \vee \Phi$ for every $x \in L$, it follows that $(x, y) \in \vartheta \vee \Phi$.

Conversely, if $(x, y) \in \vartheta \vee \Phi$ then $(\sim^2 x, \sim^2 y) \in \vartheta \vee \Phi$ whence

$$\sim^2 x = x_0 \; \vartheta \; x_1 \; \Phi \; x_2 \; \vartheta \; \cdots \; \vartheta \; x_{n-1} \; \Phi \; x_n = \sim^2 y,$$

hence

$$\sim^2 x \; \vartheta \sim^2 x_1 \; \Phi \sim^2 x_2 \; \vartheta \; \cdots \; \vartheta \sim^2 x_{n-1} \; \Phi \sim^2 y,$$

i.e.

$$\sim^2 x \; \vartheta \sim^2 x_1 = \sim^2 x_2 \; \vartheta \; \cdots \; \vartheta \sim^2 x_{n-1} = \sim^2 y.$$

Thus we have $\sim^2 x \; \vartheta \sim^2 y$. ◇

The following property is now immediate :

Theorem 8.20 *If $L \in \mathbf{K}_{1,1}$ and $\vartheta_1, \vartheta_2 \in \operatorname{Con} L$ then the following statements are equivalent :*

(1) $\vartheta_1|_{\sim L} = \vartheta_2|_{\sim L}$;
(2) ϑ_1 *is an extension of* $\vartheta_2|_{\sim L}$;
(2′) ϑ_2 *is an extension of* $\vartheta_1|_{\sim L}$;
(3) $\sim^2 a \; \vartheta_1 \sim^2 b \iff \sim^2 a \; \vartheta_2 \sim^2 b$;
(4) $\vartheta_1 \vee \Phi = \vartheta_2 \vee \Phi$. ◇

Theorem 8.21 *Let $L \in \mathbf{K}_{1,1}$ and let $\vartheta \in \operatorname{Con} L$ with $\vartheta \geqslant \Phi$. Then $\alpha \in \operatorname{Con} L$ is an extension of $\vartheta|_{\sim L}$ if and only if $\alpha \vee \Phi = \vartheta$.*

Proof This is immediate from Theorem 8.20. ◇

Corollary *Let $\vartheta \in \operatorname{Con} L$ with $\vartheta \geqslant \Phi$. Then the dual pseudocomplement of Φ in $\vartheta^{\downarrow}$ exists and is the least extension to L of $\vartheta|_{\sim L}$.*

Proof Since the meet of any family of extensions of $\vartheta|_{\sim L}$ to L is an extension of $\vartheta|_{\sim L}$, the existence of a least extension is clear. By Theorem 8.21, the least extension of $\vartheta|_{\sim L}$ to L is the smallest α such that $\alpha \vee \Phi = \vartheta$. ◇

Theorem 8.22 *If $L \in \mathbf{K}_{1,1}$ and $\varphi \in \operatorname{Con} \sim L$ then the smallest lattice congruence on L that extends φ is given by*

$$\overline{\varphi} = \bigvee_{(\sim^2 x, \sim^2 y) \in \varphi} \vartheta_{\text{lat}}(\sim^2 x, \sim^2 y).$$

Moreover, $\overline{\varphi} \in \operatorname{Con} L$.

Proof It is clear that $\overline{\varphi} \in \operatorname{Con}_{\text{lat}} L$ and extends φ. If now $\vartheta \in \operatorname{Con}_{\text{lat}} L$ extends φ then $(\sim^2 x, \sim^2 y) \in \varphi$ gives $\vartheta_{\text{lat}}(\sim^2 x, \sim^2 y) \leqslant \vartheta$ and so $\overline{\varphi} \leqslant \vartheta$.

To prove that $\overline{\varphi} \in \operatorname{Con} L$, let $(a, b) \in \overline{\varphi}$. Then there exist $x_1, \dots, x_n \in L$ such that

$$a \equiv_1 x_1 \equiv_2 x_2 \equiv_3 \cdots \equiv_n x_n \equiv_{n+1} b$$

where each $\equiv_i$ is of the form $\vartheta_{\text{lat}}(\sim^2 x, \sim^2 y)$ for some $(\sim^2 x, \sim^2 y) \in \varphi$. Observing that

$$(p, q) \in \vartheta_{\text{lat}}(\sim^2 x, \sim^2 y) \Rightarrow (\sim p, \sim q) \in \vartheta_{\text{lat}}(\sim y, \sim x)$$

and that $(\sim^2 x, \sim^2 y) \in \varphi$ gives $(\sim y, \sim x) \in \varphi$, we deduce from

$$\sim a \equiv_{\tilde{1}} \sim x_1 \equiv_{\tilde{2}} \sim x_2 \equiv_{\tilde{3}} \cdots \equiv_{\tilde{n}} \sim x_n \equiv_{\widetilde{n+1}} \sim b$$

in which, if $\equiv_i$ is $\vartheta_{\text{lat}}(\sim^2 x, \sim^2 y)$ then $\equiv_{\tilde{i}}$ is $\vartheta_{\text{lat}}(\sim y, \sim x)$, that $(\sim a, \sim b) \in \varphi$. Thus we see that $\overline{\varphi} \in \operatorname{Con} L$. ◇

Theorem 8.23 *If $L \in \mathbf{K}_{1,1}$ then the mapping $f : \operatorname{Con} \sim L \to \operatorname{Con} L$ described by $f(\varphi) = \overline{\varphi}$ is a lattice morphism.*

Proof Clearly, on the one hand, we have $\overline{\varphi} \vee \overline{\psi} \leqslant \overline{\varphi \vee \psi}$. On the other hand, $(\overline{\varphi} \vee \overline{\psi})|_{\sim L} \geqslant \varphi, \psi$ and so $(\overline{\varphi} \vee \overline{\psi})|_{\sim L} \geqslant \varphi \vee \psi$ whence $\overline{\varphi} \vee \overline{\psi} \geqslant \overline{\varphi \vee \psi}$. Thus f is a $\vee$-morphism. To show that f is also a $\wedge$-morphism, observe that clearly $\overline{\varphi \wedge \psi} \leqslant \overline{\varphi} \wedge \overline{\psi}$. Now, by Theorem 8.22, we have

$$\overline{\varphi} \wedge \overline{\psi} = \bigvee_{(\sim^2 x, \sim^2 y) \in \varphi} \vartheta_{\text{lat}}(\sim^2 x, \sim^2 y) \wedge \bigvee_{(\sim^2 a, \sim^2 b) \in \psi} \vartheta_{\text{lat}}(\sim^2 a, \sim^2 b)$$

and so, since $\operatorname{Con} L$ is a complete distributive lattice in which the infinite distributive law $\vartheta \wedge \bigvee_i \zeta_i = \bigvee_i (\vartheta \wedge \zeta_i)$ holds, in order to obtain the reverse inequality $\overline{\varphi \wedge \psi} \geqslant \overline{\varphi} \wedge \overline{\psi}$ it suffices to prove that, for all $(\sim^2 x, \sim^2 y) \in \varphi$ and all $(\sim^2 a, \sim^2 b) \in \psi$,

$$\vartheta_{\text{lat}}(\sim^2 x, \sim^2 y) \wedge \vartheta_{\text{lat}}(\sim^2 a, \sim^2 b) \leqslant \overline{\varphi \wedge \psi}.$$

Here of course we suppose that $\sim^2 x \leqslant \sim^2 y$ and $\sim^2 a \leqslant \sim^2 b$. Now

$$\vartheta_{\text{lat}}(\sim^2 x, \sim^2 y) \wedge \vartheta_{\text{lat}}(\sim^2 a, \sim^2 b) = \vartheta_{\text{lat}}\big((\sim^2 x \wedge \sim^2 b) \vee (\sim^2 y \wedge \sim^2 a),\ \sim^2 y \wedge \sim^2 b\big),$$

and $\sim^2 x \;\varphi \sim^2 y$, $\sim^2 a \;\psi \sim^2 b$ give respectively

$$(\sim^2 x \wedge \sim^2 b) \vee (\sim^2 y \wedge \sim^2 a) \;\varphi\; (\sim^2 y \wedge \sim^2 b) \vee (\sim^2 y \wedge \sim^2 a) = \sim^2 y \wedge \sim^2 b,$$
$$(\sim^2 x \wedge \sim^2 b) \vee (\sim^2 y \wedge \sim^2 a) \;\psi\; (\sim^2 x \wedge \sim^2 b) \vee (\sim^2 y \wedge \sim^2 b) = \sim^2 y \wedge \sim^2 b.$$

It follows that

$$\big((\sim^2 x \wedge \sim^2 b) \vee (\sim^2 y \wedge \sim^2 a),\ \sim^2 y \wedge \sim^2 b\big) \in \varphi \wedge \psi$$

and so $\vartheta_{\mathrm{lat}}(\sim^2 x, \sim^2 y) \wedge \vartheta_{\mathrm{lat}}(\sim^2 a, \sim^2 b) \leqslant \overline{\varphi \wedge \psi}$ as required. ◇

Theorem 8.24 *If $L \in \mathbf{K}_{1,1}$ then the relation ξ defined on* Con L *by*

$$(\vartheta_1, \vartheta_2) \in \xi \iff \vartheta_1 \vee \Phi = \vartheta_2 \vee \Phi$$

is a lattice congruence and

$$(\mathrm{Con}\, L)/\xi \simeq [\Phi, \iota].$$

Every ξ-class is of the form $[\overline{\varphi|_{\sim L}}, \varphi]$ for a unique congruence $\varphi \in [\Phi, \iota]$.

Proof The relation ξ is clearly an equivalence relation which is compatible with $\vee$. That it is also compatible with $\wedge$ follows from the distributivity of Con L. Consider the map $g : (\mathrm{Con}\, L)/\xi \to [\Phi, \iota]$ given by $g([\vartheta]\xi) = \vartheta \vee \Phi$. It is readily seen that g is an isomorphism.

For every $\pi \in \mathrm{Con}\, L$ let $\pi^\star = \pi \vee \Phi$. By Theorem 8.20, $[\pi]\xi$ is the set of all extensions to L of $\pi^\star|_{\sim L}$, has biggest element $\pi^\star$ and smallest element $\overline{\pi^\star|_{\sim L}}$. In particular, $[\omega]\xi = \Phi^{\downarrow}$ and, since Φ is dually dense, $[\iota]\xi = \{\iota\}$. ◇

Corollary *The mapping $\varphi \mapsto \overline{\varphi|_{\sim L}}$ is a residuated dual closure on* Con L.

Proof This follows immediately by [3, Theorem 15.1]. ◇

Theorem 8.25 *For every $L \in \mathbf{K}_{1,1}$ the lattices* Con L *and* Con $\sim L$ *have isomorphic centres.*

Proof Consider the mapping $f : \mathrm{Con} \sim L \to \mathrm{Con}\, L$ given by $f(\varphi) = \overline{\varphi}$. By Theorem 8.23, this is a lattice morphism. Let $\widehat{f}$ be the restriction of f to $Z(\mathrm{Con}\, L)$. Then $\widehat{f}$ is also a lattice morphism. Since $\widehat{f}(\omega) = \omega$ and $\widehat{f}(\iota) = \iota$, we see that in fact $\widehat{f}$ is a mapping into $Z(\mathrm{Con}\, L)$. Since for $\varphi \in \mathrm{Con} \sim L$ we have $\varphi = \varphi|_{\sim L}$ and therefore $\varphi = \overline{\varphi} = \widehat{f}(\varphi)$, it follows that $\widehat{f}$ is injective. To show that it is also surjective, let $\vartheta \in Z(\mathrm{Con}\, L)$ and let α be its complement. By Theorem 8.7, $\vartheta|_{\sim L}$ and $\alpha|_{\sim L}$ belong to $Z(\mathrm{Con} \sim L)$ and we have

$$\overline{\vartheta|_{\sim L}} \vee \overline{\alpha|_{\sim L}} = \iota, \quad \overline{\vartheta|_{\sim L}} \wedge \overline{\alpha|_{\sim L}} = \omega.$$

Since $\overline{\vartheta|_{\sim L}} \leqslant \vartheta$, $\overline{\alpha|_{\sim L}} \leqslant \alpha$ and complements are unique, we deduce that

$$\vartheta = \overline{\vartheta|_{\sim L}} = \widehat{f}(\vartheta|_{\sim L}). \quad ◇$$

Theorem 8.26 *If* $L \in \mathbf{K}_{1,1}$ *is finite then* Con L *is a dual Stone lattice.*

Proof If L is finite then Con L is a finite distributive lattice and is therefore pseudocomplemented. Since, by Theorem 8.14, Φ is dually dense in Con L, the congruence ξ defined on Con L as in Theorem 8.24 is the dual of the Glivenko congruence and so can be described by

$$(\vartheta_1, \vartheta_2) \in \xi \iff \vartheta_1^+ = \vartheta_2^+$$

where $^+$ denotes dual pseudocomplements. Now for every $\vartheta \in$ Con L the smallest element of $[\vartheta]\xi$ is $\vartheta^{++} = \overline{\vartheta|_{\sim L}}$. Since $\{\vartheta^{++} \mid \vartheta \in \text{Con } L\}$ is then a sublattice of Con L by Theorem 8.23, it follows that Con L is a dual Stone lattice. ◇

Example 8.1 Consider the MS-algebra L described as follows :

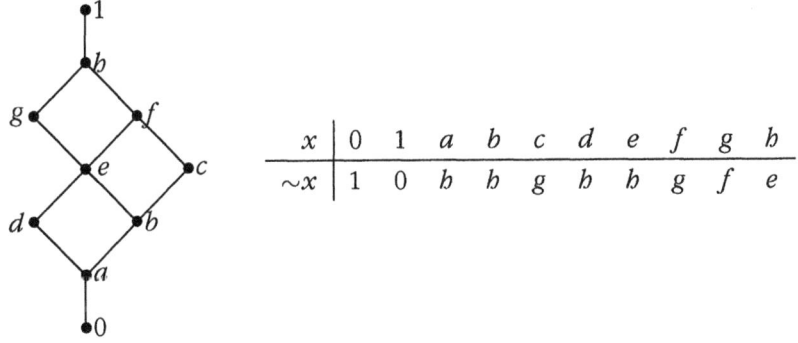

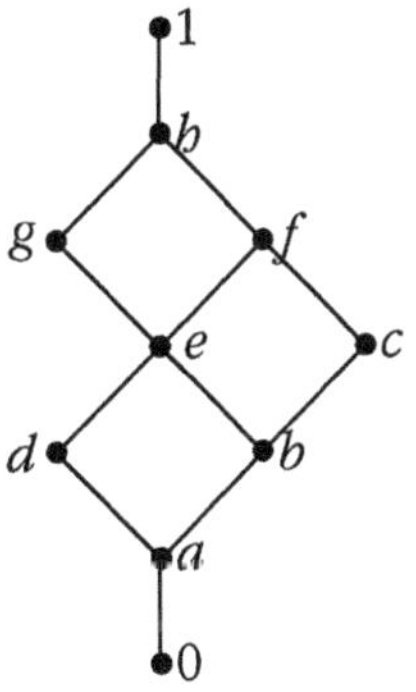

x	0	1	a	b	c	d	e	f	g	h
$\sim x$	1	0	h	h	g	h	h	g	f	e

Con L has 20 elements, namely those given by the following partitions :

$$\begin{aligned}
\iota &= \{0,1,a,b,c,d,e,f,g,h\},\\
A &= \{\{1,c,f,h\},\{0,a,b.d,e,g\}\},\\
B &= \{\{1,g,h\},\{0,a,b,c,d,e,f\}\},\\
C &= \{\{1,h\},\{c,f\},\{g\},\{0,a,b,d,e\}\},\\
D &= \{\{1\},\{a,b,c,d,e,f,g,h\},\{0\}\},\\
E &= \{\{1\},\{b,c,e,f,g,h\},\{a,d\},\{0\}\},\\
F &= \{\{1\},\{c,f,h\},\{a,b,d,e,g\},\{0\}\},\\
G &= \{\{1\},\{c,f,h\},\{b,e,g\},\{a,d\},\{0\}\},\\
H &= \{\{1\},\{d,e,f,g,h\},\{a,b,c\},\{0\}\},\\
I &= \{\{1\},\{e,f,g,h\},\{b,c\},\{d\},\{a\},\{0\}\},\\
J &= \{\{1\},\{f,h\},\{c\},\{d,e,g\},\{a,b\},\{0\}\},
\end{aligned}$$

$$
\begin{aligned}
K &= \{\{1\},\{f,h\},\{c\},\{e,g\},\{b\},\{d\},\{a\},\{0\}\},\\
L &= \{\{1\},\{g,h\},\{a,b,c,d,e,f\},\{0\}\},\\
M &= \{\{1\},\{g,h\},\{b,c,e,f\},\{a,d\},\{0\}\},\\
N &= \{\{1\},\{h\},\{c,f\},\{g\},\{a,b,d,e\},\{0\}\},\\
O &= \{\{1\},\{h\},\{c,f\},\{g\},\{b,e\},\{a,d\},\{0\}\},\\
P &= \{\{1\},\{g,h\},\{d,e,f\},\{a,b,c\},\{\{0\}\},\\
Q &= \{\{1\},\{g,h\},\{e,f\},\{b,c\},\{d\},\{a\},\{0\}\},\\
R &= \{\{1\},\{h\},\{f\},\{c\},\{g\},\{d,e\},\{a,b\},\{0\}\},\\
\omega &= \{\{1\},\{h\},\{f\},\{c\},\{g\},\{e\},\{b\},\{d\},\{a\},\{0\}\}.
\end{aligned}
$$

The congruence Φ is N and has 6 classes. The Hasse diagram for Con L is

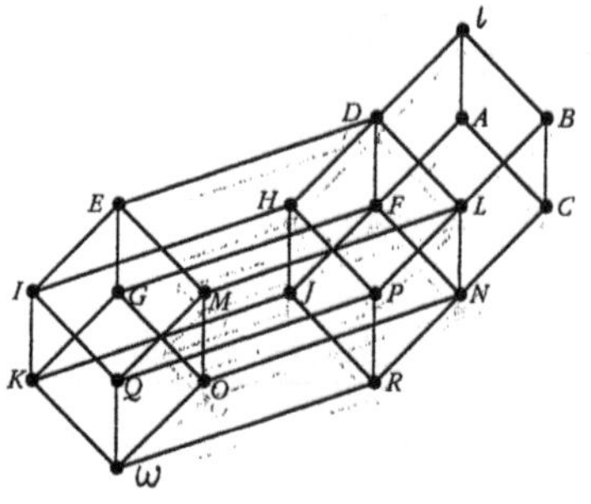

There are eight ξ-classes, namely $\{I,E,H,D\}$, $\{K,G,J,F\}$, $\{Q,M,P,L\}$, $\{\omega,O,R,N\}$, $\{A\}$, $\{B\}$, $\{C\}$, $\{\iota\}$. The principal congruences are as in the following table, where $\vartheta(x,y)$ is at the intersection of column x and row y :

0	ω									
1	ι	ω								
a	C	ι	ω							
b	C	ι	R	ω						
c	B	A	P	Q	ω					
d	C	ι	O	N	L	ω				
e	C	ι	N	O	M	R	ω			
f	B	A	L	M	O	P	Q	ω		
g	A	B	F	G	E	J	K	I	ω	
h	ι	C	D	E	G	H	I	K	Q	ω
	0	1	a	b	c	d	e	f	g	h

9 MS-spaces; fences, crowns, ...

In this chapter we shall apply both the theory of duality and the results on fixed points to a consideration of finite MS-algebras whose dual space is of a particularly simple nature, namely is connected and of length 1. As we shall see, such ordered sets are amenable to rather interesting combinatorial considerations **[50, 53]**. In order to proceed, we must first characterise the dual spaces of MS-algebras.

Definition An *MS-space* (resp. *de Morgan space*) is a Priestley space X on which there is defined a continuous antitone mapping g such that $g^2 \leqslant \mathrm{id}_X$ (resp. $g^2 = \mathrm{id}_X$).

Clearly, an MS-space (resp. a de Morgan space) is the dual space of an MS-algebra (resp. a de Morgan algebra); for $g^2 \leqslant \mathrm{id}_X$ and $g^2 = \mathrm{id}_X$ are the dual equivalents of axioms (1) and (α) respectively.

Note that if $(X; g)$ is an MS-space then $g^2[g(x)] \leqslant g(x)$ gives $g^3 \leqslant g$. But since g is antitone we also have $g \cdot g^2 \geqslant g \cdot \mathrm{id}_X$, i.e. $g^3 \geqslant g$. It therefore follows that $g^3 = g$.

Theorem 9.1 *For a Priestley space X the following statements are equivalent:*

(1) *X is the underlying set of an MS-space;*

(2) *there is a continuous dual closure map $\vartheta : X \to X$ such that* $\mathrm{Im}\,\vartheta$ *admits a continuous polarity.*

Proof (1) $\Rightarrow$ (2) : If $(X; g)$ is an MS-space, consider the mapping $\vartheta = g^2$. Clearly, ϑ is a dual closure on X and is continuous. Since g is antitone with $g^3 = g$, it is equally clear that g induces a continuous polarity on $\mathrm{Im}\,\vartheta$.

(2) $\Rightarrow$ (1) : Let $\vartheta : X \to X$ be a continuous dual closure and suppose that $\mathrm{Im}\,\vartheta$ admits a continuous polarity α. Define $g : X \to X$ by the prescription

$$(\forall x \in X) \qquad g(x) = \alpha[\vartheta(x)].$$

Then g is continuous and antitone. Since ϑ fixes the elements of $\mathrm{Im}\,\vartheta$, we have

$$(\forall x \in X) \qquad g^2(x) = \alpha\vartheta\alpha\vartheta(x) = \alpha^2\vartheta(x) = \vartheta(x)$$

whence $g^2 = \vartheta \leqslant \mathrm{id}_X$. ◇

Note that the existence of a dual closure $\vartheta : X \to X$ such that $\operatorname{Im}\vartheta$ admits a polarity is equivalent to the existence of a subset X_1 of X which admits a polarity, and a decreasing isotone retraction $\pi : X \to X_1$. In fact, it is clear that $X_1 = \operatorname{Im}\vartheta$ and that π is induced by ϑ. Observe also that, by the nature of π, the subset X_1 contains all the minimal elements of X; and that if $a, b \in X_1$ then every minimal element of the set of upper bounds of $\{a, b\}$ must also belong to X_1.

The special case where $X_1 = X$ is important. Here ϑ is necessarily id_X, so that $g^2 = \mathrm{id}_X$ and X is then a de Morgan space.

We now proceed to consider some particularly simple MS-spaces, the underlying sets of which are connected and of length 1. For each of these we shall determine the cardinality of the associated MS-algebra and that of its set of fixed points. Somewhat surprisingly, these involve the *Fibonacci numbers* and the *Lucas numbers*. In order to avoid any ambiguity, we record here that for these numbers we adopt the following definitions. The generating recurrence relation in each case being

$$x_n = x_{n-1} + x_{n-2},$$

the Fibonacci sequence $(f_n)_{n \geqslant 0}$ has $f_0 = 0$, $f_1 = 1$ and the Lucas sequence $(\ell_n)_{n \geqslant 0}$ has $\ell_0 = 2$, $\ell_1 = 1$.

Definition By an *n-fence* we shall mean an ordered set F_{2n} of the form

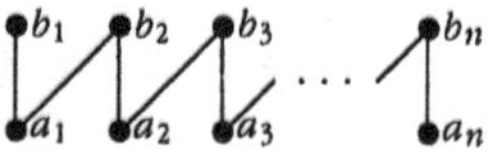

it being assumed that $n \geqslant 1$ and all the elements are distinct.

We can define two non-isomorphic fences with the same odd number $2n + 1$ of elements as follows. We let

$$F_{2n+1} = F_{2n} \cup \{b_{n+1}\}$$

with the single extra relation $a_n < b_{n+1}$; and

$$F^d_{2n+1} = F_{2n} \cup \{a_0\}$$

with the single extra relation $a_0 < b_1$. As the notation suggests, the ordered set F^d_{2n+1} is the dual of F_{2n+1}.

In what follows we shall have no interest in F^d_{2n+1}, for the following reason. If $X \simeq F^d_{2n+1}$ then by the observation following Theorem 9.1 we would require $X_1 = X$. But here X is not self-dual, so there is no appropriate g that can be defined on it.

Consider first $X \simeq F_{2n}$. Here it is clear that there is only one antitone mapping $g : X \to X$ such that $g^2 \leqslant \mathrm{id}_X$, namely that given by

$$g(a_i) = b_{n-i+1}, \quad g(b_i) = a_{n-i+1}.$$

In fact this mapping g is such that $g^2 = \mathrm{id}_X$. We can therefore deduce that there is a unique MS-algebra $L(F_{2n})$ associated with F_{2n} and that it is a de Morgan algebra. As to the size of $L(F_{2n})$, this was determined by Berman and Köhler **[29]** as an application of Theorem 5.5.

Theorem 9.2 *For* $k \geqslant 2$, $|L(F_k)| = f_{k+2}$.

Proof Applying Theorem 5.5 to F_{2n} with $x = a_n$, we obtain

$$\begin{aligned} |L(F_{2n})| &= |L(F_{2n} \setminus \{a_n\})| + |L(F_{2n-2})| \\ &= |L(F_{2n-1})| + |L(F_{2n-2})|; \end{aligned}$$

and then to F_{2n-1} with $x = b_n$, we obtain

$$|L(F_{2n-1})| = |L(F_{2n-2})| + |L(F_{2n-3})|.$$

Writing $\alpha_k = |L(F_k)|$, we therefore have the recurrence relation

$$\alpha_k = \alpha_{k-1} + \alpha_{k-2}.$$

Now $\alpha_2 = |L(F_2)| = 3 = f_4$ and $\alpha_3 = |L(F_3)| = 5 = f_5$. Hence $\alpha_k = f_{k+2}$. ◇

Example 9.1 Consider the fence F_4 :

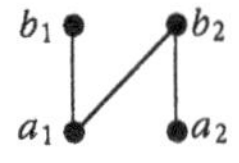

The mapping g is given by

x	a_1	a_2	b_1	b_2
$g(x)$	b_2	b_1	a_2	a_1

By Theorem 9.2, $L(F_4)$ has $f_6 = 8$ elements. Its underlying lattice is

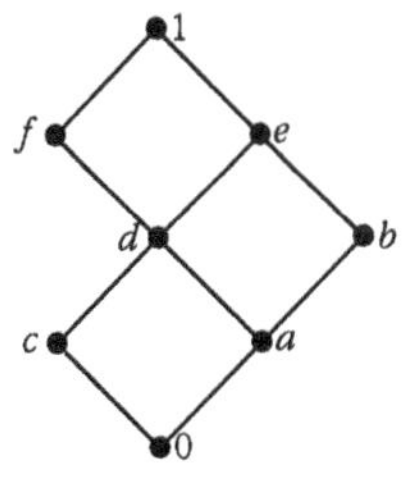

where $a = a_1^\downarrow$, $b = b_1^\downarrow$, $c = a_2^\downarrow$, $f = b_2^\downarrow$, from which the remaining elements are easily identified. Using the fact that $I^\circ = X \setminus g^{-1}(I)$, we can obtain the corresponding de Morgan negation; it is given by

x	0	a	b	c	d	e	f	1
x°	1	e	b	f	d	a	c	0

We now proceed to consider the number of fixed points of $L(F_{2n})$. By the considerations in Chapter 6, this is the number of distinguished down-sets of F_{2n}. The calculation of this is somewhat more complicated, but the outcome is pleasantly simple.

Theorem 9.3 $|\mathrm{Fix}\, L(F_{2n})| = f_{n+1}$.

Proof Consider the subsets

$$F_{2n-2} = F_{2n} \setminus \{b_1, a_n\}, \quad F_{2n-4} = F_{2n-2} \setminus \{a_1, b_n\}.$$

With g as defined above, we note that g acts inside F_{2n-2} and F_{2n-4}. Moreover, a distinguished down-set of F_{2n-2} cannot be a distinguished down-set of F_{2n-4}, and conversely. We shall establish the recurrence relation

$$|\mathrm{Fix}\, L(F_{2n})| = |\mathrm{Fix}\, L(F_{2n-2})| + |\mathrm{Fix}\, L(F_{2n-4})|.$$

For this purpose, let I be a distinguished down-set of F_{2n}. Then as $g(b_1) = a_n$ and $g(a_n) = b_1$ it is clear that I contains either b_1 or a_n.

If $b_1 \in I$ (so that $a_n \notin I$) then $I \setminus \{b_1\}$ is a distinguished down-set of F_{2n-2} which contains a_1.

If $a_n \in I$ (so that $b_1 \notin I$) and $a_1 \notin I$ then $I \setminus \{a_n\}$ is a distinguished down-set of F_{2n-2} which contains b_n.

If $\{a_1, a_n\} \subseteq I$ (so that $\{b_1, b_n\} \cap I = \emptyset$) then $I \setminus \{a_1, a_n\}$ is a distinguished down-set of F_{2n-4}.

Thus, to every distinguished down-set of F_{2n} there corresponds a distinguished down-set of either F_{2n-2} or F_{2n-4}. Clearly, this correspondence is bijective and the required relation follows.

To complete the proof, it suffices to observe from Example 9.1 that

$$|\mathrm{Fix}\, L(F_4)| = 2 = f_3. \quad \diamond$$

We now turn our attention to the fence F_{2n+1}. It is clear that if $X \simeq F_{2n+1}$ then there are precisely two antitone mappings $g : X \to X$ such that $g^2 \leqslant \mathrm{id}_X$, namely those given by

x	a_1	a_2	$\dots$	a_n	b_1	b_2	$\dots$	b_n	b_{n+1}
$g(x)$	b_n	b_{n-1}	$\dots$	b_1	a_n	a_{n-1}	$\dots$	a_1	b_1
$h(x)$	b_{n+1}	b_n	$\dots$	b_2	b_{n+1}	a_n	$\dots$	a_2	a_1

It is readily seen by relabelling X that these mappings give rise to isomorphic MS-algebras, so we shall consider only the mapping g. Since g is not surjective the corresponding MS-algebra $L(F_{2n+1})$ is not a de Morgan algebra. In fact, $\mathbf{V}(L(F_{2n+1})) = \mathbf{M}_1$ if $n > 1$, $L(F_3)$ being the subdirectly irreducible K_3. To prove this, it suffices to show that axiom (11_d) fails; or alternatively that its dual equivalent, namely $g^2 \nparallel g \vee g^0 \leqslant g^2$, fails. The latter is easier. Consider b_{n+1}; we have

$$g^2(b_{n+1}) = a_n \parallel b_1 = g(b_{n+1}), \quad b_{n+1} > a_n = g^2(b_{n+1}).$$

Example 9.2 Consider the fence F_5 :

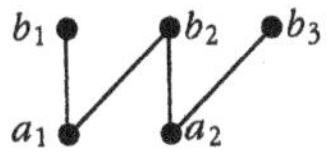

The mapping g is given by

x	a_1	a_2	b_1	b_2	b_3
$g(x)$	b_2	b_1	a_2	a_1	b_1

By Theorem 9.2, $L(F_5)$ has $f_7 = 13$ elements. Its underlying lattice is

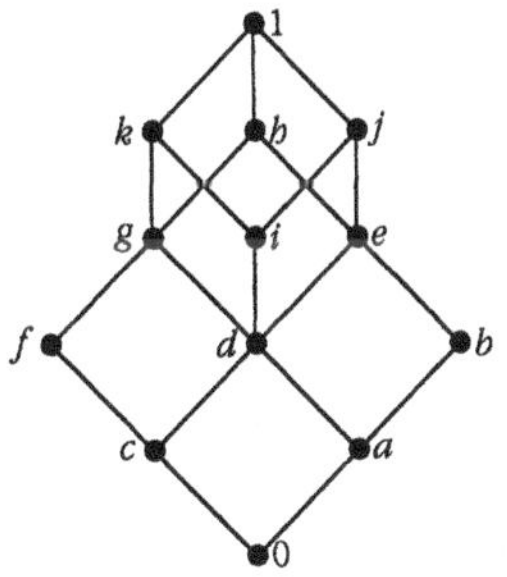

where $c = a_1^{\downarrow}$, $a = a_2^{\downarrow}$, $f = b_1^{\downarrow}$, $b = b_3^{\downarrow}$, $i = b_2^{\downarrow}$, from which the remaining elements are easily identified. The operation $^\circ$ on $L(F_5)$ is given by

x	0	a	b	c	d	e	f	g	h	i	j	k	1
x°	1	j	j	h	e	e	f	c	c	b	b	0	0

Observe from Example 9.2 that the number of fixed points of $L(F_5)$ is 2, which is the same as the number of fixed points of $L(F_4)$. In fact, we have the following result.

Theorem 9.4 $|\mathrm{Fix}\, L(F_{2n+1})| = |\mathrm{Fix}\, L(F_{2n})| = f_{n+1}$.

Proof Suppose that I is a distinguished down-set of F_{2n+1}. Then

$$g(I) \subseteq F_{2n+1} \setminus I, \quad g(F_{2n+1} \setminus I) \subseteq I.$$

If $b_{n+1} \in I$ then $I \setminus \{b_{n+1}\}$ is a distinguished down-set of F_{2n} that does not contain $g(b_{n+1}) = b_1$; whereas if $b_{n+1} \notin I$ then I is a distinguished down-set of F_{2n} that contains $g(b_{n+1}) = b_1$. It follows that the number of distinguished down-sets of $L(F_{2n+1})$ is the same as that of $L(F_{2n})$. ◇

We now turn our attention to a slightly more complicated ordered set.

Definition By an n*-crown* we shall mean an ordered set C_{2n} of the form

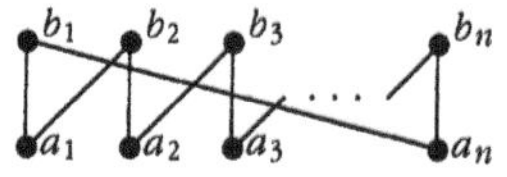

it being assumed that $n \geqslant 2$ and all the elements are distinct.

If $X \simeq C_{2n}$ then clearly $X_1 = X$. The mapping g described by

$$g(a_i) = b_{n-i+1}, \quad g(b_i) = a_{n-i+1}$$

is antitone and such that $g^2 \leqslant \mathrm{id}_X$. When n is even, this is the only such mapping. However, when n is odd there is another, namely the mapping k given by

$$k(a_i) = b_{i+\frac{1}{2}(n+1)}, \quad k(b_i) = a_{i+\frac{1}{2}(n-1)},$$

all subscripts being reduced modulo n. In fact, we have $g^2 = \mathrm{id}_X = k^2$ and so, whatever the parity of n, $L(C_{2n})$ belongs to the subvariety $\mathbf{M}$. Indeed $\mathbf{V}(L(C_{2n})) = \mathbf{M}$ if $n > 2$ since, as we shall see below, it has more than one fixed point. Note that $\mathbf{V}(L(C_4)) = \mathbf{K}$ and $L(C_4)$ has only one fixed point. As to the cardinality of $L(C_{2n})$, this rather nicely involves the Lucas numbers.

Theorem 9.5 $|L(C_{2n})| = \ell_{2n}$.

Proof Consider first the $2n$-fence F_{2n} as depicted above. If we insert a line from a_n to b_1 we obtain C_{2n}. Now by adding this line we reduce the number of down-sets. More precisely, in so doing we suppress all the down-sets of F_{2n} that contain b_1 but not a_n, that is all the down-sets of $F_{2n} \setminus \{b_n, a_1\}$ that contain b_1, hence equivalently all the down-sets of $F_{2n} \setminus \{b_1, a_1, b_n, a_n\}$. It therefore follows by Theorem 9.2 that

$$\begin{aligned} |L(C_{2n})| &= |L(F_{2n})| - |L(F_{2n-4})| \\ &= f_{2n+2} - f_{2n-2} \\ &= \ell_{2n}. \quad ◇ \end{aligned}$$

Example 9.3 The crown C_6 is

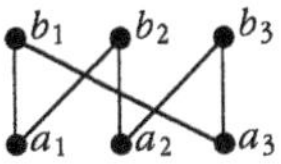

and the mappings g, k are given by

x	a_1	a_2	a_3	b_1	b_2	b_3
$g(x)$	b_3	b_2	b_1	a_3	a_2	a_1
$k(x)$	b_3	b_1	b_2	a_2	a_3	a_1

By Theorem 9.5, $L(C_6)$ has $\ell_6 = 18$ elements. Its underlying lattice is the free distributive lattice on 3 generators

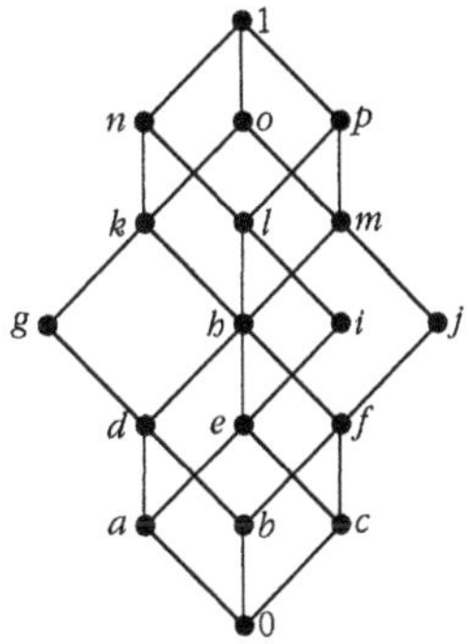

in which $a = a_1^{\downarrow}$, $b = a_2^{\downarrow}$, $c = a_3^{\downarrow}$, $g = b_2^{\downarrow}$, $i = b_1^{\downarrow}$, $j = b_3^{\downarrow}$.

On $L(C_6; g)$ the operation $^\circ$ is given by

x	0	a	b	c	d	e	f	g	h	i	j	k	l	m	n	o	p	1
x°	1	n	p	o	l	k	m	i	h	g	j	e	d	f	a	c	b	0

and on $L(C_6; k)$ it is given by

x	0	a	b	c	d	e	f	g	h	i	j	k	l	m	n	o	p	1
x°	1	n	o	p	k	l	m	g	h	i	j	d	e	f	a	b	c	0

As the following results show, the number of fixed points is governed again in a very agreeable way by the Fibonacci numbers and the Lucas numbers.

Theorem 9.6 *For every* n, $|\mathrm{Fix}\, L(C_{2n}; g)| = f_n$.

Proof Recall that g is defined on C_{2n} by

$$g(a_i) = b_{n-i+1}, \quad g(b_i) = a_{n-i+1}.$$

Observe that the subset $C_{2n} \setminus \{b_1, a_n\}$ is isomorphic to F_{2n-2} and that g acts inside it. Suppose that I is a distinguished down-set of C_{2n}. We may assume without loss of generality that I contains a_n but not b_1. Then $I \setminus \{a_n\}$ is a distinguished down-set of F_{2n-2}. Conversely, let J be a distinguished down-set of F_{2n-2}. Then $J \cup \{a_n\}$ is a distinguished down-set of C_{2n}. This correspondence between the distinguished down-sets of C_{2n} and those of F_{2n-2} is clearly bijective, so

$$|\mathrm{Fix}\, L(C_{2n}; g)| = |\mathrm{Fix}\, L(F_{2n-2})|.$$

The result now follows from Theorem 9.3. ◇

Theorem 9.7 *For n odd,* $|\mathrm{Fix}\, L(C_{2n}; k)| = \ell_n$.

Proof Recall that, for n odd, k is given by

$$k(a_i) = b_{i+\frac{1}{2}(n+1)}, \quad k(b_i) = a_{i+\frac{1}{2}(n-1)},$$

all subscripts being reduced modulo n.

Observe that a distinguished down-set I cannot contain a pair of either of the forms

$$\{b_i,\, b_{i-1+\frac{1}{2}(n+1)}\}, \quad \{b_i,\, b_{i+\frac{1}{2}(n+1)}\}.$$

In fact, if $b_i \in I$ then $\{a_{i-1}, a_i\} \subseteq I$, whence

$$b_{i-1+\frac{1}{2}(n+1)} = k(a_{i-1}) \notin I$$

and similarly

$$b_{i+\frac{1}{2}(n+1)} = k(a_i) \notin I.$$

Hence there are n forbidden pairs of the b_i, and these can be enumerated cyclically as follows :

$$\{b_1, b_{\frac{1}{2}(n+3)}\},\ \{b_{\frac{1}{2}(n+3)},\, b_2\},\ \dots,\ \{b_n, b_{\frac{1}{2}(n+1)}\},\ \{b_{\frac{1}{2}(n+1)},\, b_1\}.$$

Moreover, a distinguished down-set has cardinality n. It contains at most $\frac{1}{2}(n-1)$ of the elements b_i and among them there is of course no forbidden pair. When such elements are chosen, the elements a_j are determined since b_i and $k(b_i)$ are incomparable.

Hence we see that the number of fixed points of $L(C_{2n}; k)$ is equal to the number of subsets of the set $\{b_1, \dots, b_n\}$ that contain no forbidden pair. Our problem can therefore be restated as follows : *if S_n is a set of n points on a circle, how many subsets of S_n do not contain two neighbouring points?*

To solve this, we shall use the well-known fact that the number of subsets of $\{1, \dots, n\}$ that do not contain two consecutive integers is f_{n+2}; see, for example, [6]. Let the points of S_n be labelled consecutively $1, \dots, n$ and let t_n

be the number of subsets of S_n that do not contain two neighbouring points. If A is a subset of S_n that is counted in t_n then there are two possibilities :

(1) $1 \in A$: in this case $2 \notin A$ and $n \notin A$, so by the above fact there are f_{n-1} possibilities for A, namely $\{1\} \cup X$ where X is a subset of $\{3, \ldots, n-1\}$ that does not contain two consecutive integers.

(2) $1 \notin A$: in this case there are f_{n+1} possibilities for A, namely those subsets of $\{2, \ldots, n\}$ that do not contain two consecutive integers.

We deduce from this that

$$t_n = f_{n-1} + f_{n+1} = \ell_n,$$

whence the result follows. ◇

We now consider some more complicated ordered sets, also of length 1. For this purpose we define the sequence $(j_n)_{n \geqslant 0}$ by

$$j_0 = j_1 = 1, \quad (\forall n \geqslant 2)\ j_n = 2j_{n-1} + j_{n-2}.$$

A property of this sequence that we shall require is the following.

Theorem 9.8 $\sum_{i=0}^{n} j_i = \frac{1}{2}(j_n + j_{n+1})$.

Proof Let $x_n = \sum_{i=0}^{n} j_i$ and observe that, since $j_0 = j_1$,

$$\begin{aligned} 3x_n &= 2x_n + x_n \\ &= 2j_0 + (2j_1 + j_0) + \cdots + (2j_n + j_{n-1}) + j_n \\ &= 2j_0 + j_2 + \cdots + j_{n+1} + j_n \\ &= x_n + j_{n+1} + j_n. \end{aligned}$$

It follows that $x_n = \frac{1}{2}(j_n + j_{n+1})$. ◇

Definition By a *double fence* we shall mean an ordered set DF_{2n} of the form

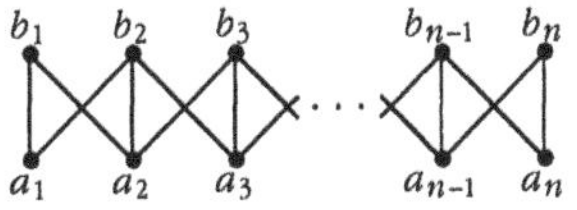

Note that $n = 1$ and $n = 2$ give respectively **2** and C_4.

On DF_{2n} there is clearly only one dual closure f with a self-dual image, namely $f = \mathrm{id}$. There are two dual isomorphisms on $\mathrm{Im} f = DF_{2n}$, namely a reflection g_1 in the horizontal, and a rotation g_2 through 180°; specifically, for each i,

$$g_1(a_i) = b_i, \quad g_1(b_i) = a_i;$$
$$g_2(a_i) = b_{n-i+1}, \quad g_2(b_i) = a_{n-i+1}.$$

Since $g_1^2 = g_2^2 = \mathrm{id}$, the MS-algebras $L(DF_{2n}; g_1)$ and $L(DF_{2n}; g_2)$ are de Morgan algebras. In fact, we can be more explicit : since $x \not\parallel g_1(x)$ we have that $L(DF_{2n}; g_1)$ is a Kleene algebra whereas, for $n \geqslant 3$, $\mathbf{V}(L(DF_{2n}; g_2)) = \mathbf{M}$ since $g_2(a_1) = b_n \parallel a_1$.

In what follows, for every ordered set X we shall denote by $\#(X)$ the number of down-sets of X, by $\#(X; a)$ the number of down-sets of X that contain the element a of X, by $\#(X; \overline{a})$ the number of down-sets of X that do not contain a, and by $\#(X; a, \overline{b})$ the number of down-sets of X that contain a but not b.

Theorem 9.9 $|L(DF_{2n})| = j_{n+1}$.

Proof Consider the element b_n of DF_{2n}. On the one hand, we have

$$\#(DF_{2n}; b_n) = \#(DF_{2n-2}; b_{n-1}) + \#(DF_{2n-4}),$$

and on the other hand

$$\begin{aligned}\#(DF_{2n}; \overline{b}_n) &= \#(DF_{2n}; a_n, \overline{b}_n) + \#(DF_{2n}; \overline{a}_n, \overline{b}_n)\\ &= |L(DF_{2n-2})| + [|L(DF_{2n-2})| - \#(DF_{2n-2}; b_{n-1})]\\ &= 2|L(DF_{2n-2})| - \#(DF_{2n-2}; b_{n-1}).\end{aligned}$$

It follows that

$$\begin{aligned}|L(DF_{2n})| &= \#(DF_{2n}; b_n) + \#(DF_{2n}; \overline{b}_n)\\ &= 2|L(DF_{2n-2})| + |L(DF_{2n-4})|.\end{aligned}$$

Writing $\alpha_n = |L(DF_{2n})|$ we therefore have

$$\alpha_n = 2\alpha_{n-1} + \alpha_{n-2}.$$

Since $\alpha_1 = |L(DF_2)| = 3 = j_2$ and $\alpha_2 = |L(DF_4)| = |L(C_4)| = 7 = j_3$, we deduce that

$$|L(DF_{2n})| = \alpha_n = j_{n+1}. \quad \diamond$$

Corollary 1 $\#(DF_{2n}; b_n) = \frac{1}{2}(j_{n-1} + j_n)$.

Proof From the first observation above we have

$$\alpha_{n-2} = |L(DF_{2n-4})| = \#(DF_{2n}; b_n) - \#(DF_{2n-2}; b_{n-1}).$$

Consequently,

$$\sum_{i=1}^{n-2} \alpha_i = \#(DF_{2n}; b_n) - \#(DF_4; b_2).$$

Note that $\#(DF_4; b_2) = 2 = j_0 + j_1$. Thus we see that

$$\begin{aligned}\#(DF_{2n}; b_n) &= \sum_{i=1}^{n-2} \alpha_i + j_0 + j_1\\ &= \sum_{i=1}^{n-2} j_{i+1} + j_0 + j_1\\ &= \sum_{i=0}^{n-1} j_i\\ &= \tfrac{1}{2}(j_{n-1} + j_n) \quad \text{by Theorem 9.8.} \quad \diamond\end{aligned}$$

Corollary 2 $\#(DF_{2n}; a_n, \overline{b}_n) = |L(DF_{2n-2})| = \alpha_{n-1} = j_n$. $\diamond$

As for fixed points, we observe that under the mapping g_1 the only distinguished down-set of $DF(2n)$ is $I = \{a_1, \ldots, a_n\}$. Consequently, we have

$$|\text{Fix } L(DF_{2n}; g_1)| = 1.$$

The situation concerning g_2 is much more complicated.

Theorem 9.10 $|\text{Fix } L(DF_{2n}; g_2)| = \begin{cases} j_{\frac{1}{2}n} & \textit{if } n \textit{ is even;} \\ j_{\frac{1}{2}(n+1)} & \textit{if } n \textit{ is odd.} \end{cases}$

Proof Consider first the case where n is even. Let $A \simeq DF_n$ be the subset (also a double fence) consisting of the elements $a_1, \ldots, a_{\frac{1}{2}n}, b_1, \ldots, b_{\frac{1}{2}n}$. For every down-set I of A let $I_\star = A' \backslash g_2(I)$ where A' denotes the complement of A in DF_{2n}. Then every distinguished down-set of DF_{2n} is of the form $I \cup I_\star$ where I is a down-set of A such that $a_{\frac{1}{2}n} \in I$ and $b_{\frac{1}{2}n} \notin I$. By Corollary 2 of Theorem 9.9, the number of such down-sets is $j_{\frac{1}{2}n}$.

Suppose now that n is odd. In this case we consider the subset B consisting of the elements $a_1, \ldots, a_{\frac{1}{2}(n+1)}, b_1, \ldots, b_{\frac{1}{2}(n-1)}$. Clearly, B is a distinguished down-set of DF_{2n}; and every distinguished down-set of DF_{2n} is of the form $I \cup I_\star$ where I is a down-set of B that contains $a_{\frac{1}{2}(n+1)}$. Clearly, this is $\#(DF_{n+1}; a_{\frac{1}{2}(n+1)}, \overline{b}_{\frac{1}{2}(n+1)})$ which, by Corollary 2 of Theorem 9.9, is $j_{\frac{1}{2}(n+1)}$. $\diamond$

Definition By a *double crown* we shall mean an ordered set DC_{2n} of the form

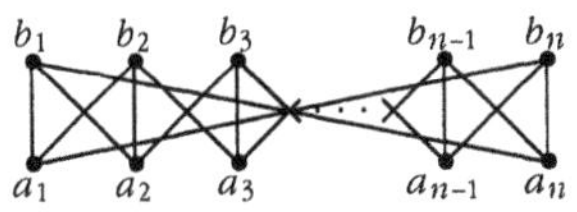

Of course, $DC_2 \simeq \mathbf{2}$ and $DC_4 \simeq DF_4$.

On DC_{2n} there is only one dual closure f with a self-dual image, namely $f = \mathrm{id}$. All antitone maps g on DC_{2n} such that $g^2 \leqslant \mathrm{id}$ are then such that $g^2 = \mathrm{id}$ and give rise to de Morgan algebras. For every value of n there are the following :

(1) the horizontal reflection g_1 given by $g_1(a_i) = b_i$, $g(b_i) = a_i$;

(2) the rotation g_2 given by $g_2(a_i) = b_{n-i+1}$, $g_2(b_i) = a_{n-i+1}$.

For odd n these are the only possibilities. For n even, however, there is also

(3) the slide-reflection k given by

$$k(a_i) = b_{i+\frac{1}{2}n},\quad k(b_i) = a_{i+\frac{1}{2}n},$$

the subscripts being reduced modulo n.

In order to determine the cardinality of $L(DC_{2n})$ we shall make use of the ordered set Z_{2n} with Hasse diagram

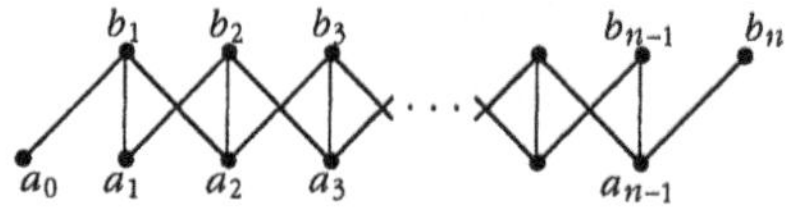

Theorem 9.11 $|L(Z_{2n})| = \frac{1}{2}(j_{n+2} - 1)$.

Proof Writing $\#(Z_{2n}) = z_n$ we have

$$\begin{aligned} z_n &= \#(Z_{2n}; a_0) + \#(Z_{2n}; \overline{a}_0) \\ &= \#(Z_{2n} \setminus \{a_0\}) + \#(Z_{2n-2}) \\ &= \#(DF_{2n+2}; b_{n+1}) + z_{n-1} \\ &= \tfrac{1}{2}(j_n + j_{n+1}) + z_{n-1}, \end{aligned}$$

the final equality following from Corollary 1 of Theorem 9.9. It follows that

$$z_n - z_0 = \frac{1}{2}\sum_{i=1}^{n}(j_i + j_{i+1}).$$

Since $z_0 = 1 = \frac{1}{2}(j_0 + j_1)$ we obtain, using Theorem 9.8,

$$\begin{aligned} z_n &= \frac{1}{2}\sum_{i=0}^{n}(j_i + j_{i+1}) \\ &= \tfrac{1}{4}(j_n + j_{n+1}) + \tfrac{1}{4}(j_{n+1} + j_{n+2}) - \tfrac{1}{2} \\ &= \tfrac{1}{2}(j_{n+2} - 1). \quad \diamond \end{aligned}$$

Theorem 9.12 $|L(DC_{2n})| = 2j_n + 1$.

Proof We obtain DC_{2n} from DF_{2n} by linking a_n with b_1, and a_1 with b_n. Consider first the effect of adding to DF_{2n} the link a_n—b_1. Clearly, this reduces the number of down-sets. More precisely, in so doing we suppress the down-sets of DF_{2n} that contain b_1 but not a_n, i.e. we suppress the down-sets of $DF_{2n}\setminus\{b_{n-1}, a_n, b_n\}$ that contain b_1; equivalently, the down-sets of $DF_{2n}\setminus\{a_1, a_2, b_1, b_{n-1}, a_n, b_n\}$; equivalently, the down-sets of Z_{2n-3}. A similar reduction in number occurs when we add the link a_1—b_n. It therefore follows by Theorems 9.9 and 9.11 that

$$\begin{aligned}|L(DC_{2n})| &= |L(DF_{2n})| - 2|Z_{2(n-3)}| \\ &= j_{n+1} - j_{n-1} + 1 \\ &= 2j_n + 1. \quad \diamond\end{aligned}$$

As for fixed points, we observe that under the mapping g_1 the only distinguished down-set of $DC(2n)$ is $I = \{a_1, \ldots, a_n\}$. Consequently,

$$|\text{Fix } L(DC_{2n}; g_1)| = 1.$$

The situation concerning g_2 is as follows.

Theorem 9.13 $|\text{Fix } L(DC_{2n}; g_2)| = \begin{cases} j_{\frac{1}{2}(n-2)} & \textit{if } n \textit{ is even;} \\ j_{\frac{1}{2}(n-1)} & \textit{if } n \textit{ is odd.} \end{cases}$

Proof Observe first that $DC\setminus\{a_1, a_n, b_1, b_n\}$ is isomorphic to DF_{2n-4} and that g_2 acts inside it.

If I is a distinguished ideal of DC_{2n} then I must contain a_1 and a_n but neither b_1 nor b_n. Consequently, $I\setminus\{a_1, a_n\}$ is a distinguished down-set of $DC\setminus\{a_1, a_n, b_1, b_n\}$. Conversely, it is clear that if J is a distinguished down-set of $DC\setminus\{a_1, a_n, b_1, b_n\}$ then $J \cup \{a_1, a_n\}$ is a distinguished down-set of DC_{2n}. This correspondence of distinguished down-sets is a bijection, so we deduce by Theorem 9.10 that

$$|\text{Fix } L(DC_{2n}; g_2)| = |\text{Fix } L(DF_{2n-2}; g_2)| = \begin{cases} j_{\frac{1}{2}(n-2)} & \text{if } n \text{ is even;} \\ j_{\frac{1}{2}(n-1)} & \text{if } n \text{ is odd.} \end{cases} \quad \diamond$$

In order to determine the number of fixed points of $L(DC_{2n}; k)$ with n even, we shall make use of the ordered set W_{2n+2} with Hasse diagram

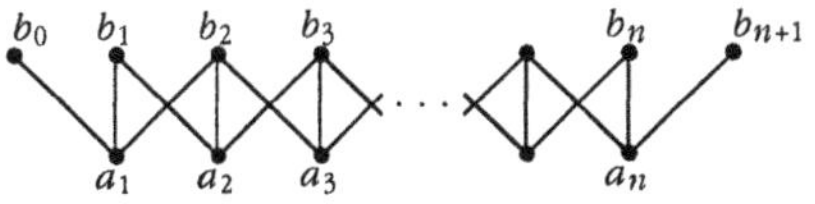

Theorem 9.14 $\#(W_{2n+2}) = \frac{1}{2}(j_{n+3} + 1)$.

Proof For every n let $\alpha_n = \#(W_{2n})$ and $\beta_n = \#(W_{2n};\overline{b}_n)$. Then we have

$$\begin{aligned}\alpha_{n+1} = \#(W_{2n+2}) &= \#(W_{2n+2};b_{n+1}) + \#(W_{2n+2};\overline{b}_{n+1})\\ &= \#(W_{2n}) + \beta_{n+1}\\ &= \alpha_n + \beta_{n+1}.\end{aligned}$$

We also have

$$\begin{aligned}\beta_n = \#(W_{2n};\overline{b}_n) &= \#(Z_{2n};\overline{a}_0)\\ &= \#(DF_{2n+2};b_{n+1})\\ &= \tfrac{1}{2}(j_n + j_{n+1}).\end{aligned}$$

Consequently,

$$\alpha_{n+1} = \alpha_n + \tfrac{1}{2}(j_{n+1} + j_{n+2}).$$

We deduce from this that

$$\alpha_{n+1} = \alpha_2 + \tfrac{1}{2}j_3 + j_4 + \cdots + j_{n+1} + \tfrac{1}{2}j_{n+2}.$$

Since $\alpha_2 = \#(W_4) = 9 = \frac{1}{2} + j_0 + j_1 + j_2 + \frac{1}{2}j_3$, we then have

$$\begin{aligned}\alpha_{n+1} &= \tfrac{1}{2} + \sum_{i=0}^{n+1} j_i + \tfrac{1}{2}j_{n+2}\\ &= \tfrac{1}{2} + \tfrac{1}{2}(j_{n+1} + j_{n+2}) + \tfrac{1}{2}j_{n+2}\\ &= \tfrac{1}{2}(1 + j_{n+3}). \quad \diamond\end{aligned}$$

Corollary $\#(DF_{2n};b_1,b_n) = \frac{1}{2}(j_{n-1}+1)$.

Proof For $n = 1,2$ the result follows by direct computation. For $n \geqslant 3$ we have

$$\#(DF_{2n};b_1,b_n) = \#(W_{2n-6}) = \tfrac{1}{2}(j_{n-1}+1). \quad \diamond$$

Theorem 9.15 *For n even,* $|\mathrm{Fix}\ L(DC_{2n};k)| = j_{\frac{1}{2}n} - 1$.

Proof Consider the subset $A = \{a_1,\dots,a_{\frac{1}{2}n},b_1,\dots,b_{\frac{1}{2}n}\}$. Let $A' = DC_{2n}\setminus A$. For every down-set I of A let $I_\star = A'\setminus k(I)$. Note that if J is a distinguished down-set of DC_{2n} then

$$b_1 \in J \Rightarrow a_{\frac{1}{2}n+1} = k(b_1) \notin J \Rightarrow b_{\frac{1}{2}n} \notin J.$$

Using the geometric nature of k it can readily be seen that a subset J of DC_{2n} is a distinguished down-set if and only if it is of the form $I \cup I_\star$ where I is a down-set of A that does not contain both b_1 and $b_{\frac{1}{2}n}$, and $I_\star$ does not contain both $b_{\frac{1}{2}n+1}$ and b_n. The latter condition is equivalent to $a_1 \in I$ and $a_{\frac{1}{2}n} \in I$. It follows that the number of fixed points of $(DC_{2n};k)$ is

$$t = \#(DF_n) - \#(DF_n;b_1,b_{\frac{1}{2}n}) - \#(DF_n;\overline{a}_1,\overline{a}_{\frac{1}{2}n}).$$

Using Theorem 9.9, the Corollary to Theorem 9.14, and the dual of Theorem 9.14, we deduce that

$$\begin{aligned} t &= j_{\frac{1}{2}n+1} - \tfrac{1}{2}(j_{\frac{1}{2}n-1}+1) - \#(W^d_{2(\frac{1}{2}n-2)+2}) \\ &= j_{\frac{1}{2}n+1} - \tfrac{1}{2}(j_{\frac{1}{2}n-1}+1) - \tfrac{1}{2}(j_{\frac{1}{2}n+1}+1) \\ &= j_{\frac{1}{2}n} - 1. \quad \diamond \end{aligned}$$

The reader will by now have realised that, even in the few cases that we have considered above, there are many problems of a combinatorial nature that arise in connection with Ockham algebras whose dual spaces are finite ordered sets of small length, the solutions to which require considerably complex arguments. There are of course many other small ordered sets to which similar considerations can be applied. As it is not our intention here to develop a 'cottage industry' in this, we simply refer the reader to **[53]** where particular ordered sets of length 2 (called *hatracks*) are considered and corresponding but quite different results are obtained, some of the algebras in question belonging to MS-subvarieties other than $\mathbf{M}$ and $\mathbf{M}_1$.

Suppose now that X is a finite ordered set and that I is a down-set of X. Then the length of I in $\mathcal{O}(X)$ is $|I|$. This is immediate from the observation that if we delete a maximal element of I then we obtain a down-set that is covered by I. In particular, consider the case where X is F_{2n} or C_{2n}. Here $\mathcal{O}(X) = L(X)$ is a de Morgan algebra of length $2n$. By a *mid-level* element we shall mean an element of length n. The question of precisely how many mid-level elements $L(X)$ has is a difficult one and involves further combinatorial arguments. In **[50]** this is answered for F_{2n} and C_{2n}. Specifically, the number of mid-level elements of $L(F_{2n})$ is

$$\sum_{m=0}^{\lfloor \frac{1}{2}n \rfloor} \binom{n-m}{m}^2$$

and the number of mid-level elements of $L(C_{2n})$ is

$$\sum_{m=0}^{\lfloor \frac{1}{2}(n-1) \rfloor} \frac{n}{n-2m} \binom{n-m-1}{m}^2.$$

Generating functions for these can also be found in **[50]**.

10 The dual space of a finite simple Ockham algebra

We recall that an n-*crown* (with $n \geqslant 2$) is an ordered set C with $2n$ elements $x_1, \ldots, x_{2n}$ whose only comparabilities are

$$x_1 < x_2, \quad x_2 > x_3, \quad x_3 < x_4, \quad \ldots, \quad x_{2n} > x_1.$$

An n-crown C is connected, has length 1, $|\text{Max } C| = |\text{Min } C| = n$, every minimal element is covered by two maximal elements, and every maximal element covers two minimal elements; so all vertices of C have degree 2.

There have been some attempts to generalise this notion. We mention two of these. In **[93]**, W. T. Trotter, Jr. defines a *crown* S_n^k as follows : for $n \geqslant 3$ and $k \geqslant 0$, S_n^k is an ordered set of length 1 with $n+k$ minimal elements $a_1, \ldots, a_{n+k}$ and $n+k$ maximal elements $b_1, \ldots, b_{n+k}$, each a_i being incomparable with $b_i, \ldots, b_{i+k}$ and being covered by the remaining $n-1$ maximal elements. Here, of course, the subscripts have to be interpreted cyclically. For example, the graph of S_4^2 is as follows :

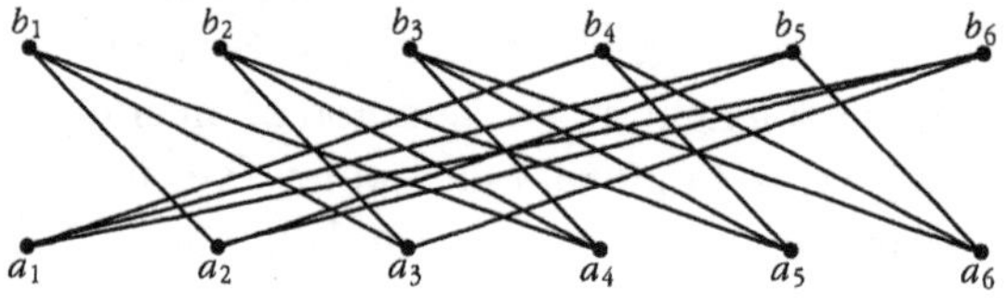

Clearly, every vertex of S_n^k has degree $n-1$ and, for every $n \geqslant 3$, an n-crown corresponds to S_3^{n-3}.

The definition of a k-*crown of order* n as given by B. Sands **[83]** is very close to Trotter's definition. A k-crown of order n (with $n \geqslant k > 0$) is an ordered set $C_k(n) = A \cup B$ where $A = \{a_1, \ldots, a_n\}$ and $B = \{b_1, \ldots, b_n\}$ are antichains and $a_i < b_j$ if and only if $j - i \in \{0, 1, \ldots, k-1\}$ modulo n. All vertices of $C_k(n)$ have degree k. All $C_k(n)$ are connected except if $k = 1$, in which case $C_1(n)$ is the disjoint union of n two-element chains. For example, $C_3(6)$ has the following graph, clearly isomorphic to S_4^2 :

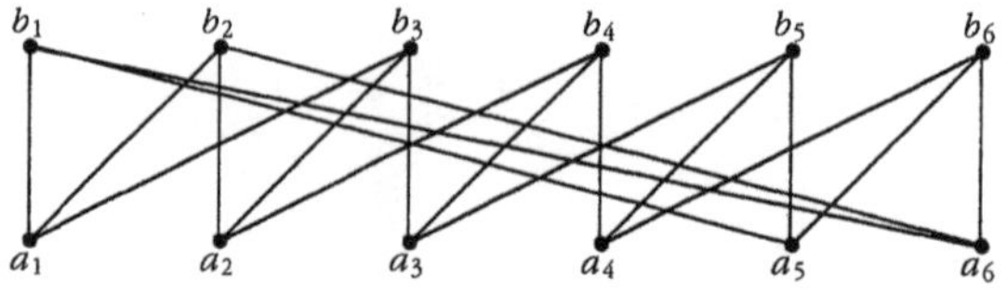

Whereas in an S_n^k every a_i is incomparable with at least one b_i, in $C_n(n)$ all a_i are comparable with all b_i. In fact, $C_n(n)$ is a complete bipartite ordered set. In every $C_k(n)$ with $k \neq n$, for two distinct minimal elements there is a maximal element that covers one of them but not the other. Except in the special case described above, S_n^k is isomorphic to $C_{n-1}(n+k)$.

For our purposes here, we introduce a more general concept.

Definition A *generalised crown* $C_{n;k}$ is an ordered set C of cardinality $2n$ (with $n \geqslant 1$) that satisfies the following conditions :

(1) C is connected;
(2) C has length 1;
(3) all vertices of C have the same degree k (with $1 \leqslant k \leqslant n$).

From this definition it follows that every $C_{n;k}$ has n minimal elements and n maximal elements. In fact, if $|\text{Min } C| = n_1$ and $|\text{Max } C| = n_2$ then the number of edges is $n_1 k = n_2 k$ whence $n_1 = n_2 = n$.

The following examples show that the conditions (1), (2), (3) are independent :

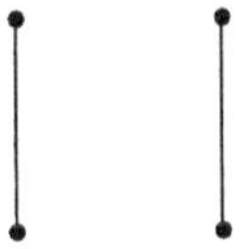

satisfies (2) and (3) but not (1);

satisfies (1) and (3) but not (2);

satisfies (1) and (2) but not (3).

The family of generalised crowns is strictly larger than that of the S_n^k. This fact is llustrated by the following $C_{6;4}$:

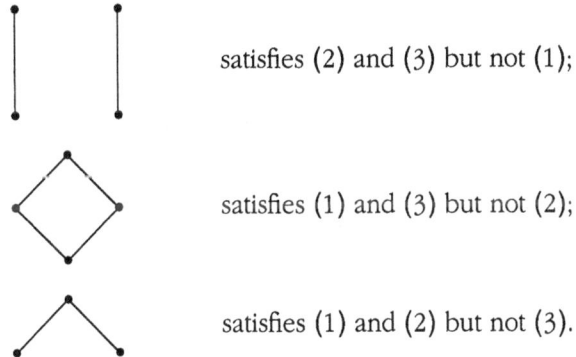

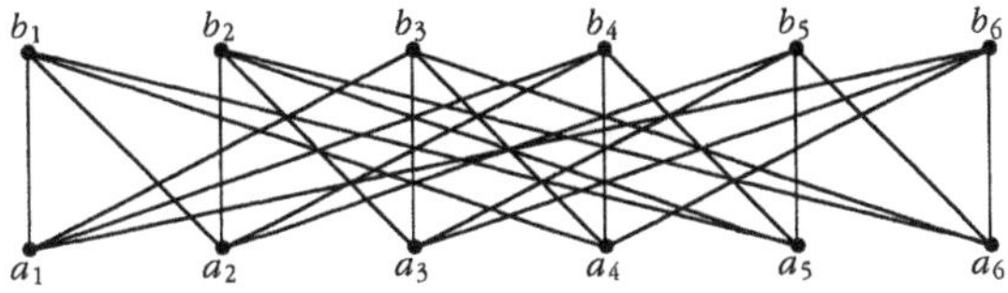

Here both a_1 and a_4 are covered by the same elements, namely b_1, b_3, b_4, b_6. Hence this ordered set is not an S_n^k. This example also shows that for some n, k there are non-isomorphic $C_{n;k}$.

Every generalised crown $C_{n;k}$ can be represented by a square $(0,1)$-matrix $[\alpha_{ij}]$ of order n in which

$$\alpha_{ij} = \begin{cases} 1 & \text{if } a_i \prec b_j; \\ 0 & \text{otherwise,} \end{cases}$$

and in which all line sums (i.e. all row sums and all column sums) are equal to k. Clearly, there is a one-one correspondence between such matrices and generalised crowns $C_{n;k}$.

For example, to the preceding $C_{6;4}$ corresponds the matrix M_1 given by

$$\begin{array}{c} \\ a_1 \\ a_2 \\ a_3 \\ a_4 \\ a_5 \\ a_6 \end{array} \begin{array}{c} \begin{array}{cccccc} b_1 & b_2 & b_3 & b_4 & b_5 & b_6 \end{array} \\ \begin{bmatrix} 1 & 0 & 1 & 1 & 0 & 1 \\ 1 & 1 & 0 & 1 & 1 & 0 \\ 0 & 1 & 1 & 0 & 1 & 1 \\ 1 & 0 & 1 & 1 & 0 & 1 \\ 1 & 1 & 0 & 1 & 1 & 0 \\ 0 & 1 & 1 & 0 & 1 & 1 \end{bmatrix} \end{array}$$

Any interchange of rows or columns yields another $C_{n;k}$ order-isomorphic to the original. For example, interchanging rows 1 and 3, and columns 2 and 6 in M_1 we obtain the matrix M_2 given by

$$\begin{array}{c} \\ a_1 \\ a_2 \\ a_3 \\ a_4 \\ a_5 \\ a_6 \end{array} \begin{array}{c} \begin{array}{cccccc} b_1 & b_2 & b_3 & b_4 & b_5 & b_6 \end{array} \\ \begin{bmatrix} 0 & 1 & 1 & 0 & 1 & 1 \\ 1 & 0 & 0 & 1 & 1 & 1 \\ 1 & 1 & 1 & 1 & 0 & 0 \\ 1 & 1 & 1 & 1 & 0 & 0 \\ 1 & 0 & 0 & 1 & 1 & 1 \\ 0 & 1 & 1 & 0 & 1 & 1 \end{bmatrix} \end{array}$$

to which corresponds the generalised crown

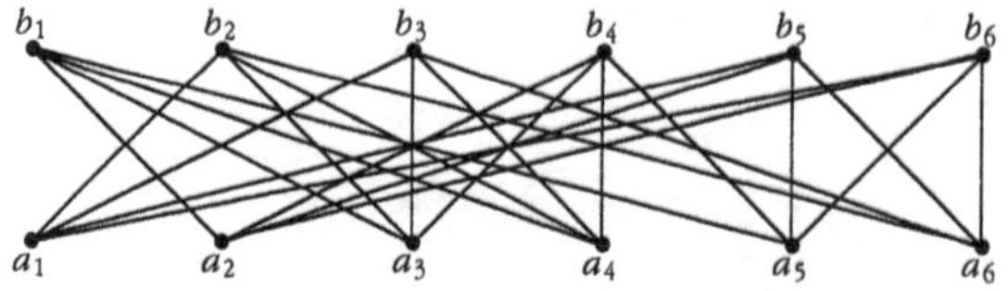

Two matrices such as M_1 and M_2 will be called *equivalent*. More precisely, M_2 is equivalent to M_1 if there are permutation matrices P and Q such that $M_2 = PM_1Q$. Clearly, two generalised crowns are order-isomorphic if and only if they have equivalent matrices.

Finally, we note that the 6×6 matrix that we associate with $C_{6;4}$ is in fact a reduced form of the 12×12 adjacency matrix of the labeled graph with 12 points as it is usually defined in graph theory **[14]**. Since $\operatorname{Min} C_{6;4}$ and $\operatorname{Max} C_{6;4}$ are totally unordered, there are only zeros in the NW and SE quarters of the adjacency matrix. Moreover, the latter is symmetric, hence the NE quarter (which is our 6×6 matrix) describes the situation unambiguously.

Our objective here is to characterise the dual space of a finite simple Ockham algebra and to determine the number of finite simple Ockham algebras in each class $\mathbf{P}_{m,n}$. For this purpose we begin by showing that every connected component of the dual space of a finite simple Ockham algebra is either a generalised crown or a singleton. In this connection, the basic result on which our investigation rests is Corollary 2 of Theorem 4.6 : *if* $(L;f) \in \mathbf{O}$ *is finite with dual space* $(X;g)$ *then* $(L;f)$ *is simple if and only if* $g^{\omega}(x) = X$ *for every* $x \in X$. It follows that the dual space $(X;g)$ of a finite simple Ockham algebra satisfies the following properties :

(1) g is surjective, hence bijective; likewise so is g^n for every n.

(2) If $|X| = N$ then for every $x \in X$ the elements $x, g(x), \dots, g^{N-1}(x)$ are distinct and $g^N(x) = x$. If we ignore the order relation, this means that X is a 'loop' and the smallest class $\mathbf{P}_{m,n}$ to which the dual algebra L belongs is $\mathbf{P}_{N,0}$. The mapping g is an order-reversing permutation of the N elements of X with a unique orbit, i.e. is an N-cycle.

(3) If N is odd then the order on X is discrete, i.e. X is an antichain. In fact, from $x < y$ we obtain $g^N(x) > g^N(y)$, giving the contradiction $x > y$.

(4) g maps every maximal element of X onto a minimal element, and conversely. In fact, if N is odd then $\operatorname{Max} X = \operatorname{Min} X$ and the result is trivial. On the other hand, if N is even, suppose that $x \in \operatorname{Max} X$ and $y < g(x)$. Then $g^{N-1}(y) > g^{N-1}(g(x)) = g^N(x) = x$, which contradicts the fact that $x \in \operatorname{Max} X$. It follows from this that $|\operatorname{Max} X| = |\operatorname{Min} X|$.

(5) The length $\ell(X)$ of X is at most one; otherwise g would act inside the set $X \backslash (\operatorname{Max} X \cup \operatorname{Min} X)$ in contradiction to property (2).

(6) All vertices of X have the same degree. Indeed, let x be an arbitrary vertex of X. Without loss of generality, we may assume that $x \in \operatorname{Min} X$. Let q be the degree of x. For every $x_i \in X$ there exists $s_i \in \mathbb{N}$ such that $x_i = g^{s_i}(x)$. As observed above, g^{s_i} is injective. If s_i is even then g^{s_i} is order-preserving, $x_i \in \operatorname{Min} X$ and is covered by q elements; if s_i is odd then g^{s_i} is order-reversing, $x_i \in \operatorname{Max} X$ and covers exactly q elements. Hence all the vertices of X have the same degree.

Thus we have proved :

Theorem 10.1 *Every connected component of the dual space of a finite simple Ockham algebra is either a generalised crown or a singleton.* ◇

In what follows, $(X; g)$, or simply X, will always denote the dual space of a finite simple Ockham algebra. We shall write the elements of X as $0, 1, \ldots, N-1$ with $0 \in \operatorname{Min} X$, and $g(0) = 1$, $g(1) = 2$, ..., $g(N-1) = 0$. With this simplification of notation, $g^r(0) = r$ modulo N and, more generally, $g^r(i) = i + r$ modulo N. We point out, once and for all, that such equalities have to be understood modulo N. Since we are using integers to denote the elements of X we shall denote the order on X by $\preccurlyeq$. Note that if $i \prec j$ then we have $i + r \prec j + r$ if r is even, and $i + r \succ j + r$ if r is odd.

We have already seen that the case where N is odd is uninteresting. Suppose then that N is even, say $N = 2n$. Note that in this case, if $i \prec i + r$ then $i \prec i + 2n - r$; in fact, since $2n - r$ must be odd, we have

$$i \prec i + r \Rightarrow i + 2n - r \succ i + r + 2n - r = i + 2n = i.$$

In particular, if $0 \prec j$ then $0 \prec 2n - j$. Note also that

$$r \prec s \iff 2n - r \prec 2n - s.$$

We define the subset $\Gamma(0)$ by

$$\Gamma(0) = \{x \in X \mid 0 \prec x,\ x \leqslant n\},$$

noting that all the elements of $\Gamma(0)$ are odd. If $x, y \in X$ are connected then we write $x \bowtie y$. If t is the number of connected components of X then for $0 \leqslant i \leqslant t - 1$ we denote by C_i the component that contains i.

The connected component that contains 0, namely $C_0 = \{x \in X \mid x \bowtie 0\}$ has the following important property. If r is the smallest non-zero element of $\Gamma(0)$ then, r being odd, we have $0 \prec r$, $r \succ 2r$, $2r \prec 3r$, ..., so C_0 is the subgroup $\langle r \rangle$ of the group $\mathbb{Z}_{2n}$, the remaining connected components being the cosets of C_0.

The following two properties appear in **[69]**.

Theorem 10.2 *C_0 is self-dual.*

Proof Let p be a fixed element of $\Gamma(0)$ and define $\varphi_p : C_0 \to X$ by

$$\varphi_p(q) = 2n - p - q.$$

If $q \in C_0$ then, since $0 \prec p$ implies $0 \prec 2n - p$, we have $q \bowtie 2n - p$ so

$$0 = 2n = 2n - q + q \bowtie 2n - q + 2n - p = 2n - p - q = \varphi_p(q).$$

Thus $\operatorname{Im} \varphi_p \subseteq C_0$. If now $r \in C_0$ then $2n - r - p \in C_0$ and

$$\varphi_p(2n - r - p) = 2n - p - 2n + r + p = r.$$

Thus $\operatorname{Im}\varphi_p = C_0$. Since p is odd, we also have

$$\varphi_p(r) \preceq \varphi_p(s) \iff 2n - r - p \preceq 2n - s - p \iff r \succeq s.$$

Hence φ_p is a dual order embedding and $C_0 = \operatorname{Im}\varphi_p$ is self-dual. ◇

If X is not connected then we have $1 = g(0) \notin C_0$; for $1 \in C_0$ gives $0 \bowtie 1$ whence $1 = g(0) \bowtie g(1) = 2$, and so on, whence X would be connected.

Theorem 10.3 *Two connected components of X are either isomorphic or dually isomorphic. More precisely,*

$$C_0 \overset{d}{\simeq} C_1 \overset{d}{\simeq} C_2 \overset{d}{\simeq} \cdots \overset{d}{\simeq} C_{t-1} \overset{d}{\simeq} C_0$$

and the number t of connected components is necessarily odd.

Proof It suffices to prove that $C_0 \overset{d}{\simeq} C_1$. For this purpose, consider the map $\psi : C_0 \to X$ defined by the prescription $\psi(q) = q + 1$. Since $0 \bowtie q$ implies $1 \bowtie q + 1$ we have $\operatorname{Im}\psi \subseteq C_1$. But if $r \in C_1$ then $r \bowtie 1$ gives $r - 1 = r + 2n - 1 \bowtie 0$, so that $r - 1 \in C_0$ with $\psi(r-1) = r$. Hence $\operatorname{Im}\psi = C_1$. Since

$$\begin{aligned} \psi(r) \preceq \psi(s) &\iff r + 1 \preceq s + 1 \\ &\iff r = r + 1 + 2n - 1 \succeq s + 1 + 2n - 1 = s \end{aligned}$$

it follows that ψ is a dual order-embedding and $C_0 \overset{d}{\simeq} C_1$. Finally, that t must be odd is clear from the sequence of dual isomorphisms. ◇

Corollary *Every connected component of X is self-dual.* ◇

We now proceed to consider the number of connected components. The simplest case arises when $\Gamma(0)$ is a singleton.

Theorem 10.4 *If $\Gamma(0) = \{r\}$ then X has $t = \gcd\{n, r\}$ connected components each of which is an $\frac{n}{t}$-crown.*

Proof In the cyclic group $\mathbb{Z}_{2n}$ the order of r is $\frac{2n}{t}$ where $t = \gcd\{2n, r\}$ (equivalently, for r odd, $t = \gcd\{n, r\}$). Since the elements of C_0 constitute the subgroup $\langle r \rangle$ of $\mathbb{Z}_{2n}$ it follows that $C_0 = \{0, r, 2r, \ldots, (\frac{2n}{t} - 1)r\}$ and we have

$$0 \prec r, \quad r \succ 2r, \quad \ldots, \quad (\tfrac{2n}{t} - 1)r \succ 0.$$

Consequently we see that C_0 is an $\frac{n}{t}$-crown. Now each connected component of X is a coset of C_0. The number of connected components is therefore $|\mathbb{Z}_{2n}|/|C_0| = t$. ◇

Corollary *X is an n-crown if and only if $\Gamma(0) = \{r\}$ with r, n coprime.* ◇

Example 10.1 Let $n = 15$ and $\Gamma(0) = \{9\}$. Then we have $t = 3$ and the connected components are the following 5-crowns :

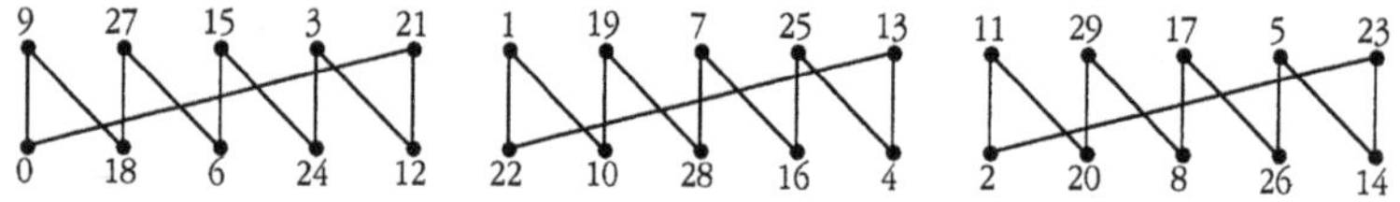

Theorem 10.4 can be generalised as follows.

Theorem 10.5 *Let* $\Gamma(0) = \{r_1, \dots, r_k\}$. *Then the number of connected components of* X *is* $t = \gcd\{n, r_1, \dots, r_k\}$.

Proof As in the proof of Theorem 10.4, consider the cyclic group $\mathbb{Z}_{2n}$. Here

$$\langle r_1 \rangle \vee \cdots \vee \langle r_k \rangle = \langle u \rangle$$

where $u = \gcd\{r_1, \dots, r_k\}$, and the order of $\langle u \rangle$ is $2n/\gcd\{2n, u\}$. It follows that the number of components of X is the number of cosets of $\langle u \rangle$ in $\mathbb{Z}_{2n}$, namely $t = |\mathbb{Z}_{2n}|/|\langle u \rangle| = \gcd\{2n, u\} = \gcd\{n, r_1, \dots, r_k\}$. ◇

Corollary *If* $\Gamma(0) = \{r_1, \dots, r_k\}$ *then* X *is connected if and only if the integers* $n, r_1, \dots, r_k$ *are coprime.* ◇

Example 10.2 Let $n = 15$ and $\Gamma(0) = \{3, 9\}$. Then $t = 3$ and X is

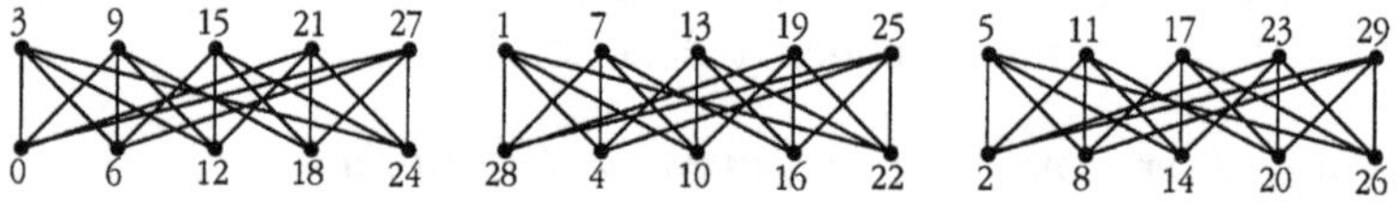

Theorem 10.6 *Let* X *be of cardinality* $2n$ *and have* t *compoents. Then*

(1) *when* n *is even the degree of each vertex can take any even value between* 2 *and* n/t;

(2) *when* n *is odd the degree of each vertex can take any value between* 1 *and* n/t.

Proof All vertices of X have the same degree, so it suffices to establish the property for the vertex 0.

If n is even then $\operatorname{Max} C_0$ has even cardinality and, since $0 \prec j$ implies $0 \prec 2n - j$, the degree of 0 is $2|\Gamma(0)|$ with $1 \leqslant |\Gamma(0)| \leqslant n/2t$.

If n is odd then $\operatorname{Max} C_0 = \{t, 3t, \dots, n, \dots, 2n{-}3t, 2n{-}t\}$ and has odd cardinality, and the degree of 0 is $2|\Gamma(0)|$ if $n \notin \Gamma(0)$ and $2|\Gamma(0)| - 1$ if $n \in \Gamma(0)$. It follows that the degree of each vertex can take any value between 1 and n/t. It is 1 if $n = t$ (i.e. X is the disjoint union of n two-element chains), and it is n if $t = 1$ (i.e. X is connected, complete bipartite). ◇

Corollary *There are generalised crowns that cannot be made into the dual space of a finite simple Ockham algebra.*

Proof Consider the generalised crown

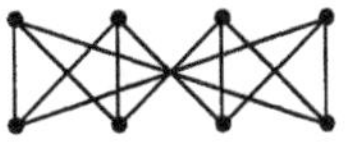

Here we have $n = 4$ and all vertices have degree 3. By the above, X cannot be made into the dual space of a finite simple Ockham algebra. ◇

We can now characterise those generalised crowns that are the connected components of X. We require the notion of a *circulant matrix* **[10]**, i.e. an $n \times n$ matrix $[\alpha_{ij}]$ with the property that, modulo n, $\alpha_{ij} = \alpha_{i+1,j+1}$. In a circulant $(0,1)$-matrix we shall say that the i-th row is *symmetric* if

$$\alpha_{ij} = 1 \iff \alpha_{i,n-j} = 1.$$

Theorem 10.7 *A generalised crown $C_{n;k}$ is a connected component of X if and only if it can be represented by a circulant $(0,1)$-matrix of order n in which the first row is symmetric and has k entries 1.*

Proof Let M be a matrix satisfying the conditions. We show that M can represent the connected component C_0 of X. Since the first row of M is symmetric, the entries 1 of this row yield the vertices that cover 0. The number t of connected components of X can be chosen arbitrarily, the only restriction being that t must be odd. Then the vertices that cover 0 are

$$k_1t,\ k_2t,\ \ldots,\ 2n-k_2t,\ 2n-k_1t,$$

all k_i being odd, and $\operatorname{Min} C_0 = \{0, 2t,\ \ldots,\ 2(n-t)\}$. Since M is circulant, its second row yields the vertices that cover $2t$ and, more generally, the m-th row yields the vertices that cover $2(m-1)t$. Thus we obtain a $C_{n;k}$ which is a connected component of several X.

Conversely, let X be the dual space of a finite simple Ockham algebra and suppose that X has t connected components. By Theorem 10.1 we know that if C_0 is not a singleton then it is a generalised crown $C_{n;k}$, so it can be represented by a square $(0,1)$-matrix M of order n. The row of M that corresponds to the vertex 0 contains k entries 1 and is necessarily symmetric since $0 \prec r$ implies $0 \prec 2n-r$. By permuting two rows, this row can become the first row. It remains to show that the resulting matrix is equivalent to a circulant matrix. Observe now that if 0 is covered by

$$k_1t,\ k_2t,\ \ldots,\ 2n-k_2t,\ 2n-k_1t$$

then $2t$ is covered by

$$(k_1+2)t,\ (k_2+2)t,\ \ldots,\ 2n-(k_2-2)t,\ 2n-(k_1-2)t,$$

i.e. there is in M a row whose elements are identical to those of the first row but are moved one place to the right. This row can then become the second row. By repeating this procedure, we obtain a circulant matrix with the required properties. ◇

At this point we draw the reader's attention to the fact that two circulant $(0,1)$-matrices which are equivalent can give rise to non-isomorphic dual spaces, hence to non-isomorphic finite simple Ockham algebras. The simplest example is provided by the matrices

$$A=\begin{bmatrix}1&0&0&1\\1&1&0&0\\0&1&1&0\\0&0&1&1\end{bmatrix}\qquad\text{and}\qquad B=\begin{bmatrix}0&1&1&0\\0&0&1&1\\1&0&0&1\\1&1&0&0\end{bmatrix}$$

with which are associated the graphs

G_A : 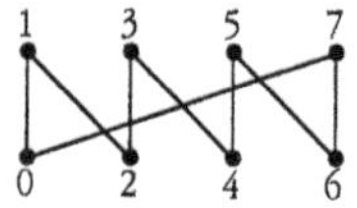

and G_B : 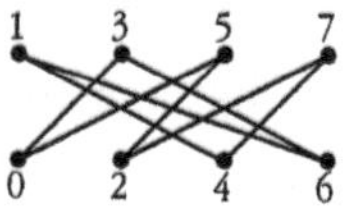

In G_A we have $x \prec g(x)$ or $x \succ g(x)$, whereas in G_B we have $x \parallel g(x)$.

Theorem 10.8 *For every even $n \geqslant 6$ there is at least one finite simple Ockham algebra whose dual space is a generalised crown that is not an S_n^k.*

Proof Taking $\Gamma(0)=\{r_1,\ldots,r_k,n-r_k,\ldots,n-r_1\}$ we see that 0 and n are both covered by all the elements of

$$S=\{r_1,\ldots,r_k,n-r_k,\ldots,n-r_1,n+r_1,\ldots,n+r_k,2n-r_k,\ldots,2n-r_1\}.$$

Since $n \geqslant 6$, we can choose the index k small enough to obtain for S a proper subset of Max X. ◇

We can express Theorem 10.8 in matrix terms : for every even $n \geqslant 6$, say $n=2n'$, there is at least one circulant $(0,1)$-matrix of order n that represents a finite simple Ockham algebra and in which the first row is *doubly symmetric* in the sense that $\alpha_{1j}=1 \Leftrightarrow \alpha_{1,n-j}=1$ and $(\forall j \leqslant n')\ \alpha_{1j}=1 \Leftrightarrow \alpha_{1,n'-j}=1$.

Example 10.3 Let $n=6$ and take $\Gamma(0)=\{1,5\}$. Then 0 and 6 are both covered by 1, 5, 7, 11. It follows that 2 and 8 are covered by 1, 3, 7, 9, whereas 4 and 10 are covered by 3, 5, 9, 11 . The corresponding matrix is

$$\begin{bmatrix} 1 & 0 & 1 & 1 & 0 & 1 \\ 1 & 1 & 0 & 1 & 1 & 0 \\ 0 & 1 & 1 & 0 & 1 & 1 \\ 1 & 0 & 1 & 1 & 0 & 1 \\ 1 & 1 & 0 & 1 & 1 & 0 \\ 0 & 1 & 1 & 0 & 1 & 1 \end{bmatrix}.$$

Now the question to solve is : how to recognise when a square $(0,1)$-matrix all of whose line sums are equal is equivalent to a circulant matrix the first row of which is symmetric? In order to examine this question we require the following notions. Let $M = [\alpha_{ij}]$ be a square $(0,1)$-matrix of order n. For every pair (i,j) denote by $\rho_{i,j}(M)$ the number of values of k for which $\alpha_{ik} = \alpha_{jk} = 1$. In particular, $\rho_{i,i+1}(M)$ does not depend on i if M is circulant. Let also $\mathrm{P}_{i,j}(M) = \{\rho_{i,j}(M)\ ;\ i,j \in \{1,\ldots,n\}\}$. Clearly, $\mathrm{P}_{i,j}(M)$ is invariant under arbitrary permutations of the columns and the rows. The symbols $\rho_{i,i+1,i+2,\ldots}(M)$ and $\mathrm{P}_{i,j,k,\ldots}(M)$ have obvious meanings.

Example 10.4 The $C_{5;3}$ whose matrix M is

$$\begin{bmatrix} 1 & 1 & 0 & 0 & 1 \\ 0 & 1 & 1 & 1 & 0 \\ 1 & 0 & 1 & 0 & 1 \\ 0 & 1 & 1 & 1 & 0 \\ 1 & 0 & 0 & 1 & 1 \end{bmatrix}$$

cannot be the dual space of a finite simple Ockham algebra. In fact, M has to be equivalent either to $A = \mathrm{Circ}\{0,1,1,1,0\}$ or to $B = \mathrm{Circ}\{1,0,1,0,1\}$. But $\mathrm{P}_{i,j}(A) = \mathrm{P}_{i,j}(B) = \{1,2\}$ whereas $\mathrm{P}_{i,j}(M) = \{1,2,3\}$. The fact that n is odd here provides another direct proof : in A and B the rows are distinct, whereas in M rows 2 and 4 are identical.

Example 10.5 The $C_{6;4}$ whose (circulant) matrix M is

$$\begin{bmatrix} 1 & 1 & 0 & 1 & 0 & 1 \\ 1 & 1 & 1 & 0 & 1 & 0 \\ 0 & 1 & 1 & 1 & 0 & 1 \\ 1 & 0 & 1 & 1 & 1 & 0 \\ 0 & 1 & 0 & 1 & 1 & 1 \\ 1 & 0 & 1 & 0 & 1 & 1 \end{bmatrix}$$

cannot be the dual space of a finite simple Ockham algebra. Indeed, M has to be equivalent to one of

$$A=\mathrm{Circ}\{0,1,1,1,1,0\},\ B=\mathrm{Circ}\{1,1,0,0,1,1\},\ C=\mathrm{Circ}\{1,0,1,1,0,1\}.$$

Here we have $\mathrm{P}_{i,j}(A) = \mathrm{P}_{i,j}(B) = \{2,3\}$ and $\mathrm{P}_{i,j}(C) = \{2,4\}$. Since $\mathrm{P}_{i,j}(M) = \{2,3\}$ we must go further into the analysis of the first two cases. Therefore we consider $\mathrm{P}_{i,j,k}$ and note that $\mathrm{P}_{i,j,k}(M) = \{1,3\}$ and $\rho_{i,i+1,i+2}(A) = \rho_{i,i+1,i+2}(B) = 2$.

Example 10.6 Suppose that we are given the matrix

$$M = \begin{bmatrix} 0 & 1 & 1 & 0 & 1 & 1 \\ 1 & 0 & 0 & 1 & 1 & 1 \\ 1 & 1 & 1 & 1 & 0 & 0 \\ 1 & 1 & 1 & 1 & 0 & 0 \\ 1 & 0 & 0 & 1 & 1 & 1 \\ 0 & 1 & 1 & 0 & 1 & 1 \end{bmatrix}.$$

Since both $\mathrm{Circ}\{0,1,1,1,1,0\}$ and $\mathrm{Circ}\{1,1,0,0,1,1\}$ have distinct rows, we need only decide whether or not M is equivalent to $\mathrm{Circ}\{1,0,1,1,0,1\}$ in which, like M, the rows are identical in pairs. To make the third row of M symmetric, we permute columns 2 and 6. Rearranging the rows, we obtain $\mathrm{Circ}\{1,0,1,1,0,1\}$. Consequently the $C_{6;4}$ represented by M is the dual space of a finite simple Ockham algebra.

We are of course aware of the weakness of the above procedure. Given a $C_{n;k}$ it is necessary to consider all circulant $(0,1)$-matrices of order n whose first row is symmetric and has k entries 1. If n is large, such an examination can be long. An algorithmic method would be welcome.

We have already seen that if $|X| = N$ is odd then there is only one finite simple Ockham algebra in the class $\mathbf{P}_{N,0}$. The situation is quite different if N is even.

Theorem 10.9 *The number α_n of non-isomorphic finite simple Ockham algebras that belong properly to $\mathbf{P}_{2n,0}$ is given by*

$$\alpha_n = \begin{cases} 2^{\frac{1}{2}n} & \text{if } n \text{ is even;} \\ 2^{\frac{1}{2}(n+1)} & \text{if } n \text{ is odd.} \end{cases}$$

This includes ϑ_n algebras whose dual spaces are not connected, where ϑ_n is the number of subsets $\{r_1,\dots,r_k\}$ of $\Gamma(0)$ for which $\gcd\{n,r_1,\dots,r_k\} \neq 1$.

Proof The Ockham space X is completely determined by its cardinality $2n$ and the subset $\Gamma(0)$. If n is even then $0 \leqslant |\Gamma(0)| \leqslant \frac{1}{2}n$ and the number of possibilities is

$$\binom{\frac{1}{2}n}{0} + \binom{\frac{1}{2}n}{1} + \cdots + \binom{\frac{1}{2}n}{\frac{1}{2}n} = 2^{\frac{1}{2}n}.$$

When n is odd we have $0 \leqslant |\Gamma(0)| \leqslant \frac{1}{2}(n+1)$ and the number of possibilities becomes $2^{\frac{1}{2}(n+1)}$. Moreover, if $|X| = |X'|$ with $\Gamma(0) = \{r_1, \ldots, r_k\}$ in X and $\Gamma(0') = \{r'_1, \ldots, r'_k\}$ in X' then we have

$$X \simeq X' \iff \{r_1, \ldots, r_k\} = \{r'_1, \ldots, r'_k\}.$$

The final assertion is an immediate consequence of the Corollary to Theorem 10.5. ◇

Example 10.7 Let $n = 15$. Then there are 256 non-isomorphic finite simple Ockham algebras that belong properly to $\mathbf{P}_{30,0}$. Here $\vartheta_{15} = 10$ and the dual spaces that are not connected are the following (in which it suffices to describe in each case only the component C_0 and to indicate the number t of components).

For $\Gamma(0) = \emptyset$ we have $C_0 = \{0\}$ and $t = 30$.

For $|\Gamma(0)| = 1$ the possibilities are

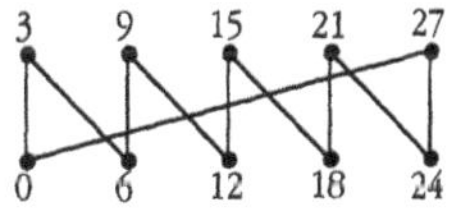

Γ(0)={3}
t=3

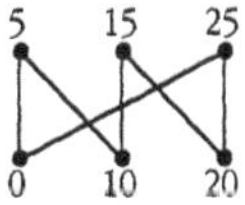

Γ(0)={5}
t=5

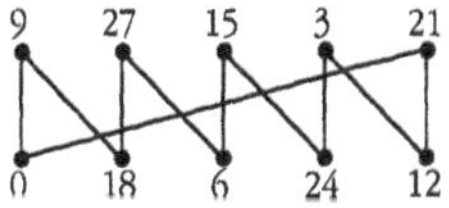

Γ(0)={9}
t=3

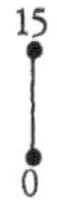

Γ(0)={15}
t=15

For $|\Gamma(0)| = 2$ the possibilities are

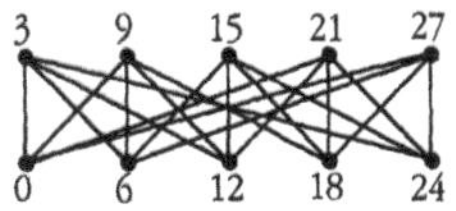

Γ(0)={3,9}
t=3

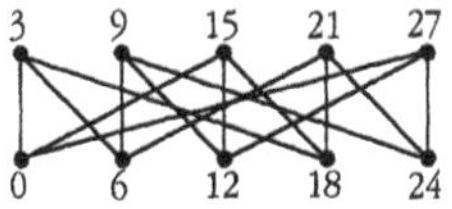

Γ(0)={3,15}
t=3

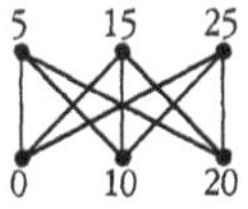

Γ(0)={5,15}
t=5

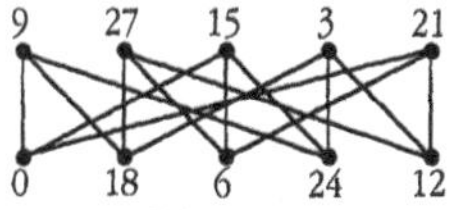

Γ(0)={9,15}
t=3

For $|\Gamma(0)| = 3$ we have

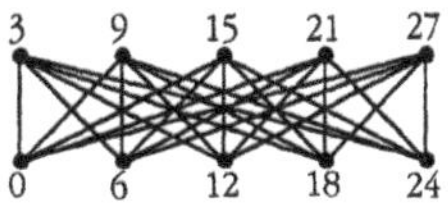

Γ(0)={3,9,15}
t=3

Remark The following table gives the first 35 values of ϑ_n.

n	ϑ_n	n	ϑ_n	n	ϑ_n	n	ϑ_n	n	ϑ_n
1	1	8	1	15	10	22	2	29	2
2	1	9	4	16	1	23	2	30	38
3	2	10	2	17	2	24	16	31	2
4	1	11	2	18	8	25	8	32	1
5	2	12	4	19	2	26	2	33	66
6	2	13	2	20	4	27	32	34	2
7	2	14	2	21	18	28	4	35	22

There appears to be no obvious rule of formation here. However, the following observations are useful :

(1) if $n = 2^a$ then $\vartheta_n = 1$;

(2) if $n = p^r$ with p an odd prime then $\vartheta_n = 2^{\frac{1}{2}(p^{r-1}+1)}$. In particular, if $n = p$ then $\vartheta_n = 2$. Indeed, the integers not exceeding n that are not prime to n are the elements of $S = \{p, 3p, 5p, \ldots, p^r\}$. Now S has cardinality $\frac{1}{2}(p^{r-1} + 1)$, and ϑ_n is the cardinality of the power set of S.

Theorem 10.10 *Let L be a finite simple Ockham algebra and let X be its dual space. Then*

(1) *L is fixed point free if and only if $|X|$ is odd*;

(2) *L has two fixed points if and only if X is an antichain of even cardinality*;

(3) *L has a single fixed point if and only if X is a generalised crown or a disjoint union of generalised crowns.*

Proof In each case we apply Theorem 6.1.

(1) If $|X|$ is odd then the order on X is discrete and there is no bipartition of the graph of g.

(2) If X is the antichain $\{0, 1, \ldots, 2n-1\}$ then the partition

$$\{\{0, 2, \ldots, 2n-2\}, \{1, 3, \ldots, 2n-1\}\}$$

gives two fixed points since both the blocks are down-sets.

(3) If X is a generalised crown, or a disjoint union of generalised crowns, then the partition $\{\text{Min } X, \text{Max } X\}$ gives the unique fixed point.

Since these cases exhaust the possibilities, the result follows. ◇

If $\leqslant$ and $\leqslant_1$ are orders on the same set X and if $\leqslant_1$ is an extension of $\leqslant$ then a mapping $g : X \to X$ can be order-reversing on $(X; \leqslant)$ but not on $(X; \leqslant_1)$, and conversely, as shown by the following examples in which $g(n) = g(n+1)$ modulo $|X|$.

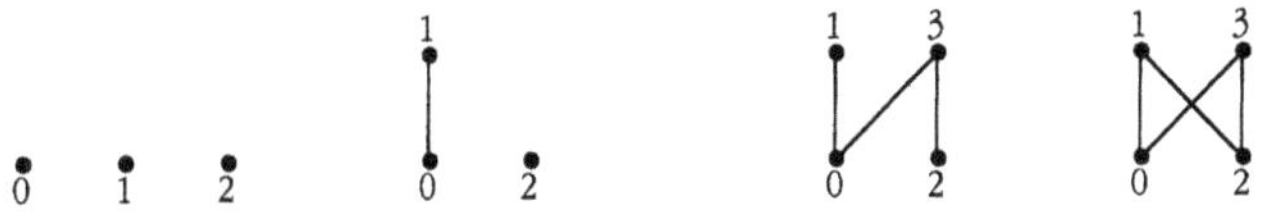

$(X_1; \leqslant)$ $(X_1; \leqslant_1)$ $(X_2; \leqslant)$ $(X_2; \leqslant_1)$

g is order-reversing on $(X_1; \leqslant)$ but not on $(X_1; \leqslant_1)$, and is not order-reversing on $(X_2; \leqslant)$ but is on $(X_2; \leqslant_1)$.

The following result, although trivial, is of interest in this context.

Lemma *Let $(X; \leqslant, g)$ be a finite Ockham space. If $\leqslant_1$ extends $\leqslant$ in such a way that g remains order-reversing on $(X; \leqslant_1)$ then*

(1) *$(X; \leqslant_1, g)$ is also an Ockham space;*

(2) *the dual algebra $(L_1; f)$ of $(X; \leqslant_1, g)$ is a subalgebra of the dual algebra $(L; f)$ of $(X; \leqslant, g)$;*

(3) *the number of fixed points of (L_1, f) is at most equal to that of $(L; f)$.*

Proof It suffices to observe that the operation f on $\mathcal{O}(X; \leqslant_1, g)$ is the restriction of f on $\mathcal{O}(X; \leqslant, g)$, that every down-set of $(X; \leqslant_1)$ is a down-set of $(X; \leqslant)$, and that every fixed point of $(\mathcal{O}(X; \leqslant_1, g); f)$ is a fixed point of $(\mathcal{O}(X; \leqslant, g); f)$. ◇

In fact, the following result is now clear.

Theorem 10.11 *Every finite simple Ockham algebra whose dual space has cardinality $2n$ is a subalgebra of the finite simple Ockham algebra whose dual space is the antichain $\{0, 1, \dots, 2n-1\}$.* ◇

We illustrate this property for $n = 3$. By Theorem 10.9, there are four non-isomorphic finite simple Ockham algebras that belong properly to $\mathbf{P}_{6,0}$, namely $L_0 = (\mathbf{2}^6; f)$ whose dual space is the antichain $\{0, 1, \dots, 5\}$ and the subalgebras L_1, L_2, L_3 whose dual spaces are

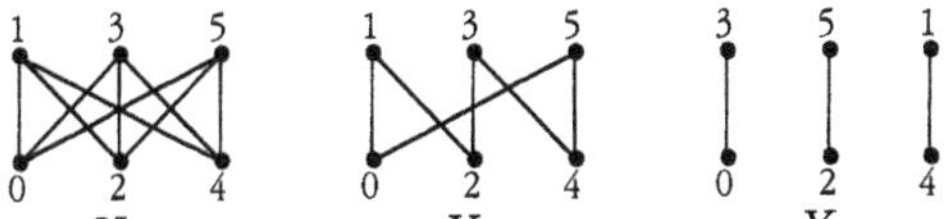

X_1 X_2 X_3

In the diagrams that follow we omit, for visual reasons, a great many of the lines. As a lattice, $L_0 \simeq \mathbf{2}^6$. The algebra L_0 has two fixed points, corresponding to the down-sets $\{0, 2, 4\}$ and $\{1, 3, 5\}$. Each of L_1, L_2, L_3 has a single fixed point. As a lattice, $L_1 \simeq \mathbf{2}^3 \mathbin{\overline{\oplus}} \mathbf{2}^3$. The elements of L_2 are those of L_1 together with the three elements that correspond to the down-sets $\{0,1,2\},\{2,3,4\},\{0,4,5\}$. The elements of L_3 are those of L_1 together with twelve others. Amongst these are the mid-level elements that correspond to the down-sets $\{0,2,3\},\{0,3,4\},\{0,1,4\},\{1,2,4\},\{2,4,5\},\{0,2,5\}$. By Theorem 10.11, L_1 is a subalgebra of both L_2 and L_3.

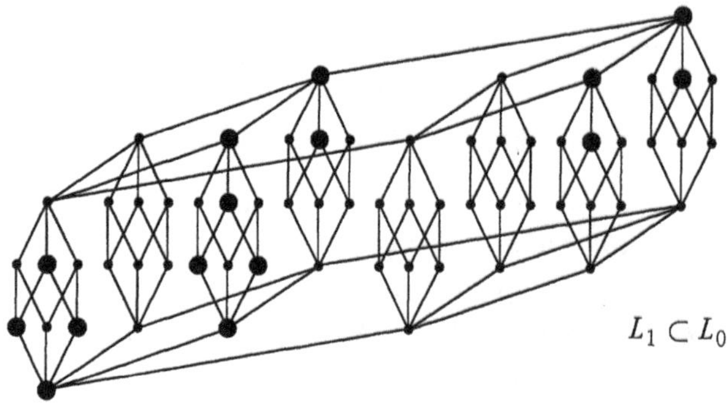

$L_1 \subset L_0$

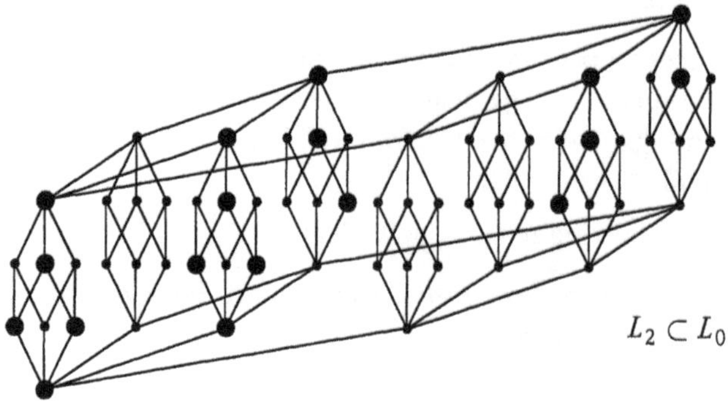

$L_2 \subset L_0$

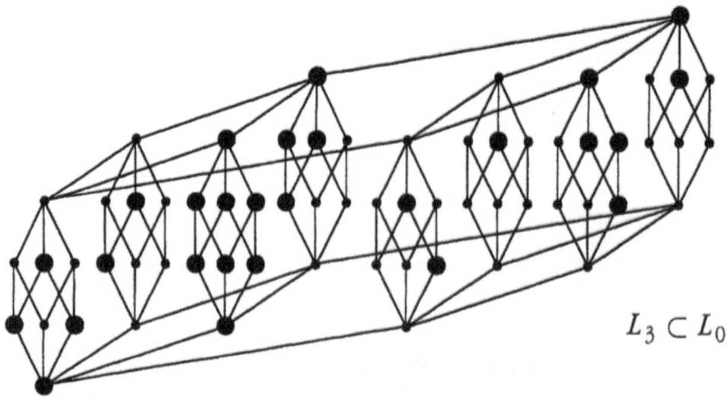

$L_3 \subset L_0$

11 Relative Ockham algebras

In Chapter 1 we gave an affirmative answer to the question of whether every bounded distributive lattice L can be made into an Ockham algebra $(L; \sim)$. If the subvariety $\mathbf{V}$ of $\mathbf{O}$ to which $(L; \sim)$ has to belong is prescribed, then the answer is far from being affirmative, even when L is finite. For example, as we observed in **[41]**, the 5-element distributive lattice with Hasse diagram

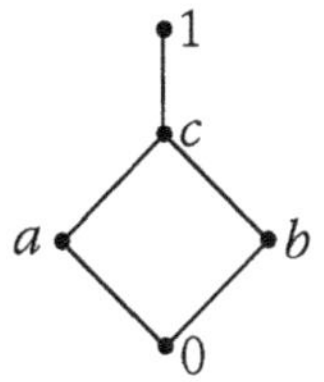

cannot be made into an $\mathbf{M}_1$-algebra since otherwise we would have

$$1 = {\sim}0 = {\sim}(a \wedge b) = {\sim}a \vee {\sim}b$$

whence either ${\sim}a = 1$ or ${\sim}b = 1$, so that either ${\sim}^2 a = 0$ or ${\sim}^2 b = 0$ from which it follows by axiom (1) that either $a = 0$ or $b = 0$, a contradiction.

So the following general problem arises : a subvariety $\mathbf{V}$ of $\mathbf{O}$ being given, characterise the (finite) distributive lattices that can be made into an algebra in $\mathbf{V}$. A particularly simple first step in the solution of this problem is given by

Theorem 11.1 *Let* $\mathbf{V} \in \Lambda(\mathbf{O})$ *with* $\mathbf{V} \supseteq \mathbf{P}_{2,1}$. *Then every finite distributive lattice can be made into a* $\mathbf{V}$*-algebra.*

Proof Let L be a finite distributive lattice, let X be its dual space, and let $y \in X$. Define $g : X \to X$ by $g(x) = y$ for every $x \in X$. Clearly, g is order-reversing and $g^2 = g$. Thus X has been made into a $\mathbf{P}_{2,1}$-space. ◇

In **[59]** G. Bordalo and H. A. Priestley established the converse of Theorem 11.1, thus proving

Theorem 11.2 *Let* $\mathbf{V} \in \Lambda(\mathbf{O})$. *Then every finite distributive lattice is a reduct of a* $\mathbf{V}$*-algebra if and only if* $\mathbf{V} \supseteq \mathbf{P}_{2,1}$. ◇

In the same spirit, they solved **[58]** for the five subvarieties $\mathbf{B}$, $\mathbf{K}$, $\mathbf{S}$, $\overline{\mathbf{S}}$, $\widetilde{\mathbf{S}}$ the following interesting question :

A subvariety $\mathbf{V}$ of $\mathbf{O}$ being given, what are the finite distributive lattices all of whose closed intervals can be made into a $\mathbf{V}$-algebra?

We shall follow their work very closely, reporting the main results (with or without proof) and show that in fact their solution is far from being limited to the above five subvarieties. We begin with some definitions.

Definition Let $\mathbf{V}$ be a subvariety of $\mathbf{O}$ and let L be a finite distributive lattice. Then L is a *relative* $\mathbf{V}$*-algebra* if every interval $[a, b]$ of L can be given the structure of a $\mathbf{V}$-algebra.

We shall denote the class of finite relative $\mathbf{V}$-algebras by $\mathbf{V}_r$. On $\Lambda(\mathbf{O})$ we can define an equivalence relation $\approx$ by

$$\mathbf{V} \approx \mathbf{W} \iff \mathbf{V}_r = \mathbf{W}_r.$$

The $\approx$-class of $\mathbf{V}$ will be denoted by $[\mathbf{V}]$.

We shall use duality throughout, but only in the finite case : all lattices and ordered sets involved will be finite. If P and Q are ordered sets then we shall say that P has Q as a *subposet* if there is an order-embedding of Q into P; and that P has Q as a *convex poset* if there is an order embedding $\alpha : Q \to P$ such that $\alpha(Q)$ is a convex subset of P.

For example, the ordered set

contains  as a subposet but not as a convex subposet.

A class $\mathcal{E}$ of ordered sets is said to be *convex-closed* if, whenever $P \in \mathcal{E}$, every convex subposet of P also belongs to $\mathcal{E}$.

In Chapter 5 the reader can find equational bases for the five subvarieties we are interested in, as well as the dual equivalents of the axioms involved. The following lemma is therefore straightforward.

Lemma 11.1 *Let* $(L; f) \in \mathbf{O}$ *and let* $(X; g)$ *be its dual space. Then*

$L \in \mathbf{B}$ *if and only if* $g = g^0$;
$L \in \mathbf{K}$ *if and only if* $g^0 = g^2$ *and* $g^0 \nparallel g$;
$L \in \mathbf{S}$ *if and only if* $g^0 \geqslant g$;
$L \in \overline{\mathbf{S}}$ *if and only if* $g^0 \leqslant g$;
$L \in \widetilde{\mathbf{S}}$ *if and only if* $g = g^2$ *and* $g^0 \nparallel g$. ◇

For every (finite) $L \in \mathbf{D}$ we denote by X_L the dual space of X, and for every (finite) ordered set P we denote by L_P the lattice of down-sets of P. The following theorem establishes an interesting correspondence between the intervals of L and the convex subposets of X_L.

Theorem 11.3 (1) *Let L be a finite distributive lattice and let $M = [a, b]$ be a closed interval of L. Then X_M is a convex subposet of X_L.*

(2) *Let Q be a convex subset of the finite ordered set P. Then there is a closed interval M of L_P such that X_M is order-isomorphic to Q.*

Proof (1) : If the down-sets A, B of X_L are such that $A \subseteq B$ then $B \setminus A$ is a convex subset of X_L. In fact, let $p, q \in B \setminus A$ with $p \leqslant x \leqslant q$. Since B is decreasing and $q \in B$, we have $x \in B$. Since A', the complement of A, is increasing and $p \in A'$, we have $x \in A'$. Hence $x \in B \setminus A$.

We now show that $\mathcal{O}(B \setminus A) \simeq M = [a, b]$. For each $P \in \mathcal{O}(B \setminus A)$ define $f(P) = P \cup A$. Clearly, $f(P)$ is a down-set and $A \subseteq f(P) \subseteq B$, whence $f(P) \in M$. Moreover, f is injective, $f(\mathcal{O}(B \setminus A)) = M$, and f is a lattice morphism.

(2) : Let Q be a convex subset of P. Then

$$\begin{aligned} Q &= Q^{\downarrow} \cap Q^{\uparrow} \\ &= Q^{\downarrow} \setminus (Q^{\downarrow} \setminus Q) \\ &= Q^{\downarrow} \setminus (Q^{\downarrow} \setminus (Q^{\downarrow} \cap Q^{\uparrow})) \\ &= Q^{\downarrow} \setminus (Q^{\downarrow} \cap (X_L \setminus Q^{\uparrow})). \end{aligned}$$

Let $Q^{\downarrow} = B$ and $Q^{\downarrow} \cap (X_L \setminus Q^{\uparrow}) = A$. Both A and B are down-sets. Hence Q is order-isomorphic to X_M where $M = [A, B]$. ◇

Corollary *A finite distributive lattice L is a relative $\mathbf{V}$-algebra if and only if every convex subposet of X_L can be endowed with an order-reversing map g which makes it into a $\mathbf{V}$-space.* ◇

This latter property enables us to define the 'strategy' which will be successful in the case of the subvarieties **B**, **K**, **S**, and $\overline{\mathbf{S}}$.

- Seek a family $(P_i)_{i \in I}$ of ordered sets (a family that is as 'small' as possible) such that no P_i can be made into a **V**-space.
- Consider the class $\mathcal{E}$ of ordered sets which have no P_i as a convex poset.
- If every member of $\mathcal{E}$ can be made into a **V**-space and if $\mathcal{E}$ is convex-closed, then the finite relative **V**-algebras are exactly those L for which $X_L \in \mathcal{E}$.

Moreover, the lattices for which $X_L \in \mathcal{E}$ are the finite relative **W**-algebras for any $\mathbf{W} \supseteq \mathbf{V}$ which is such that no P_i can be made into a **W**-space.

Lemma 11.2 *Let P be a finite ordered set. Then the following are equivalent:*

(1) *P is an antichain;*

(2) *P does not contain*  *as a subposet;*

(3) *P does not contain*  *as a convex subposet.*

The class of all finite antichains is convex-closed. ◇

An ordered set P is a *tree* if, for every $x \in P$, $x^{\downarrow}$ is a chain.

Lemma 11.3 *Let P be a finite ordered set. Then the following are equivalent:*

(1) *P is a disjoint union of trees;*

(2) *P does not contain*  *as a subposet;*

(3) *P does not contain*  *as a convex subposet.*

The class of all disjoint unions of finite trees is convex-closed.

Proof Only (3) $\Rightarrow$ (2) is non-trivial. Suppose that$\{x, u, v\} \subseteq P$ with $x > u$, $x > v$, and $u \parallel v$. Let x' be a minimal element of the set $x^{\downarrow} \cap u^{\uparrow} \cap v^{\uparrow}$. Then $x' > u$, $x' > v$, and there is no y such that $x' > y > u$ and $x' > y > v$. Take u' to be a maximal element of $(x'^{\downarrow} \setminus \{x'\}) \cap u^{\uparrow}$ and v' to be a maximal element of $(x'^{\downarrow} \setminus \{x'\}) \cap v^{\uparrow}$. Then $Q = \{x', u', v'\}$ is convex by the definition of u' and v'. Moreover, $u' \parallel v'$ since, for example, $u' \geqslant v'$ would give $u' \in x'^{\downarrow} \cap u^{\uparrow} \cap v^{\uparrow}$, which is impossible by the minimality of x'. Hence Q is isomorphic to . ◇

Combining Lemma 11.3 with its dual version, we obtain

Lemma 11.4 *Let P be a finite ordered set. Then the following are equivalent:*

(1) *P is a disjoint union of chains;*

(2) *P contains neither*  *nor*  *as a subposet;*

(3) *P contains neither*  *nor*  *as a convex subposet.*

The class of all disjoint unions of finite chains is convex-closed. ◇

Theorem 11.4 *The $\approx$-classes of $\mathbf{B}, \mathbf{K}, \mathbf{S}, \overline{\mathbf{S}}$ are as follows:*

(1) $[\mathbf{B}] = \{\mathbf{V} \in \Lambda(\mathbf{O}) \mid \mathbf{V} \supseteq \mathbf{B},\ \mathbf{V} \not\supseteq \mathbf{K},\ \mathbf{V} \not\supseteq \mathbf{S};\ \mathbf{V} \not\supseteq \overline{\mathbf{S}}\}$;
(2) $[\mathbf{K}] = \{\mathbf{V} \in \Lambda(\mathbf{O}) \mid \mathbf{V} \supseteq \mathbf{K},\ \mathbf{V} \not\supseteq \mathbf{S};\ \mathbf{V} \not\supseteq \overline{\mathbf{S}}\}$;
(3) $[\mathbf{S}] = \{\mathbf{V} \in \Lambda(\mathbf{O}) \mid \mathbf{V} \supseteq \mathbf{S};\ \mathbf{V} \not\supseteq \overline{\mathbf{S}}\}$;
(4) $[\overline{\mathbf{S}}] = \{\mathbf{V} \in \Lambda(\mathbf{O}) \mid \mathbf{V} \supseteq \overline{\mathbf{S}};\ \mathbf{V} \not\supseteq \mathbf{S}\}$.

Proof We restrict ourselves to establishing (1) and illustrate the 'strategy' described above. The family $\{P_i\}_{i \in I}$ is formed by the . In fact, there are only three distinct g-maps on the two-element chain $x < y$, namely those described as follows :

α	x	y
$g_1(\alpha)$	y	x
$g_2(\alpha)$	x	x
$g_3(\alpha)$	y	y

The maps g_1, g_2, g_3 make  into a **K**-, **S**-, $\overline{\mathbf{S}}$-algebra respectively. By Lemma 11.2, the class $\mathcal{E}$ is the class of finite antichains. Finally, every antichain is a **B**-space if it is endowed with the identity map as a g-map. The conclusion follows easily. ◇

The next result gives an algebraic description of $\mathbf{V}_r$ where $\mathbf{V}$ is in any of the equivalence classes $[\mathbf{B}]$, $[\mathbf{K}]$, $[\mathbf{S}]$, $[\overline{\mathbf{S}}]$.

Theorem 11.5 *Let $\mathbf{V} \in \Lambda(\mathbf{O})$ and let L be a finite distributive lattice. Then*

(α) *if $\mathbf{V} \in [\mathbf{B}]$ then $L \in \mathbf{V}_r$ if and only if L is boolean;*

(β) *if $\mathbf{V} \in [\mathbf{K}]$ then $L \in \mathbf{V}_r$ if and only if it satisfies any of the following equivalent conditions:*

(1) *L is a direct product of chains;*

(2) *L contains neither $\mathbf{2}^2 \oplus \mathbf{1}$ nor $\mathbf{1} \oplus \mathbf{2}^2$ as an interval;*

(3) *L has neither $\mathbf{2}^2 \oplus \mathbf{1}$ nor $\mathbf{1} \oplus \mathbf{2}^2$ as a homomorphic image.*

(γ) *If $\mathbf{V} \in [\mathbf{S}]$ (resp. if $\mathbf{V} \in [\overline{\mathbf{S}}]$) then $\mathbf{V}_r$ is defined in the following way:*

(a) *the trivial algebra and the 2-element chain are in $\mathbf{V}_r$;*

(b) *any finite direct product of elements of $\mathbf{V}_r$ is in $\mathbf{V}_r$;*

(c) *if $L \in \mathbf{V}_r$ then $\mathbf{1} \oplus L \in \mathbf{V}_r$ (resp. $L \oplus \mathbf{1} \in \mathbf{V}_r$);*

(d) *any element of $\mathbf{V}_r$ can be constructed by repeated application of the properties (a), (b), (c).*

Further, $L \in \mathbf{V}_r$ if and only if it satisfies either of the following equivalent conditions:

(1) *L does not contain $\mathbf{2}^2 \oplus \mathbf{1}$ (resp. $\mathbf{1} \oplus \mathbf{2}^2$) as an interval;*

(2) *L does not have $\mathbf{2}^2 \oplus \mathbf{1}$ (resp. $\mathbf{1} \oplus \mathbf{2}^2$) as a homomorphic image.* ◇

If we apply Theorem 11.5 to the subvariety $\mathbf{M}_1$ (see the diagram on page 92), we obtain $[\mathbf{B}] = \{\mathbf{B}\}$, $[\mathbf{K}] = \{\mathbf{K}, \mathbf{M}, \mathbf{K}_1, \mathbf{M} \vee \mathbf{K}_1\}$, and $[\mathbf{S}]$ consists of all the other subvarieties of $\mathbf{M}_1$. So parts (β) and (γ) of Theorem 11.5 generalise results previously obtained by Varlet **[102]** and Bordalo **[57]** for **M** and **S** respectively.

Whereas the relative **V**-algebras for $\mathbf{V} \in \{\mathbf{B}, \mathbf{K}, \mathbf{S}, \overline{\mathbf{S}}\}$ can be characterised by the exclusion of a set of intervals, for $\mathbf{V} = \widetilde{\mathbf{S}}$ the procedure fails and a more sophisticated method has to be used.

Lemma 11.5 *A non-empty finite ordered set P can be made into an $\widetilde{\mathbf{S}}$-space if and only if every connected component of P has a node.*

Proof $\Leftarrow$: Suppose that $P = \bigcup_{1 \leqslant i \leqslant k} C_{x_i}$. Define g by $g(x) = x_i$ whenever $x \not\parallel x_i$. Then g is order-reversing, $g^2 = g$, and $g \not\parallel g^0$; so $(P; g)$ is an $\widetilde{\mathbf{S}}$-space.

$\Rightarrow$: Let $(P;g)$ be the dual space of an $\widetilde{\mathbf{S}}$-algebra. Since $g^2 = g$, we have that $\mathcal{Y} = g(P)$ is an antichain; in fact, if $g(y_1) = g^2(y_1) \leqslant g^2(y_2) = g(y_2)$ then $g^2(y_1) \geqslant g^2(y_2)$ and so $g(y_1) = g(y_2)$. Since $g \nparallel g^0$ we have $P = \mathcal{Y}^{\downarrow} \cup \mathcal{Y}^{\uparrow}$. ◇

Lemma 11.6 *Let $(X;g)$ be the dual space of a finite relative $\widetilde{\mathbf{S}}$-algebra. Then X has no subposet isomorphic to the fence* .

Proof Suppose that X contains the subposet $\{u, v, x, y\}$ with

$$u < x,\ y < x,\ y < v,\ u \parallel v,\ u \parallel y,\ x \parallel v.$$

We shall show that X has a convex subposet $Q = \{a, b\} \cup [d, c]$ such that

$$a \prec c,\ d \prec b,\ d < c,\ a \parallel b,\ a \parallel d,\ c \parallel b.$$

First, consider

$$X_1 = u^{\uparrow} \cap y^{\uparrow} \cap (X \setminus v^{\uparrow}).$$

This set contains x but no element t below v (otherwise $u \leqslant t < v$ which contradicts $u \parallel v$). Let x_1 be a minimal element of X_1. Then $\{u, x_1, y, v\}$ is isomorphic to .

Next, consider

$$X_2 = x_1^{\downarrow} \cap (X \setminus y^{\uparrow}) \cap (X \setminus y^{\downarrow}) \cap u^{\uparrow}.$$

This set contains u. Let u_1 be a maximal element of X_2. Then we have that $u_1 \parallel v$; for $u_1 \leqslant v$ gives the contradiction $u \leqslant v$, and $u_1 \geqslant v$ gives the contradiction $x_1 \geqslant v$. Moreover, $u_1 \parallel y$.

We now show that $u_1 \prec x_1$. Suppose in fact that there existed u_2 such that $u_1 < u_2 < x_1$. Then $u_2 \geqslant u$; and $u_2 \ngeqslant v$ since otherwise we have the contradiction $x_1 \geqslant v$. By the minimality of x_1 we then have $u_2 \ngeqslant y$. We also have $u_2 \nleqslant y$, for otherwise we have the contradiction $u_1 \leqslant y$. Consequently, $u_2 \in X_2$. But this contradicts the maximality of u_1. Hence we have that $u_1 \prec x_1$.

Similarly, we can show the existence of an element v_1 that covers y and thus we obtain the subposet Q .

Finally, we show that no map g can be defined on Q in such a way that it becomes an $\widetilde{\mathbf{S}}$-space. Suppose in fact that such a g existed. Then since $g \nparallel g^0$ we have $g(a) \in \{a, c\}$. Now

(1) if $g(a) = a$ then $g(c) = a$ and $g(d) = c$, But then $g^2(d) = a$ and we have the contradiction $g^2(d) \neq g(d)$;

(2) if $g(a) = c$ then $g(c) = c$ and $g(d) = c$. But then $g(b) \leqslant c$ and therefore $g(b) \in \{a, c\}$, which is impossible since $b \parallel a$ and $b \parallel c$. ◇

The lattice dual of the fence  is the lattice L_8 with Hasse diagram

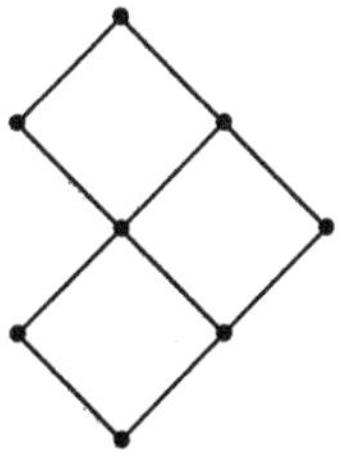

It is known that the finite distributive lattices which do not have L_8 as a homomorphic image are precisely those whose duals do not contain the fence  as a subposet.

The latter class of ordered sets coincides with the class of series-parallel posets **[58,71]**, defined as follows.

A finite (non-empty) ordered set is *series-parallel* if it can be constructed from singleton sets using the operations of disjoint union and linear sum. For instance, all trees are series-parallel. By Lemma 11.6, the dual spaces of $\widetilde{\mathbf{S}}_r$-algebras are series-parallel.

The four-element crown

is series-parallel, being the linear sum of the antichains $\{a, b\}$ and $\{c, d\}$. By Lemma 11.6 it cannot be made into an $\widetilde{\mathbf{S}}_r$-space.

The non-trivial finite distributive lattices whose duals are series-parallel posets are those that can be built up from 2-element chains using direct product and vertical sum.

Lemmas 11.5 and 11.6 lead rather easily to the following

Theorem 11.6 *Let P be a finite series-parallel poset. Then the following statements are equivalent:*

(1) *every connected component of P has a node;*

(2) *no convex subset of P is* ;

(3) *in the construction of P from singletons there is a restriction on linear sums, namely that $P_1 \oplus P_2$ is permitted only when P_1 has a biggest element or P_2 has a smallest element (or both).* ◇

The translation of Theorem 11.6 into algebraic terms is provided by the following result.

Theorem 11.7 *Let L be a finite distributive lattice. Then the following statements are equivalent :*

(1) $L \in \widetilde{\mathbf{S}}_r$;

(2) *L is a member of the class C defined as follows :*

(*a*) *the trivial algebra and the 2-element chain are in C;*

(*b*) *the direct product of two members of C is in C;*

(*c*) *if L_1 and L_2 are in $C \cup \{\emptyset\}$ then $L_1 \oplus \mathbf{1} \oplus L_2$ is in C;*

(*d*) *every member of C is obtained in a finite number of steps using the properties* (*a*), (*b*), (*c*).

(3) *L does not have L_8 as a homomorphic image and L contains no interval isomorphic to $\mathbf{2}^2 \overline{\oplus} \mathbf{2}^2$.* ◇

12 Double MS-algebras

The notion of a *double Stone algebra* is well known. This is an algebra $(L; \vee, \wedge, ^*, ^+, 0, 1)$ of type $(2,2,1,1,0,0)$ such that both $(L; ^*)$ and $(L^{\mathrm{op}}; ^+)$ are Stone algebras with the unary operations $a \mapsto a^*$ and $a \mapsto a^+$ linked by the properties

$$a^{+*} = a^{++} \leqslant a \leqslant a^{**} = a^{*+}.$$

Our objective now is to consider the following natural generalisation.

Definition A *double MS-algebra* is an algebra $(L; \vee, \wedge, ^\circ, ^+, 0, 1)$ of type $(2,2,1,1,0,0)$ such that $(L; ^\circ)$ is an MS-algebra, $(L; ^+)$ is a dual MS-algebra, and the unary operations are linked by the properties

(D1) $(\forall a \in L) \quad a^{+\circ} = a^{++}$;

(D2) $(\forall a \in L) \quad a^{\circ +} = a^{\circ\circ}$.

A double MS-algebra will generally be denoted by $(L; ^\circ, ^+)$. The double MS-algebras form a variety that we shall denote by **DMS**.

Example 12.1 If B is a boolean algebra and $B^{[2]} = \{(a,b) \in B^2 \mid a \leqslant b\}$ then $B^{[2]}$ is a double Stone algebra in which $(a,b)^\circ = (b', b')$ and $(a,b)^+ = (a', a')$.

Example 12.2 Every de Morgan algebra $(M; ^-)$ is a double MS-algebra; it suffices to define $a^\circ = a^+ = \overline{a}$ for every $a \in M$.

Example 12.3 If $(M; ^-)$ is a de Morgan algebra and $(S; ^*, ^+)$ is a double Stone algebra then $(M \times S; ^\circ, ^\oplus)$ is a double MS-algebra where $(a,b)^\circ = (\overline{a}, b^*)$ and $(a,b)^\oplus = (\overline{a}, b^+)$.

Immediate consequences of (D1) and (D2) are the following.

Theorem 12.1 *If* $(L; ^\circ, ^+)$ *is a double MS-algebra then*

(1) $(\forall a \in L) \quad a^\circ \leqslant a^+$;

(2) $L^{\circ\circ} = L^{++}$.

Proof (1) : Writing a° for a in (D2), we obtain $a^{\circ\circ +} = a^{\circ\circ\circ} = a^\circ$; and from $a \leqslant a^{\circ\circ}$ we have, since $a \mapsto a^+$ is antitone, $a^{\circ\circ +} \leqslant a^+$.

(2) : (D1) gives $L^{++} \subseteq L^{\circ\circ}$; and (D2) gives the reverse inclusion. ◇

The *skeleton* $S(L)$ of a double MS-algebra $(L; ^\circ, ^+)$ is defined in the same way as for any Ockham algebra, namely by

$$S(L) = L^\circ = \{x^\circ \mid x \in L\}.$$

By Theorem 12.1 we also have

$$S(L) = L^+ = \{x^+ \mid x \in L\}$$

and clearly $a \in S(L)$ if and only if $a = a^{\circ\circ} = a^{++}$. Every double MS-algebra has a de Morgan skeleton.

Fundamental to the study of double MS-algebras is the notion of a *residuated mapping*. We recall here from **[3]** the following basic facts concerning such mappings. If E and F are ordered sets then a mapping $f : E \to F$ is said to be *residuated* if the pre-image under f of every principal down-set $y{\downarrow}$ of F is a principal down-set of E. By [**3**, Theorem 2.5], $f : E \to F$ is residuated if and only if it is isotone and there is an isotone mapping $g : F \to E$ such that $f \circ g \leqslant \mathrm{id}_F$ and $g \circ f \geqslant \mathrm{id}_E$. Such a mapping g is necessarily unique, is called the *residual* of f, and is written as f^+. It is readily seen that

$$(\forall y \in F) \qquad f^+(y) = \max\{x \in E \mid f(x) \leqslant y\}.$$

For a residuated mapping f we have

$$f \circ f^+ \circ f = f \quad \text{and} \quad f^+ \circ f \circ f^+ = f^+.$$

Moreover, by [**3**, Theorem 2.10], the following statements are equivalent :

$$f \text{ is a closure;} \quad f^+ \text{ is a dual closure;} \quad f = f^+ \circ f; \quad f^+ = f \circ f^+.$$

The importance of residuated mappings in our discussion can be seen in the theory of duality that applies to double MS-algebras, which we now proceed to describe. If we consider a double MS-algebra $(L; {}^\circ, {}^+)$ then the Ockham algebras $(L; {}^\circ)$ and $(L; {}^+)$ give rise to the Ockham spaces $(X; g)$ and $(X; h)$ respectively, each of the mappings g and h being both antitone and continuous.

The following table translates properties in L to properties on X.

$a \leqslant a^{\circ\circ}$	$a \geqslant a^{++}$	$a^{\circ +} = a^{\circ\circ}$	$a^{+\circ} = a^{++}$	$a^\circ \leqslant a^+$
$g^2 \leqslant \mathrm{id}_X$	$h^2 \geqslant \mathrm{id}_X$	$g \circ h = g^2$	$h \circ g = h^2$	$g \leqslant h$

By way of example, for each $x \in I_p(L)$ we have

$$g(x) = \{a \mid a^\circ \notin x\}, \qquad h(x) = \{a \mid a^+ \notin x\}$$

and so

$$gh(x) = \{a \mid a^\circ \notin h(x)\} = \{a \mid a^{\circ +} \in x\}$$
$$g^2(x) = \{a \mid a^\circ \notin g(x)\} = \{a \mid a^{\circ\circ} \in x\},$$

whence $a^{\circ +} = a^{\circ\circ}$ gives $g \circ h = g^2$; and

$$a \notin h(x) \Rightarrow a^+ \in x \Rightarrow a^\circ \in x \Rightarrow a \notin g(x)$$

so $g(x) \subseteq h(x)$ whence $g \leqslant h$.

These considerations lead naturally to the following notion.

Definition A *double MS-space* $(X; g, h)$ is a Priestley space X on which there are defined two continuous antitone mappings g, h such that

$$g \circ h = g^2 \leqslant \mathrm{id}_X \leqslant h^2 = h \circ g.$$

Theorem 12.2 *For a Priestley space X the following statements are equivalent:*

(1) *X is the underlying set of a double MS-space;*

(2) *there is a residuated dual closure map $\vartheta : X \to X$ such that both ϑ and its residual ϑ^+ are continuous, and* Im ϑ *admits a continuous polarity.*

Proof (1) $\Rightarrow$ (2) : If $(X; g, h)$ is a double MS-space, consider the mapping $\vartheta = g^2$. Clearly, ϑ is a dual closure on X and is continuous. Since g is antitone and $g^3 = g$, it is equally clear that g induces a polarity on Im ϑ. Now

$$g^2 \circ h^2 = g \circ g \circ h \circ h = g^3 \circ h = g \circ h = g^2 \leqslant \mathrm{id}_X,$$

and similarly $h^2 \circ g^2 = h^2 \geqslant \mathrm{id}_X$. Consequently, ϑ is residuated with residual $\vartheta^+ = h^2$ which is continuous.

(2) $\Rightarrow$ (1) : Let ϑ be a residuated dual closure on X. Suppose that ϑ and ϑ^+ are continuous, and that Im ϑ has a continuous polarity α. Then $\vartheta = \vartheta \circ \vartheta^+$ and $\vartheta^+ = \vartheta^+ \circ \vartheta$ with ϑ^+ a closure on X. Define $g, h : X \to X$ by

$$(\forall x \in X) \qquad g(x) = \alpha\vartheta(x), \quad h(x) = \vartheta^+ g(x).$$

Then g, h are antitone and continuous. Now ϑ fixes the elements of Im ϑ, so we have

$$g^2(x) = \alpha\vartheta\alpha\vartheta(x) = \alpha^2\vartheta(x) = \vartheta(x);$$
$$h^2(x) = \vartheta^+\alpha\vartheta\vartheta^+\alpha\vartheta(x) = \vartheta^+\alpha\vartheta\alpha\vartheta(x) = \vartheta^+\vartheta(x) = \vartheta^+(x),$$

whence $g^2 = \vartheta \leqslant \mathrm{id}_X$ and $h^2 = \vartheta^+ \geqslant \mathrm{id}_X$. Also,

$$(g \circ h)(x) = \alpha\vartheta\vartheta^+\alpha\vartheta(x) = \alpha\vartheta\alpha\vartheta(x) = g^2(x),$$
$$(h \circ g)(x) = \vartheta^+\alpha\vartheta\alpha\vartheta(x) = h^2(x),$$

whence $g \circ h = g^2$ and $h \circ g = h^2$. ◊

Precisely when a Priestley space X is a double MS-space can also be determined using equivalence relations. For this purpose, we recall that an equivalence relation Θ on an ordered set E is *strongly lower regular* if

$$z \leqslant x\Theta y \;\Rightarrow\; (\exists z' \in E) \quad z\Theta z' \leqslant y;$$

strongly upper regular if

$$z \geqslant y\Theta x \;\Rightarrow\; (\exists z' \in E) \quad z\Theta z' \geqslant x;$$

and *strongly regular* if it is both.

Theorem 12.3 *Let X be an ordered set and let Θ be an equivalence relation on X. Then the following statements are equivalent:*

(1) *there is a dual closure $\vartheta : X \to X$ with* $\operatorname{Ker} \vartheta = \Theta$;

(2) Θ *is strongly lower regular and every Θ-class is bounded below.*

Proof (1) $\Rightarrow$ (2) : If (1) holds then clearly every Θ-class is bounded below, the smallest element of $[x]\Theta$ being $\vartheta(x)$. To see that Θ is strongly lower regular, suppose that $z \leqslant x\Theta y$. Then $\vartheta(z) \leqslant \vartheta(x) = \vartheta(y) \leqslant y$ and so $z\Theta\vartheta(z) \leqslant y$.

(2) $\Rightarrow$ (1) : Suppose now that (2) holds. For every $x \in X$ define

$$\vartheta(x) = \min [x]\Theta.$$

Then clearly $\vartheta = \vartheta^2 \leqslant \mathrm{id}_X$ and $\operatorname{Ker} \Theta = \vartheta$. Since Θ is strongly lower regular it follows from $z \leqslant y\Theta\vartheta(y)$ that there exists $z' \in X$ such that $z\Theta z' \leqslant \vartheta(y)$ whence $\vartheta(z) = \vartheta(z') \leqslant \vartheta^2(y) = \vartheta(y)$. Consequently ϑ is also isotone and hence is a dual closure. ◇

Theorem 12.4 *Let X be an ordered set and let $\vartheta : X \to X$ be a dual closure. Then the following statements are equivalent:*

(1) ϑ *is residuated*;

(2) $\Theta = \operatorname{Ker} \vartheta$ *is strongly upper regular and every Θ-class is bounded above.*

Proof (1) $\Rightarrow$ (2) : If the dual closure ϑ is residuated then ϑ^+ is a closure on X. Moreover, from the relations $\vartheta = \vartheta^+ \circ \vartheta$ and $\vartheta^+ = \vartheta \circ \vartheta^+$ we deduce that $\operatorname{Ker} \vartheta^+ = \operatorname{Ker} \vartheta$. The dual of Theorem 12.3 now gives (2).

(2) $\Rightarrow$ (1) : Suppose now that (2) holds and define $\varphi : X \to X$ by

$$\varphi(x) = \max [x]\Theta.$$

Then on the one hand we have

$$\begin{aligned} \vartheta(y) \leqslant x \quad &\Rightarrow \quad y\Theta\vartheta(y) \leqslant x \\ &\Rightarrow \quad (\exists z \in X) \quad y \leqslant z\Theta x \\ &\Rightarrow \quad y \leqslant \max [x]\Theta = \varphi(x); \end{aligned}$$

and on the other, since by definition x and $\varphi(x)$ are in the same Θ-class,

$$y \leqslant \varphi(x) \;\Rightarrow\; \vartheta(y) \leqslant \vartheta\varphi(x) = \vartheta(x) \leqslant x.$$

These observations show that ϑ is residuated with $\vartheta^+ = \varphi$. ◇

Suppose now that E is an ordered set and that Θ is an equivalence relation on E. If $A \subseteq E$ then we shall denote by A^Θ the union of all the Θ-classes that contain an element of A, so that

$$A^\Theta = \bigcup\{[a]\Theta \mid a \in A\}.$$

Now the topology on a Priestley space has as a sub-basis the clopen decreasing sets and the clopen increasing sets. We shall say that Θ is *lower saturated* on X if

$$U \text{ decreasing clopen } \Rightarrow U^\Theta \text{ clopen;}$$

upper saturated if

$$U \text{ increasing clopen } \Rightarrow U^\Theta \text{ clopen;}$$

and *saturated* if both hold.

Theorem 12.5 *Let X be a Priestley space and let Θ be an equivalence relation on X. Suppose that there is a dual closure $\vartheta : X \to X$ with* Ker $\vartheta = \Theta$. *Then ϑ is continuous if and only if Θ is lower saturated.*

Proof First observe that if U is a decreasing subset then $\vartheta^{-1}(U) = U^\Theta$. Indeed, if $x \in U^\Theta$ then $x\Theta u \in U$ so $\vartheta(x) = \vartheta(u) \leqslant u \in U$ and therefore $x \in \vartheta^{-1}(U)$; and, conversely, if $x \in \vartheta^{-1}(U)$ then $\vartheta(x) \in U$ and therefore, since $x\Theta\vartheta(x)$, we have $x \in U^\Theta$.

$\Rightarrow$: It is immediate from the above observation that if ϑ is continuous then Θ is lower saturated.

$\Leftarrow$: Suppose now that Θ is lower saturated. If U is clopen and decreasing then $\vartheta^{-1}(U) = U^\Theta$ is clopen; and if U is clopen and increasing then $\vartheta^{-1}(U) = -(-U)^\Theta$ is also clopen. It follows that ϑ is continuous. ◇

The above results give the following characterisation of Priestley spaces that admit the structure of a double MS-space.

Theorem 12.6 *Let X be a Priestley space. Then X is the underlying set of a double MS-space if and only if there can be defined on X an equivalence relation Θ such that*

(1) *Θ has bounded classes;*
(2) *Θ is strongly regular;*
(3) *X/Θ admits a continuous polarity;*
(4) *Θ is saturated.*

Proof $\Rightarrow$: By Theorem 12.2 there is a residuated dual closure ϑ on X such that ϑ and ϑ^+ are continuous and Im ϑ admits a continuous polarity. By

Theorems 12.3 and 12.4, Θ = Ker ϑ is strongly regular and every Θ-class is bounded, so (1) and (2) hold. As for (3), this follows from

$$\text{Im } \vartheta \simeq X/\text{Ker } \vartheta = X/\Theta.$$

Finally, (4) follows from Theorem 12.5 and its dual.

$\Leftarrow$: If the conditions hold then by (1), (2), and Theorems 12.3, 12.4 there is a residuated dual closure ϑ on X with Ker $\vartheta = \Theta$. By (4), Theorem 12.5 and its dual, both ϑ and ϑ^+ are continuous. Finally, by (3) and Theorem 12.2, X is the underlying set of a double MS-space. ◇

As the following two examples show, the four conditions of Theorem 12.6 are independent.

Example 12.4 Consider the ordered set X with Hasse diagram

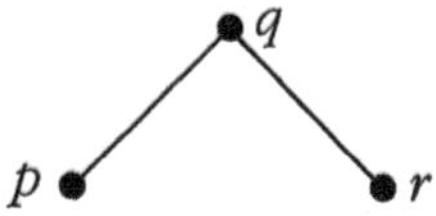

There are five equivalence relations on X, namely $\Theta_1 = \omega, \Theta_2 = \iota$, and

$$\Theta_3 \equiv \{\{p,r\},\{q\}\}, \quad \Theta_4 \equiv \{\{p,q\},\{r\}\}, \quad \Theta_5 \equiv \{\{p\},\{q,r\}\}.$$

Since X is finite the discrete topology gives a Priestley space, and in this situation condition (4) of Theorem 12.6 is redundant. Of the remaining properties, it is readily verified that Θ_1 satisfies all but (3); Θ_2 satisfies all but (1); Θ_3 satisfies all but (1); Θ_4 and Θ_5 satisfy all but (2).

Example 12.5 Let $X = \mathbb{N} \oplus \mathbf{3} \oplus \mathbb{N}^{op}$ where $\mathbf{3}$ is the chain $p < q < r$. Endow X with the interval topology, i.e. that generated by the sets $\{x \mid x < a\}$ and $\{x \mid x > a\}$ for every $a \in X$. Then X is a Priestley space. Consider the dual closure $\vartheta : X \to X$ given by

$$\vartheta(x) = \begin{cases} x & \text{if } x \neq q; \\ p & \text{if } x = q. \end{cases}$$

Here Θ = Ker ϑ clearly satisfies conditions (1), (2), (3) of Theorem 12.6. However, it does not satisfy condition (4). For example, $U = q^{\uparrow}$ is clopen but $U^{\Theta} = p^{\uparrow}$ is not open.

Note that in this example ϑ is residuated; in fact, we have

$$\vartheta^+(x) = \begin{cases} x & \text{if } x \neq p; \\ q & \text{if } x = p. \end{cases}$$

Although ϑ is continuous, ϑ^+ is not; for example, $q^{\uparrow}$ is open but $(\vartheta^+)^{-1}(q^{\uparrow}) = p^{\uparrow}$ is not.

For a finite ordered set X, regarded as a Priestley space under the discrete topology, the number of double MS-spaces definable on X depends on the number of polarities on X/Θ for each appropriate equivalence relation Θ on X. This is illustrated in the following example.

Example 12.6 If X has Hasse diagram

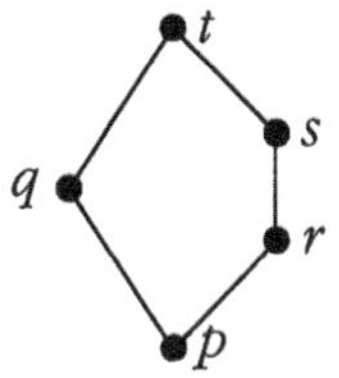

then on X there are five equivalence relations that satisfy the conditions of Theorem 12.6, namely

$$\Theta_1 = \omega, \quad \Theta_2 = \iota, \quad \Theta_3 \equiv \{\{p,r,s\},\{q,t\}\},$$
$$\Theta_4 \equiv \{\{r,s,t\},\{p,q\}\}, \quad \Theta_4 \equiv \{\{r,s\},\{p\},\{q\},\{t\}\}.$$

There are therefore five corresponding dual closures ϑ_i such that Ker $\vartheta_i = \Theta_i$, namely

x	p	q	r	s	t
$\vartheta_1(x)$	p	q	r	s	t
$\vartheta_2(x)$	p	p	p	p	p
$\vartheta_3(x)$	p	q	p	p	q
$\vartheta_4(x)$	p	p	r	r	r
$\vartheta_5(x)$	p	q	r	r	t

Since X/Θ_5 is the four-element boolean lattice, which admits two distinct polarities, there are in all six distinct antitone mappings g (these being given as in the proof of Theorem 12.2 by $g(x) = \alpha\vartheta(x)$ where α is a polarity on Im ϑ), namely

x	p	q	r	s	t
$g_1(x)$	t	q	s	r	p
$g_2(x)$	p	p	p	p	p
$g_3(x)$	q	p	q	q	p
$g_4(x)$	r	r	p	p	p
$g_5(x)$	t	q	r	r	p
$g'_5(x)$	t	r	q	q	p

In an entirely similar manner, we can compute the 6 corresponding antitone mappings h.

Now $L = \mathcal{O}(X)$ has Hasse diagram

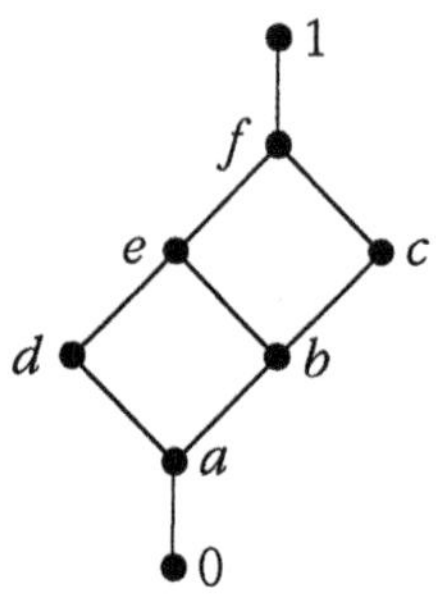

and on L we can define six non-isomorphic double MS-algebras, namely

x	0	a	b	c	d	e	f	1	
x°	1	f	e	d	c	b	a	0	L_1
x^+	1	f	e	d	c	b	a	0	
x°	1	0	0	0	0	0	0	0	L_2
x^+	1	1	1	1	1	1	1	0	
x°	1	c	c	c	0	0	0	0	L_3
x^+	1	1	1	c	1	1	c	0	
x°	1	d	0	0	d	0	0	0	L_4
x^+	1	1	1	1	d	d	d	0	
x°	1	f	d	d	c	a	a	0	L_5
x^+	1	f	f	d	c	c	a	0	
x°	1	f	c	c	d	a	a	0	L_6
x^+	1	f	f	c	d	d	a	0	

Note that L_1 is a Kleene algebra and L_2 is a double Stone algebra.

We shall now consider the following purely algebraic question : given an MS-algebra $(L; {}^\circ)$, precisely when can this be made into a double MS-algebra $(L; {}^\circ, {}^+)$? For this purpose, we recall that a non-empty subset M of an ordered set E is said to be *bicomplete* if, for every $x \in E$, the set $x^{\downarrow} \cap M$ has a biggest element and the set $x^{\uparrow} \cap M$ has a smallest element. In an MS-algebra $(L; {}^\circ)$ the smallest element of $x^{\uparrow} \cap L^{\circ\circ}$ is $x^{\circ\circ}$. By [**3**, Theorem 20.1], there is a bijection between the set of residuated closure maps on E and the set of bicomplete subsets of E; if $\vartheta : E \to E$ is a residuated closure then $\mathrm{Im}\,\vartheta = \mathrm{Im}\,\vartheta^+$ is bicomplete and ϑ^+ is given by

$$(\forall x \in E) \quad \vartheta^+(x) = \max(x^{\downarrow} \cap L^{\circ\circ}).$$

Theorem 12.7 *An MS-algebra $(L; {}^\circ)$ can be made into a double MS-algebra $(L; {}^\circ, {}^+)$ if and only if the closure $a \mapsto a^{\circ\circ}$ is residuated and its residual preserves suprema.*

Proof $\Rightarrow$: If $(L; {}^\circ, {}^+)$ is a double MS-algebra then for every $a \in L$ we have, by (D1) and (D2),

$$a^{\circ\circ++} = a^{\circ+++} = a^{\circ+} = a^{\circ\circ} \geqslant a;$$
$$a^{++\circ\circ} = a^{+\circ\circ\circ} = a^{+\circ} = a^{++} \leqslant a.$$

Consequently the closure map $a \mapsto a^{\circ\circ}$ is residuated with residual the dual closure $a \mapsto a^{++}$, which clearly preserves suprema.

$\Leftarrow$: Suppose conversely that $\vartheta : a \mapsto a^{\circ\circ}$ is residuated and that ϑ^+ preserves suprema. Then $\mathrm{Im}\,\vartheta = L^{\circ\circ}$ is bicomplete and

$$(\forall a \in E) \quad \vartheta^+(a) = \max(a^{\downarrow} \cap L^{\circ\circ}).$$

For every $a \in L$ define

$$a^+ = [\vartheta^+(a)]^\circ = [\max(a^{\downarrow} \cap L^{\circ\circ})]^\circ.$$

Then we have $0^+ = 1$, $1^+ = 0$ and

$$(a \vee b)^+ = [\vartheta^+(a \vee b)]^\circ = [\vartheta^+(a) \vee \vartheta^+(b)]^\circ = [\vartheta^+(a)]^\circ \wedge [\vartheta^+(b)]^\circ = a^+ \wedge b^+$$

so that $(L, {}^+)$ is a dual MS-algebra. Moreover,

$$a^{\circ+} = [\max(a^{\circ\downarrow} \cap L^{\circ\circ})]^\circ = a^{\circ\circ},$$

and, since $L^{\circ\circ} = \mathrm{Im}\,\vartheta = \mathrm{Im}\,\vartheta^+ = L^{++}$,

$$a^{++} = [\max(a^{+\downarrow} \cap L^{\circ\circ})]^\circ = a^{+\circ}.$$

Thus (D1) and (D2) hold, whence $(L, {}^\circ, {}^+)$ is a double MS-algebra. ◇

Corollary 1 *If an MS-algebra $(L; {}^\circ)$ can be made into a double MS-algebra $(L; {}^\circ, {}^+)$ then this can be done in only one way, namely with*

$$(\forall a \in L) \quad a^+ = [\max(a^{\downarrow} \cap L^{\circ\circ})]^\circ.$$

Proof Suppose that $(L; {}^\circ, {}^\oplus)$ is also a double MS-algebra. Then $a \mapsto a^{\oplus\oplus}$ is also the residual of $a \mapsto a^{\circ\circ}$ and so $a^{\oplus\oplus} = a^{++}$. By (D1) and (D2) we have

$$a^\oplus = a^{\oplus\oplus\oplus} = a^{++\oplus} = a^{+\circ\oplus} = a^{+\circ\circ} = a^{+++} = a^+.$$

Hence the operations ${}^\oplus$ and ${}^+$ coincide. ◇

Corollary 2 *If an MS-algebra* $(L;^{\circ})$ *can be made into a double MS-algebra then every* $a \in L^{\circ\circ}$ *that is* $\vee$*-reducible in* L *must be* $\vee$*-reducible in* $L^{\circ\circ}$.

Proof Suppose that $a \in L^{\circ\circ}$ is such that $a = b \vee c$ where $b, c < a$. Then $a = a^{++} = b^{++} \vee c^{++}$. Since $b^{++} = a$ gives the contradiction $a \leqslant b$, and likewise for c, it follows that $b^{++}, c^{++} < a$. ◇

Example 12.7 Consider the MS-algebra K_3. Here we have that 1 is $\vee$-reducible in K_3 but is not $\vee$-reducible in $K_3^{\circ\circ}$. Hence, by Corollary 2 of Theorem 12.7, K_3 cannot be made into a double MS-algebra.

Although we have considered here only double MS-algebras, it is possible to consider other double Ockham algebras. For example, M. Sequeira **[91]** defines an $\mathbf{O}_2$*-algebra* to be an algebra $(L; \wedge, \vee, f, g, 0, 1)$ of type $(2, 2, 1, 1, 0, 0)$ such that $(L; \wedge, \vee, f, 0, 1)$ and $(L; \wedge, \vee, g, 0, 1)$ are Ockham algebras. In particular, she considers the notion of a *double* MS_n*-algebra*, namely an algebra $(L; f, g) \in \mathbf{O}_2$ such that

$$fg = g^{2n} \leqslant \mathrm{id} \leqslant f^{2n} = gf.$$

Clearly, this generalises the notion of a double MS-algebra.

13 Subdirectly irreducible double MS-algebras

In order to determine the subdirectly irreducible double MS-algebras we can extend the discussion given in Chapter 4 for MS-algebras. Corresponding to the notion of a *g-subset* in the dual space $(X; g)$ of an MS-algebra $(L; {}^{\circ})$, we define a *$\{g, h\}$-subset* in the dual space $(X; g, h)$ of a double MS-algebra $(L; {}^{\circ}, {}^{+})$ to be a subset of X that is both a g-subset and an h-subset. Such a subset, for example, is $g^{\omega}\{x\} \cup h^{\omega}\{x\}$. Working with this we see that, corresponding to Corollary 1 of Theorem 4.6, a finite double MS-algebra $(L; {}^{\circ}, {}^{+})$ is subdirectly irreducible if and only if there exists $x \in X$ such that $X = g^{\omega}\{x\} \cup h^{\omega}\{x\}$.

In examining for a double MS-space $(X; g, h)$ the corresponding situation to that in Example 4.8 we have to consider a 'double noose'

in which the arrows to the right of p indicate the (partial) effect of g and those to the left of p that of h. Here the inequalities $g \circ h = g^2 \leqslant \mathrm{id}_X \leqslant h^2 = h \circ g$ force, for example, $g(s) = g[h(p)] = g^2(p) = r \leqslant p$. We are therefore led to consider the double MS-space $(X; g, h)$ described as follows, in which the complete actions of g and h on the chains $r < p < t$ and $q < s$ are as indicated :

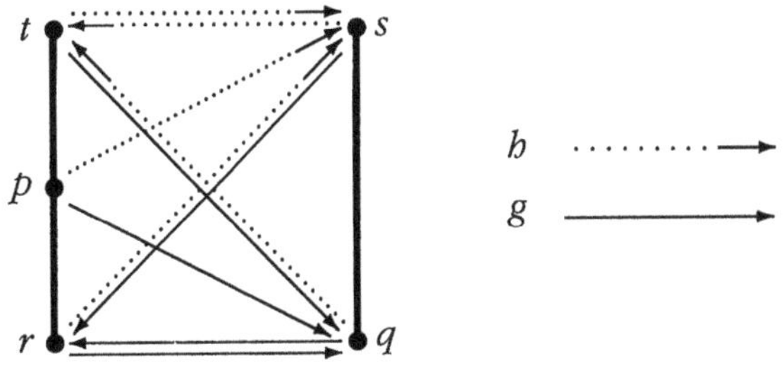

Note that here $X = g^{\omega}\{p\} \cup h^{\omega}\{p\}$ so the corresponding double MS-algebra is subdirectly irreducible. Using arguments similar to those in Chapter 4, we can see that the subdirectly irreducible double MS-algebras are precisely the subalgebras of this algebra. Up to isomorphism, there are 21 in all. We arrange them in decreasing order of cardinality and label them $\mathrm{SID}_{21}, \ldots, \mathrm{SID}_1$.

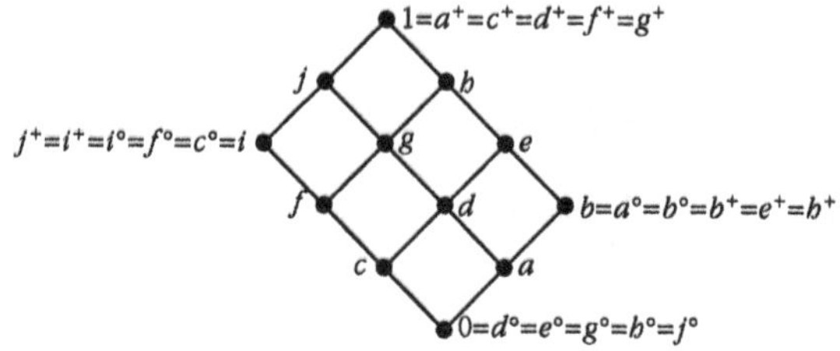

SID_{21}

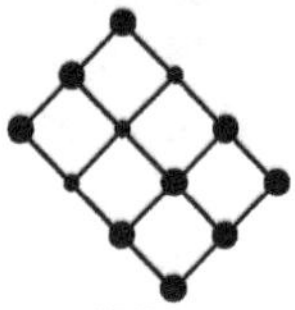
SID_{20}

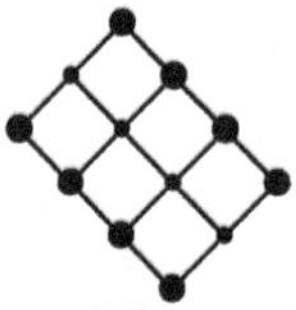
SID_{19}

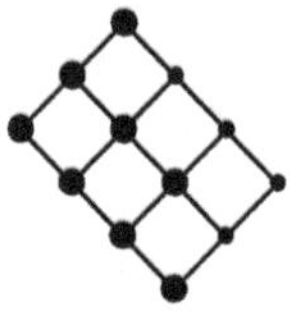
SID_{18}

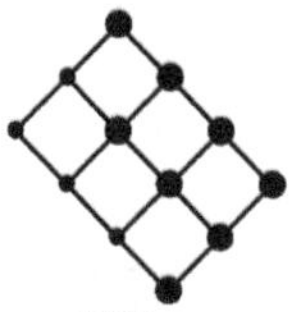
SID_{17}

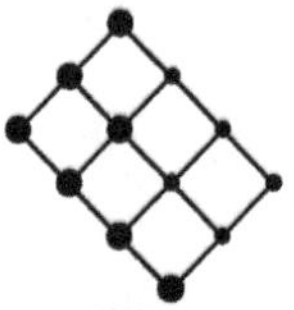
SID_{16}

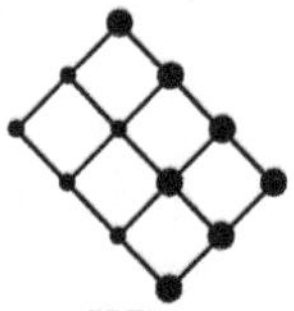
SID_{15}

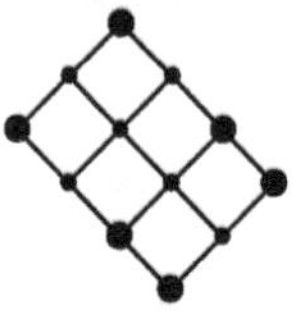
SID_{14}

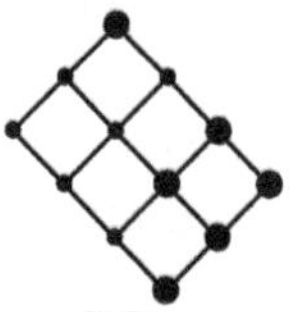
SID_{13}

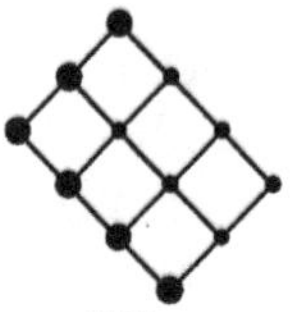
SID_{12}

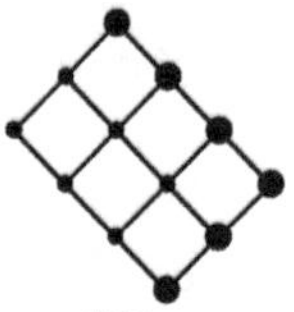
SID_{11}

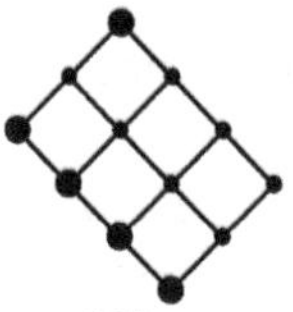
SID_{10}

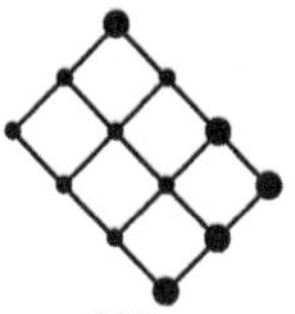
SID_9

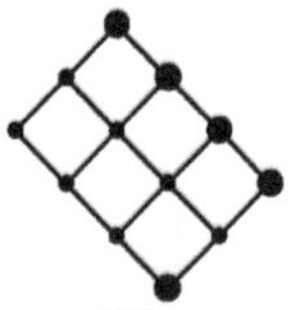
SID_8

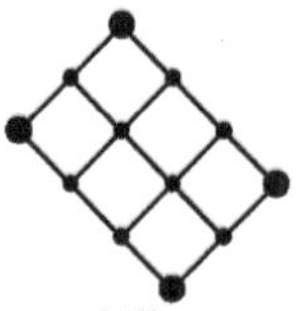
SID_7

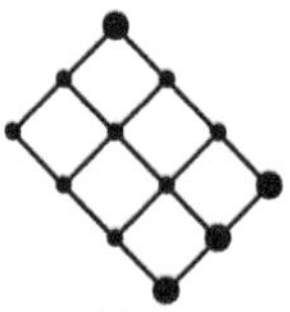
SID_6

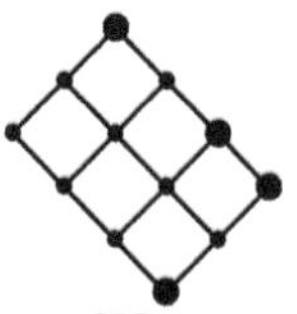
SID_5

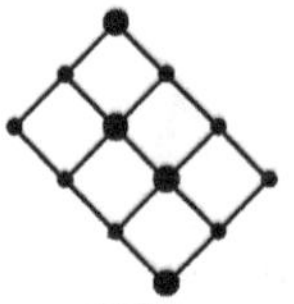
SID_4

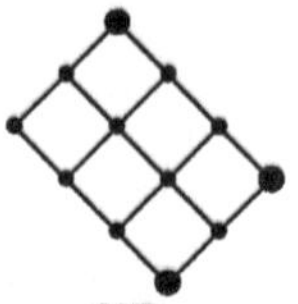
SID_3

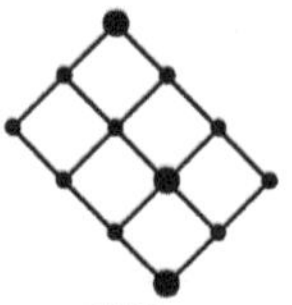
SID_2

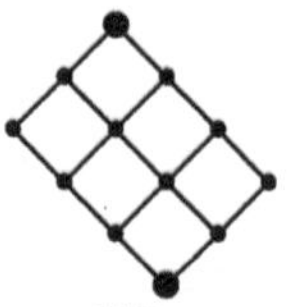
SID_1

These algebras were originally determined in **[38]** without using duality.

In a double MS-algebra $(L;\ ^\circ,\ ^+)$ we define the congruence Φ_+° by

$$(a,b)\in\Phi_+^\circ \iff (a^\circ = b^\circ \text{ and } a^+ = b^+).$$

From the above description, we then have immediately :

Theorem 13.1 *A double MS-algebra L is subdirectly irreducible if and only if* Con L *reduces to the chain* $\omega \preceq \Phi_+^\circ \prec \iota$. ◇

Since in SID_{21} the non-trivial Φ_+°-classes are $\{c,f\}$, $\{d,g\}$, $\{e,h\}$ we also have :

Theorem 13.2 *Of the* 21 *non-isomorphic subdirectly irreducible double MS-algebras* 11 *are non-simple, namely*

$$\mathrm{SID}_4,\ \mathrm{SID}_8,\ \mathrm{SID}_{10},\ \mathrm{SID}_{11},\ \mathrm{SID}_{12},\ \mathrm{SID}_{15},\ \mathrm{SID}_{16},\ \mathrm{SID}_{17},\ \mathrm{SID}_{18},\ \mathrm{SID}_{19},\ \mathrm{SID}_{21};$$

and 10 *are simple, namely*

$$\mathrm{SID}_1,\ \mathrm{SID}_2,\ \mathrm{SID}_3,\ \mathrm{SID}_5,\ \mathrm{SID}_6,\ \mathrm{SID}_7,\ \mathrm{SID}_9,\ \mathrm{SID}_{13},\ \mathrm{SID}_{14},\ \mathrm{SID}_{20}.\ \diamond$$

We note also that

$$\mathrm{SID}_1,\ \mathrm{SID}_2,\ \mathrm{SID}_3,\ \mathrm{SID}_4,\ \mathrm{SID}_7,\ \mathrm{SID}_9,\ \mathrm{SID}_{13},\ \mathrm{SID}_{14},\ \mathrm{SID}_{19},\ \mathrm{SID}_{20},\ \mathrm{SID}_{21}$$

are self-dual, whereas we have dual isomorphisms

$$\mathrm{SID}_5 \overset{d}{\simeq} \mathrm{SID}_6,\ \mathrm{SID}_8 \overset{d}{\simeq} \mathrm{SID}_{10},\ \mathrm{SID}_{11} \overset{d}{\simeq} \mathrm{SID}_{12},\ \mathrm{SID}_{15} \overset{d}{\simeq} \mathrm{SID}_{16},\ \mathrm{SID}_{17} \overset{d}{\simeq} \mathrm{SID}_{18}.\ \diamond$$

Theorem 13.3 *There are* 3 *non-isomorphic subdirectly irreducible double Stone algebras, namely* SID_1, SID_2, SID_4. *Moreover only one of these, namely* SID_4, *is non-simple.* ◇

We can order the subdirectly irreducible double MS-algebras by writing $A \leqslant B$ if and only if A is isomorphic to a subalgebra of B. In so doing, we obtain the following Hasse diagram, in which n denotes SID_n :

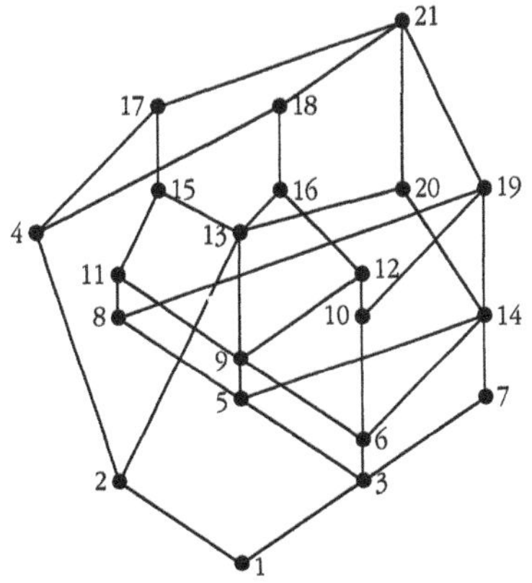

The lattice of subvarieties of double MS-algebras can, in theory, be obtained from this by applying the theorem of Davey in precisely the same way as we did in Chapter 5 to obtain the lattice of subvarieties of MS-algebras. However, in the case of double MS-algebras this lattice is rather large and so we shall concentrate on some important ideals of it. Specifically, if $\mathbf{SID}_n$ denotes the subvariety generated by SID_n, we describe the lattices of subvarieties of $\mathbf{SID}_{20}$, $\mathbf{SID}_{19}$, and $\mathbf{SID}_{17} \vee \mathbf{SID}_{18}$. We also obtain equational bases for the subvarieties generated by the subdirectly irreducible double MS-algebras.

(1) *Semisimple double MS-algebras*

An algebra is *semisimple* if it is isomorphic to a subdirect product of simple algebras. A variety $\mathbf{V}$ is semisimple if every member of $\mathbf{V}$ is semisimple. It is well known that a variety $\mathbf{V}$ is semisimple if and only if every subdirectly irreducible member of $\mathbf{V}$ is simple. Now from Theorem 13.2 there are 10 simple double MS-algebras. As can be seen, they constitute the down-set $20^{\downarrow}$ in the above Hasse diagram. So a double MS-algebra is semisimple if and only if it is a subdirect product of copies of SID_{20}. The variety $\mathbf{SID}_{20}$ can be characterised as follows.

Theorem 13.4 *On a double MS-algebra* $(L; {}^\circ, {}^+)$ *the following conditions are equivalent*:

(A_0) *L is semisimple*;
(A) $(\forall a, b \in L)\ a \wedge b^{\circ\circ} \leqslant a^{++} \vee b$;
(A_1) $\Phi_+^\circ = \omega$.

Proof $(A_0) \Rightarrow (A)$: It suffices to observe that SID_{20} satisfies (A).

$(A) \Rightarrow (A_0)$: Examination of each of the subdirectly irreducible double MS-algebras reveals that only those that are simple satisfy (A). Thus, if L satisfies (A) then by Birkhoff's theorem L is a subdirect product of copies of SID_{20}.

$(A) \Rightarrow (A_1)$: Suppose that L satisfies (A) and that $a, b \in L$ are such that $a^\circ = b^\circ$ and $a^+ = b^+$. Then (A) gives $a = a \wedge a^{\circ\circ} = a \wedge b^{\circ\circ} \leqslant a^{++} \vee b = b^{++} \vee b = b$. Similarly, $b \leqslant a$ and so $a = b$.

$(A_1) \Rightarrow (A)$: Suppose that $\Phi_+^\circ = \omega$ and consider the elements

$$p = a \wedge b^{\circ\circ}, \quad q = a \wedge b^{\circ\circ} \wedge (a^{++} \vee b).$$

We have

$$q^\circ = a^\circ \vee b^\circ \vee (a^+ \wedge b^\circ) = a^\circ \vee b^\circ = p^\circ;$$
$$q^+ = a^+ \vee b^\circ \vee (a^+ \wedge b^+) = a^+ \vee b^\circ = p^+.$$

It follows that $q = p$, whence (A) follows. ◇

We can construct the lattice of subvarieties of semisimple double MS-algebras by applying Davey's theorem. Its size can of course be predicted by Theorem 5.5 applied to the down-set $\mathrm{SID}_{20}^{\downarrow}$. We leave to the reader the verification that it has 30 elements and, with $\mathbf{n} = \mathbf{SID}_n$, is

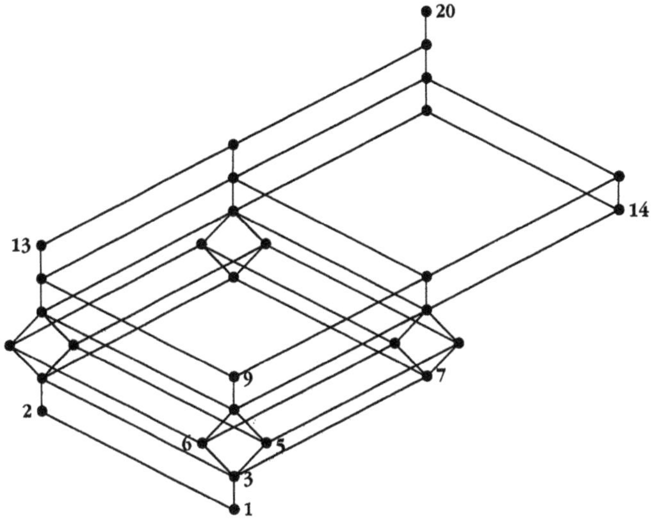

(2) *Locally convex skeletons*

The *skeleton* of a double MS-algebra L is the set

$$S(L) = \{a \in L \mid a^{++} = a = a^{\circ\circ}\}.$$

Clearly, $S(L)$ is a de Morgan algebra. We shall say that L has a *locally convex skeleton* if every interval of the form $[a^\circ, a^+]$ belongs to $S(L)$.

We can characterise the subvariety $\mathbf{SID}_{19}$ as follows.

Theorem 13.5 *On a double MS-algebra* $(L; {}^\circ, {}^+)$ *the following conditions are equivalent*:

(B_0) *L has a locally convex skeleton*;

(B) $(\forall a, b \in L)\ a^{\circ\circ} \wedge b^+ \leqslant a^{++} \vee b^\circ$;

(B_1) $L \in \mathbf{SID}_{19}$.

Proof $(B_0) \Leftrightarrow (B)$: It is clear that $z \in [b^\circ, b^+]$ if and only if z is of the form $(a \vee b^\circ) \wedge b^+$, and that this is in $S(L)$ if and only if

$$(a^{\circ\circ} \vee b^\circ) \wedge b^+ = (a^{++} \vee b^\circ) \wedge b^+,$$

which is equivalent to (B).

$(B) \Leftrightarrow (B_1)$: It suffices to check that (B) is satisfied by SID_{19} but is not satisfied by either SID_9 or SID_2. ◇

Using the same technique as before, we can construct the lattice of subvarieties of $\mathbf{SID}_{19}$. This has 24 elements and is

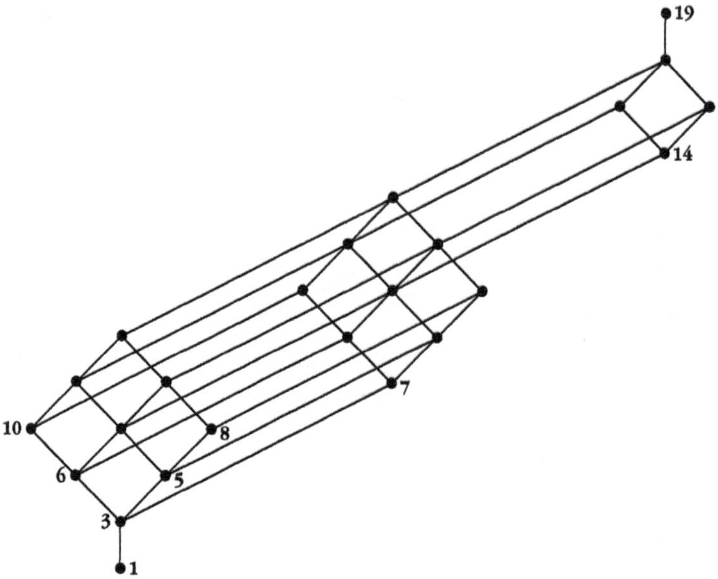

(3) *Kleene skeletons*

We now turn our attention to the subvariety $\mathbf{SID}_{17} \vee \mathbf{SID}_{18}$.

Theorem 13.6 *The following inequalities are equivalent*:

$(i)\ f \leqslant b^{++} \vee b^{+}$, $(ii)\ f \leqslant b \vee b^{+}$, $(iii)\ f \leqslant b^{\circ\circ} \vee b^{+}$, $(iv)\ f \leqslant b^{\circ\circ} \vee b^{\circ}$.

Likewise, so are the inequalities

$(j)\ a^{\circ\circ} \wedge a^{\circ} \leqslant g$, $(jj)\ a \wedge a^{\circ} \leqslant g$, $(jjj)\ a^{++} \wedge a^{\circ} \leqslant g$, $(jw)\ a^{++} \wedge a^{+} \leqslant g$.

Proof Clearly, $(i) \Rightarrow (ii) \Rightarrow (iii)$. For $(iii) \Rightarrow (i)$ and $(iv) \Rightarrow (i)$, write b^{++} for b; and for $(i) \Rightarrow (iv)$, write $b^{\circ\circ}$ for b. ◇

Theorem 13.7 *The following inequalities are equivalent*

$(v)\ b \vee b^{\circ} \geqslant f \in S(L)$, $(vi)\ b^{++} \vee b^{\circ} \geqslant f \in S(L)$;
$(w)\ a \wedge a^{+} \leqslant g \in S(L)$, $(wj)\ a^{\circ\circ} \wedge a^{+} \leqslant g \in S(L)$. ◇

Theorem 13.8 *On a double MS-algebra* $(L; {}^{\circ}, {}^{+})$ *the following conditions are equivalent:*

(C_0) *L has a Kleene skeleton;*

(C) $(\forall a, b \in L)\ a \wedge a^{\circ} \leqslant b \vee b^{+}$;

(C_1) $L \in \mathbf{SID}_{17} \vee \mathbf{SID}_{18}$.

Proof (C_0) holds if and only if $a^{\circ\circ} \wedge a^{\circ} \leqslant b^{\circ\circ} \vee b^{\circ}$ and by Theorem 13.6 this is equivalent to (C). The equivalence of (C) and (C_1) results from the fact that SID_{17} and SID_{18} satisfy (C) whereas SID_7 does not. ◇

The lattice of subvarieties of double MS-algebras with a Kleene skeleton can be constructed using the same technique as before. The lattice in question has 98 elements and can be visualised as follows. It consists of the three 'layers' shown below, the second projecting down onto the first with $\mathbf{SID}_2$ directly above $\mathbf{SID}_1$, and the third projecting down onto the second with $\mathbf{SID}_4$ directly above $\mathbf{SID}_2$.

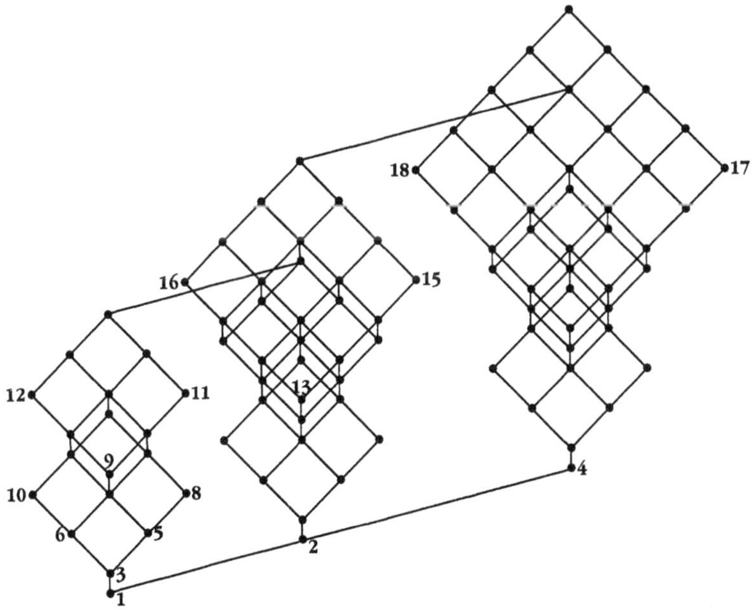

Equational bases for $\mathbf{SID}_n$ are not in general unique. In what follows we shall obtain equational bases each of which involves at most three relations. We first add to the results of Theorems 13.4, 13.5, 13.8 further single-term equational bases.

Definition For each axiom (n) we define $(n^\star)$ to be the axiom obtained from (n) by replacing $\vee, \wedge, {}^\circ, {}^+, 0, 1, \leqslant$ respectively by $\wedge, \vee, {}^+, {}^\circ, 1, 0, \geqslant$. When (n) is equivalent to $(n^\star)$ we shall say that (n) is *self-dual in* **DMS**.

Observe for example that although the axiom (2), namely $a \vee a^\circ = 1$, is not self-dual in **MS** it is self-dual in **DMS**. In fact, $a \vee a^\circ = 1$ gives $a^+ \wedge a^{\circ\circ} = 0$, whence $a^+ \wedge a = 0$. Conversely, $a^+ \wedge a = 0$ gives $a^{++} \vee a^\circ = a^{+\circ} \vee a^\circ = 1$, whence $a \vee a^\circ = 1$.

The axioms $(A), (B), (C)$ above are self-dual in **DMS**, as are (E), (F), (G), (2_d), (α), (γ) in the following result.

Theorem 13.9 *The following subvarieties of double MS-algebras have the equational bases indicated:*

$\mathbf{SID}_{18} \vee \mathbf{SID}_{20}$	(D)	$a \wedge b^{\circ\circ} \leqslant a^{++} \vee b \vee b^+$
$\mathbf{SID}_{17} \vee \mathbf{SID}_{20}$	$(D^\star)$	$a \wedge a^\circ \wedge b^{\circ\circ} \leqslant a^{++} \vee b$
$\mathbf{SID}_{15} \vee \mathbf{SID}_{16} \vee \mathbf{SID}_{20}$	(E)	$a \wedge a^+ \wedge b^{\circ\circ} \leqslant a^{++} \vee b \vee b^\circ$
$\mathbf{SID}_{11} \vee \mathbf{SID}_{12}$	(F)	$a^{\circ\circ} \wedge a^+ \leqslant b^{++} \vee b^\circ$
$\mathbf{SID}_{2} \vee \mathbf{SID}_{11} \vee \mathbf{SID}_{12}$	(G)	$a \wedge a^+ \leqslant b \vee b^\circ$
$\mathbf{SID}_{7}$	(α)	$a = a^{\circ\circ}$ $(\equiv a^\circ = a^+)$
$\mathbf{SID}_{4} \vee \mathbf{SID}_{11} \vee \mathbf{SID}_{12}$	(γ)	$a \wedge a^\circ \leqslant b \vee b^\circ$
$\mathbf{SID}_{4}$	(2_d)	$a \wedge a^\circ = 0$
$\mathbf{SID}_{1}$	(2)	$a \vee a^\circ = 1$
$\mathbf{SID}_{7} \vee \mathbf{SID}_{10}$	(15)	$a \vee a^\circ = a^{\circ\circ} \vee a^\circ$
$\mathbf{SID}_{7} \vee \mathbf{SID}_{8}$	$(15^\star)$	$a \wedge a^+ = a^{++} \wedge a^+$

Proof This uses the same principle that we employed in Chapter 5.

(D) is satisfied by SID_{18} and SID_{20}, but not by SID_{8}.
$(D^\star)$ is satisfied by SID_{17} and SID_{20}, but not by SID_{10}.
(E) is satisfied by SID_{15}, SID_{16}, and SID_{20}, but not by SID_{4} or SID_{19}.
(F) is satisfied by SID_{11} and SID_{12}, but not by SID_{2} or SID_{7}.
(G) is satisfied by SID_{2}, SID_{11}, and SID_{12} but not by SID_{4}, SID_{13}, or SID_{7}.
(α) is satisfied by SID_{7} but not by SID_{2}, SID_{5}, or SID_{6}.
(γ) is satisfied by SID_{4}, SID_{11}, and SID_{12} but not by SID_{7} or SID_{13}.
(2_d) is satisfied by SID_{4} but not by SID_{3}.
(2) is satisfied by SID_{1} but not by SID_{2} or SID_{3}.
(15) is satisfied by SID_{7} and SID_{10}, but not by SID_{2} or SID_{5}.
$(15^\star)$ is satisfied by SID_{7} and SID_{8}, but not by SID_{2} or SID_{6}. ◇

The above axioms are linked by the following implications.

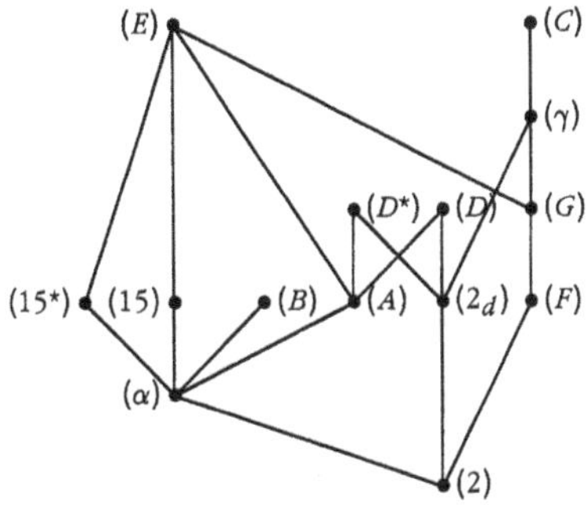

Theorem 13.10 *An equational basis for* $\mathbf{SID}_{15} \vee \mathbf{SID}_{16}$ *is* (C, E).

Proof Writing $\mathbf{SID}_n$ as $\mathbf{n}$, we have that $\mathbf{20} \wedge \mathbf{17} = \mathbf{13} = \mathbf{20} \wedge \mathbf{18}$ and so

$$\begin{aligned} \mathbf{15} \vee \mathbf{16} &= \mathbf{15} \vee \mathbf{16} \vee \mathbf{13} \\ &= \mathbf{15} \vee \mathbf{16} \vee [\mathbf{20} \wedge (\mathbf{17} \vee \mathbf{18})] \\ &= (\mathbf{15} \vee \mathbf{16} \vee \mathbf{20}) \wedge (\mathbf{17} \vee \mathbf{18}) \end{aligned}$$

whence the result follows by Theorems 13.8 and 13.9. ◇

In the following result we shall use our knowledge of the various parts of the lattice $\Lambda(\mathbf{DMS})$.

Theorem 13.11 *Equational bases for* $\mathbf{n} = \mathbf{SID}_n$ $(n = 1, \ldots, 20)$ *are*:

20	(A)	**15**	$(C, D^\star, E)$	**10**	$(F, 15)$	**5**	$(A, F, 15^\star)$
19	(B)	**14**	(A, B)	**9**	(A, F)	**4**	(2_d)
18	(C, D)	**13**	(A, C)	**8**	$(F, 15^\star)$	**3**	(γ, α)
17	$(C, D^\star)$	**12**	(D, F)	**7**	(α)	**2**	$(A, 2_d)$
16	(C, D, E)	**11**	$(D^\star, F)$	**6**	$(A, F, 15)$	**1**	(2)

Proof Theorems 13.4, 13.5, 13.9 yield equational bases for **20**, **19**, **7**, **4**, **1**.

Now

$$\mathbf{18} = \mathbf{18} \vee \mathbf{13} = \mathbf{18} \vee (\mathbf{20} \wedge \mathbf{17}) = (\mathbf{18} \vee \mathbf{20}) \wedge (\mathbf{18} \vee \mathbf{17})$$

whence, by Theorems 13.9 and 13.8, **18** has equational basis (D, C).

From the lattice in Theorem 13.8, we have $\mathbf{16} = (\mathbf{15} \vee \mathbf{16}) \wedge \mathbf{18}$ and so, by Theorem 13.10 and the above, an equational basis for **16** is (C, D, E).

Since $\mathbf{14} = \mathbf{20} \wedge \mathbf{19}$, an equational basis for **14** is (A, B).

Since $\mathbf{13} = \mathbf{20} \wedge (\mathbf{17} \vee \mathbf{18})$, an equational basis for **13** is (A, C).

From the lattice of Theorem 13.8 we have $\mathbf{12} = (\mathbf{11} \vee \mathbf{12}) \wedge \mathbf{16}$ so, by Theorem 13.9, the above, and the fact that $(F) \Rightarrow (C, E)$, an equational basis for **12** is (D, F).

Since $\mathbf{12} \wedge \mathbf{7} = \mathbf{3} = \mathbf{11} \wedge \mathbf{7}$, we have

$$\mathbf{10} = \mathbf{10} \vee \mathbf{3} = \mathbf{10} \vee [\mathbf{7} \wedge (\mathbf{11} \vee \mathbf{12})] = (\mathbf{10} \vee \mathbf{7}) \wedge (\mathbf{11} \vee \mathbf{12}).$$

Consequently, by Theorem 13.9, an equational basis for **10** is $(F, 15)$.

Since $\mathbf{9} = (\mathbf{11} \vee \mathbf{12}) \wedge \mathbf{13}$, an equational basis for **9** is (A, F).

Since $\mathbf{6} = \mathbf{9} \wedge \mathbf{10}$, an equational basis for **6** is $(A, F, 15)$.

Since $\mathbf{3} = \mathbf{7} \wedge (\mathbf{4} \vee \mathbf{11} \vee \mathbf{12})$, an equational basis for **3** is (α, γ).

Since $\mathbf{2} = \mathbf{20} \wedge \mathbf{4}$, an equational basis for **2** is $(A, 2_d)$.

Finally, equational bases for the remaining five subvarieties can be deduced from the above using the dual isomorphisms on page 199. ◇

Of the above classes, some are worthy of especial mention, namely :

$\mathbf{SID}_1$: the class of boolean algebras;
$\mathbf{SID}_3$: the class of Kleene algebras;
$\mathbf{SID}_7$: the class of de Morgan algebras;
$\mathbf{SID}_4$: the class of double Stone algebras.

A further class worth mentioning is the class $\mathbf{SID}_2$, characterised by the properties (A) and (2_d). From the above, the members of this class are the double Stone algebras that are semisimple. Such algebras are called *trivalent Lukasiewicz algebras*. In fact, these algebras can be characterised by $(G, 2_d)$ since, from the lattice of Theorem 13.8, we have $\mathbf{2} = (\mathbf{2} \vee \mathbf{11} \vee \mathbf{12}) \wedge \mathbf{4}$. For further considerations of trivalent Lukasiewicz algebras we refer the reader to **[96, 98]**.

14 Congruences on double MS-algebras

We now turn our attention to congruences on double MS-algebras. Here also an important rôle is played by the principal congruences $\vartheta(a,b)$ with $a \leqslant b$. It is clear from Theorem 2.1 that if $L \in \mathbf{DMS}$ and $a, b \in L$ are such that $a \leqslant b$ then we have

$$\vartheta(a,b) = \vartheta_{\mathrm{lat}}(a,b) \vee \vartheta_{\mathrm{lat}}(b^\circ, a^\circ) \vee \vartheta_{\mathrm{lat}}(a^{\circ\circ}, b^{\circ\circ}) \vee \vartheta_{\mathrm{lat}}(b^+, a^+) \vee \vartheta_{\mathrm{lat}}(a^{++}, b^{++}).$$

Arguing precisely as in Theorem 8.1, we can characterise $\vartheta(a,b)$ as follows.

Theorem 14.1 *If $L \in \mathbf{DMS}$ and $a, b \in L$ with $a \leqslant b$ then $(x,y) \in \vartheta(a,b)$ if and only if*

(1) $x \wedge a^{++} \wedge b^\circ = y \wedge a^{++} \wedge b^\circ$;
(2) $(x \wedge a \wedge b^\circ) \vee b^{++} = (y \wedge a \wedge b^\circ) \vee b^{++}$;
(3) $(x \vee a^\circ) \wedge a^{++} \wedge b^+ = (y \vee a^\circ) \wedge a^{++} \wedge b^+$;
(4) $[(x \vee a^\circ) \wedge a \wedge b^+] \vee b^{++} = [(y \vee a^\circ) \wedge a \wedge b^+] \vee b^{++}$;
(5) $(x \vee a^+) \wedge a^{++} = (y \vee a^+) \wedge a^{++}$;
(6) $(x \wedge a) \vee a^+ \vee b^{++} = (y \wedge a) \vee a^+ \vee b^{++}$;
(7) $(x \vee b) \wedge b^\circ \wedge a^{\circ\circ} = (y \vee b) \wedge b^\circ \wedge a^{\circ\circ}$;
(8) $(x \wedge b^\circ) \vee b^{\circ\circ} = (y \wedge b^\circ) \vee b^{\circ\circ}$;
(9) $(x \vee b \vee a^\circ) \wedge a^{\circ\circ} \wedge b^+ = (y \vee b \vee a^\circ) \wedge a^{\circ\circ} \wedge b^+$;
(10) $(x \vee b \vee a^+) \wedge a^{\circ\circ} = (y \vee b \vee a^+) \wedge a^{\circ\circ}$;
(11) $(x \vee a^\circ \vee b^{\circ\circ}) \wedge b^+ = (y \vee a^\circ \vee b^{\circ\circ}) \wedge b^+$;
(12) $x \vee a^+ \vee b^{\circ\circ} = y \vee a^+ \vee b^{\circ\circ}$. ◇

In general, these conditions are difficult to handle, but in the case of double Stone algebras a considerable simplification occurs.

Theorem 14.2 *In a double Stone algebra $(x,y) \in \vartheta(a,b)$ if and only if*

(13) $x \wedge a \wedge (a^{++} \vee b^+) = y \wedge a \wedge (a^{++} \vee b^+)$;
(14) $(x \vee b) \wedge (a^{\circ\circ} \vee b^\circ) \wedge b^+ = (y \vee b) \wedge (a^{\circ\circ} \vee b^\circ) \wedge b^+$.

Proof Using the identity $a \wedge a^\circ = 0$, which has the following five equivalent forms

$$a^{\circ\circ} \wedge a^\circ = 0, \quad a^{++} \wedge a^+ = 0, \quad a^{\circ\circ} \vee a^\circ = 1, \quad a^{++} \vee a^+ = 1, \quad a \vee a^+ = 1,$$

and the fact that $a \leqslant b$, we see that the equations (1), (2), (3), (6), (7), (10), (11), (12) are trivial. As for (4), this reduces to

$$(x \wedge a \wedge b^+) \vee b^{++} = (y \wedge a \wedge b^+) \vee b^{++}.$$

Since $x \wedge a \wedge b^+ \wedge b^{++} = 0 = y \wedge a \wedge b^+ \wedge b^{++}$, it follows by distributivity that (4) becomes

$$(13') \qquad x \wedge a \wedge b^+ = y \wedge a \wedge b^+.$$

Clearly, (5), (8), (9) become respectively

$$(13'') \quad x \wedge a^{++} = y \wedge a^{++};$$
$$(14') \quad x \vee b^{\circ\circ} = y \vee b^{\circ\circ};$$
$$(14'') \quad (x \vee b) \wedge a^{\circ\circ} \wedge b^+ = (y \vee b) \wedge a^{\circ\circ} \wedge b^+.$$

Now it is clear that (13′) and (13″) together are equivalent to (13). Since L is a double Stone algebra, (14′) is equivalent to

$$x \wedge b^\circ = y \wedge b^\circ.$$

This, together with (14″), gives (14). Conversely, taking the join of each side of (14) with $b^{\circ\circ}$, we obtain (14′); and taking the meet of each side with $a^{\circ\circ}$ we obtain (14″). ◇

As with MS-algebras, there is a strong relationship between principal congruences on double MS-algebras and principal lattice congruences.

Theorem 14.3 *Let $L \in$ **DMS** and let $a, b \in L$ be such that*

$$a \leqslant b, \quad a \wedge a^\circ = b \wedge b^\circ, \quad a \vee a^+ = b \vee b^+.$$

Then we have

$$\begin{aligned}\vartheta(a,b) &= \vartheta_{\text{lat}}(a,b) \vee \vartheta_{\text{lat}}(b^\circ, a^\circ) \vee \vartheta_{\text{lat}}(b^+, a^+)\\ &= \vartheta_{\text{lat}}\big((a \vee b^\circ) \wedge b^+, (b \vee a^\circ) \wedge a^+\big).\end{aligned}$$

Proof As observed in the proof of Theorem 8.2, the equality $a \wedge a^\circ = b \wedge b^\circ$ gives $\vartheta_{\text{lat}}(b^\circ, a^\circ) = \vartheta_{\text{lat}}(a^{\circ\circ}, b^{\circ\circ})$. This, together with its dual, gives the first equality.

As for the second equality, we note first that $\vartheta(a,b)$ is a lattice congruence that identifies each of the pairs

$$(a,b), \quad (b^\circ, a^\circ), \quad (b^+, a^+).$$

It follows that $\big((a \vee b^\circ) \wedge b^+, (b \vee a^\circ) \wedge a^+\big) \in \vartheta(a,b)$ and so

$$\vartheta_{\text{lat}}\big((a \vee b^\circ) \wedge b^+, (b \vee a^\circ) \wedge a^+\big) \leqslant \vartheta(a,b).$$

As for the reverse inequality, observe that

$$\begin{aligned} a \wedge (a \vee b^\circ) \wedge b^+ = a \wedge b^+ &= [a \vee (a \wedge a^\circ)] \wedge b^+ \\ &= (a \wedge b^+) \vee (a \wedge a^\circ \wedge b^+) \\ &= (a \wedge b^+) \vee (b \wedge b^\circ \wedge b^+) \\ &= b \wedge (a \vee b^\circ) \wedge b^+ \end{aligned}$$

and similarly

$$b \vee [(b \vee a^\circ) \wedge a^+] = a \vee [(b \vee a^\circ) \wedge a^+],$$

so that we have

$$(a, b) \in \vartheta_{\text{lat}}\big((a \vee b^\circ) \wedge b^+, (b \vee a^\circ) \wedge a^+\big).$$

Likewise, we can show that $\vartheta_{\text{lat}}\big((a \vee b^\circ) \wedge b^+, (b \vee a^\circ) \wedge a^+\big)$ identifies (a, b). Then, by the first equality, the reverse inequality follows. ◊

Corollary *If L is a double Stone algebra then every principal congruence is a principal lattice congruence; specifically,*

$$\vartheta(a, b) = \vartheta_{\text{lat}}\big((a \vee b^\circ) \wedge b^+, (b \vee a^\circ) \wedge a^+\big). \quad \diamond$$

We can in fact say more.

Theorem 14.4 *The class* $\mathbf{SID}_4$ *of double Stone algebras is the largest subvariety of* **DMS** *in which every principal congruence is a principal lattice congruence.*

Proof By considering the ordered set of subdirectly irreducible double MS-algebras (see page 199), we see that it is enough to show that the stated property fails in the subvariety $\mathbf{SID}_3$ of Kleene algebras. For this purpose, consider the 4-element chain $0 < a < b < 1$ on which the operation $^\circ$ is defined by

$$0^\circ = 1, \quad a^\circ = b, \quad b^\circ = a, \quad 1^\circ = 0.$$

This defines a Kleene algebra on which the principal congruence $\vartheta(0, a)$ has the partition

$$\{\{0, a\}, \{b, 1\}\}$$

and is not a principal lattice congruence. ◊

Our objective now is to determine precisely when, for a given $L \in$ **DMS**, the lattice Con L is boolean. For this purpose, we embark on an investigation of those principal congruences that are complemented. Given $L \in$ **DMS** and $a, b \in L$ with $a \leqslant b$, define

$$\begin{aligned} \varphi_1(a, b) &= \vartheta(0, b^\circ \wedge a^{\circ\circ}) \vee \vartheta(b^{\circ\circ} \wedge b^\circ, b^\circ) \vee \vartheta(a^\circ, a^{\circ\circ} \vee a^\circ), \\ \varphi_2(a, b) &= \vartheta(0, b^+ \wedge a^+) \vee \vartheta(b^{++} \wedge b^+, b^+) \vee \vartheta(a^+, a^{++} \vee a^+). \end{aligned}$$

If we write

$$\varphi(a,b) = \varphi_1(a,b) \wedge \varphi_2(a,b),$$

then by Theorem 8.9 and its dual we have immediately

$$\vartheta(a,b) \vee \varphi(a,b) = \iota.$$

Theorem 14.5 *If $\vartheta(a,b)$ is complemented then its complement is $\varphi(a,b)$.*

Proof Suppose that $\vartheta(a,b)$ is complemented and denote its complement by $\vartheta(a,b)^c$. Then from the above equality it follows that

$$\vartheta(a,b)^c \leqslant \varphi(a,b).$$

To establish the reverse inequality, observe that

$$\vartheta(a,b) = \vartheta_\circ(a,b) \vee \vartheta_+(a,b)$$

where

$$\vartheta_\circ(a,b) = \vartheta_{\text{lat}}(a,b) \vee \vartheta_{\text{lat}}(b^\circ, a^\circ) \vee \vartheta_{\text{lat}}(a^{\circ\circ}, b^{\circ\circ})$$

is an MS-congruence and $\vartheta_+(a,b)$ is the corresponding dual MS-congruence. Now, by applying ${}^{\circ\circ}$ to the six equations in Corollary 1 of Theorem 8.1, we see that

$$(x^{\circ\circ}, y^{\circ\circ}) \in \vartheta_\circ(a,b) \iff (x^{\circ\circ}, y^{\circ\circ}) \in \vartheta_\circ(a^{\circ\circ}, b^{\circ\circ}).$$

Using this, and its dual, we deduce that in the skeleton $S(L)$ we have

$$\begin{aligned} \vartheta(a,b)|_{S(L)} &= \vartheta_\circ(a,b)|_{S(L)} \vee \vartheta_+(a,b)|_{S(L)} \\ &= \vartheta_\circ(a^{\circ\circ}, b^{\circ\circ}) \vee \vartheta_+(a^{++}, b^{++}) \\ &= \vartheta(a^{\circ\circ}, b^{\circ\circ})|_{S(L)} \vee \vartheta(a^{++}, b^{++})|_{S(L)} \end{aligned}$$

and consequently, as in Theorem 8.9,

$$\begin{aligned} \vartheta(a,b)^c|_{S(L)} &= (\vartheta(a,b)|_{S(L)})^c \\ &= \vartheta(a^{\circ\circ}, b^{\circ\circ})^c|_{S(L)} \wedge \vartheta_+(a^{++}, b^{++})^c|_{S(L)} \\ &= \varphi_1(a,b)|_{S(L)} \wedge \varphi_2(a,b)|_{S(L)}. \end{aligned}$$

It follows that $\vartheta(a,b)^c|_{S(L)} = \varphi(a,b)|_{S(L)}$ and so $\vartheta(a,b)^c$ is an extension to L of $\varphi(a,b)|_{S(L)}$.

Now it is clear that $\varphi(a,b)|_{S(L)}$ identifies each of the pairs

$$\begin{gathered} (0, b^\circ \wedge a^{\circ\circ}),\ \ (b^{\circ\circ} \wedge b^\circ, b^\circ),\ \ (a^\circ, a^{\circ\circ} \vee a^\circ), \\ (0, b^+ \wedge a^{++}),\ \ (b^{++} \wedge b^+, b^+),\ \ (a^+, a^{++} \vee a^+) \end{gathered}$$

of elements of $S(L)$, and hence so does every extension to L of $\varphi(a,b)|_{S(L)}$. In particular, $\vartheta(a,b)^c$ identifies each of these pairs, and consequently we have $\varphi(a,b) \leqslant \vartheta(a,b)^c$ as required. ◇

This knowledge of $\vartheta(a,b)^c$, when it exists, allows us to determine precisely under what conditions it does exist.

Theorem 14.6 *$\vartheta(a,b)$ is complemented if and only if*

(1) $b \wedge b^\circ \wedge a^{\circ\circ} \leqslant b^{++} \vee a$;

(2) $b \wedge a^{\circ\circ} \leqslant b^{++} \vee a \vee a^+$;

(3) $b \wedge b^+ \wedge a^{\circ\circ} \leqslant b^{++} \vee a \vee a^\circ$.

Proof By Theorem 14.5, $\vartheta(a,b)$ is complemented if and only if

$$\vartheta(a,b) \wedge \varphi(a,b) = \omega.$$

Now

$$\vartheta(a,b) \wedge \varphi(a,b) = \vartheta(a,b) \wedge \varphi_1(a,b) \wedge \varphi_2(a,b)$$

and each of these components can be expressed as a join of lattice congruences. Using the identity

$$\vartheta_{\text{lat}}(a,b) \wedge \vartheta_{\text{lat}}(c,d) = \vartheta_{\text{lat}}\big(b \wedge d \wedge (a \vee c),\, b \wedge d\big),$$

we can then express $\vartheta(a,b) \wedge \varphi(a,b)$ as a join of lattice congruences. We leave to the reader the task of verifying that $\vartheta(a,b) \wedge \varphi(a,b)$ can be expressed as the join of the lattice congruences

$$\vartheta_{\text{lat}}\big(b \wedge b^\circ \wedge a^{\circ\circ} \wedge (b^{++} \vee a),\, b \wedge b^\circ \wedge a^{\circ\circ}\big),$$
$$\vartheta_{\text{lat}}\big(b \wedge (a^\circ \vee a^{\circ\circ}) \wedge (a \vee a^+ \vee b^{++}),\, b \wedge (a^\circ \vee a^{\circ\circ})\big),$$
$$\vartheta_{\text{lat}}\big(b \wedge b^+ \wedge (a^\circ \vee a^{\circ\circ}) \wedge (b^{++} \vee a \vee a^\circ),\, b \wedge b^+ \wedge (a^\circ \vee a^{\circ\circ})\big).$$

Clearly, each of these is ω if and only if (1), (2), (3) hold. ◇

Again, in the case of a double Stone algebra there is a simplification; and in this case we can identify the complement as a principal lattice congruence.

Theorem 14.7 *In a double Stone algebra the following are equivalent*:

(1) *$\vartheta(a,b)$ is complemented*;

(2) $b \wedge b^+ \leqslant a \vee a^\circ$;

(3) *$a \wedge b^+$ is complemented in $[0, b \wedge b^+]$.*

Moreover, when it exists, the complement of $\vartheta(a,b)$ is given by

$$\vartheta(a,b)^c = \vartheta_{\text{lat}}\big((a^\circ \wedge b^{\circ\circ}) \vee (a^+ \wedge b^{++}),\, 1\big).$$

Proof (1) $\Leftrightarrow$ (2) : When L is a double Stone algebra the conditions of Theorem 14.6 reduce to

$$b \wedge b^+ \wedge a^{\circ\circ} \leqslant b^{++} \vee a \vee a^\circ$$

which is equivalent to

$$(\star)\qquad b \wedge b^+ \wedge a^{\circ\circ} = b \wedge b^+ \wedge a^{\circ\circ} \wedge (b^{++} \vee a \vee a^\circ) = b \wedge b^+ \wedge a.$$

This implies that

$$b \wedge b^+ = b \wedge b^+ \wedge (a^\circ \vee a^{\circ\circ}) = b \wedge b^+ \wedge (a^\circ \vee a) \leqslant a^\circ \vee a,$$

which is (2). Conversely, if (2) holds then

$$b \wedge b^+ \wedge a^{\circ\circ} \leqslant (a \vee a^\circ) \wedge a^{\circ\circ} = a,$$

whence $(\star)$ follows.

(2) $\Rightarrow$ (3) : If (2) holds then we have

$$b \wedge b^+ = b \wedge b^+ \wedge (a \vee a^\circ) = (a \wedge b^+) \vee (b \wedge b^+ \wedge a^\circ).$$

Since also $(a \wedge b^+) \wedge (b \wedge b^+ \wedge a^\circ) = 0$, it follows that $b \wedge b^+ \wedge a^\circ$ is the complement of $a \wedge b^+$ in $[0, b \wedge b^+]$.

(3) $\Rightarrow$ (2) : Let x be the complement of $a \wedge b^+$ in $[0, b \wedge b^+]$. Then we have $a \wedge b^+ \wedge x = 0$ and $(a \wedge b^+) \vee x = b \wedge b^+$. Since $(L; {}^\circ)$ is a Stone algebra, it follows that $x \leqslant (a \wedge b^+)^\circ = a^\circ \vee b^{++}$ and hence that $b \wedge b^+ \leqslant (a \wedge b^+) \vee a^\circ \vee b^{++}$. Consequently,

$$\begin{aligned} b \wedge b^+ &= b \wedge b^+ \wedge [(a \wedge b^+) \vee a^\circ \vee b^{++}] \\ &= (a \wedge b^+) \vee (b \wedge b^+ \wedge a^\circ) \\ &\leqslant a \vee a^\circ. \end{aligned}$$

As for the final statement, suppose that $\vartheta(a, b)^c$ exists. Then by Theorem 14.5 we have, primes denoting complements in $\mathrm{Con}_{\mathrm{lat}}\, L$,

$$\begin{aligned} \vartheta(a, b)^c &= [\vartheta(0, b^\circ) \vee \vartheta(a^\circ, 1)] \wedge [\vartheta(0, b^+) \vee \vartheta(a^+, 1)] \\ &= [\vartheta_{\mathrm{lat}}(0, b^\circ) \vee \vartheta_{\mathrm{lat}}(b^{\circ\circ}, 1) \vee \vartheta_{\mathrm{lat}}(a^\circ, 1) \vee \vartheta_{\mathrm{lat}}(0, a^{\circ\circ})] \\ &\qquad \wedge [\vartheta_{\mathrm{lat}}(0, b^+) \vee \vartheta_{\mathrm{lat}}(b^{++}, 1) \vee \vartheta_{\mathrm{lat}}(a^+, 1) \vee \vartheta_{\mathrm{lat}}(0, a^{++})] \\ &= [(\vartheta_{\mathrm{lat}}(b^\circ, a^\circ))' \vee (\vartheta_{\mathrm{lat}}(a^{\circ\circ}, b^{\circ\circ}))'] \\ &\qquad \wedge [(\vartheta_{\mathrm{lat}}(b^+, a^+))' \vee (\vartheta_{\mathrm{lat}}(a^{++}, b^{++}))'] \\ &= [\vartheta_{\mathrm{lat}}(b^\circ, a^\circ) \wedge \vartheta_{\mathrm{lat}}(a^{\circ\circ}, b^{\circ\circ})]' \wedge [\vartheta_{\mathrm{lat}}(b^+, a^+) \wedge \vartheta_{\mathrm{lat}}(a^{++}, b^{++})]' \\ &= (\vartheta_{\mathrm{lat}}(0, a^\circ \wedge b^{\circ\circ}))' \wedge (\vartheta_{\mathrm{lat}}(0, a^+ \wedge b^{++}))' \\ &= \vartheta_{\mathrm{lat}}(a^\circ \wedge b^{\circ\circ}, 1) \wedge \vartheta_{\mathrm{lat}}(a^+ \wedge b^{++}, 1) \\ &= \vartheta_{\mathrm{lat}}((a^\circ \wedge b^{\circ\circ}) \vee (a^+ \wedge b^{++}), 1). \qquad \diamond \end{aligned}$$

These results provide the following characterisation of semisimple double MS-algebras.

Theorem 14.8 *The subvariety* $\mathbf{SID}_{20}$ *of semisimple double MS-algebras is the largest subvariety of* **DMS** *in which every principal congruence is complemented.*

Proof The subvariety $\mathbf{SID}_{20}$ of semisimple double MS-algebras is characterised by the axiom

$$(A) \qquad a \wedge b^{\circ\circ} \leqslant a^{++} \vee b.$$

The conditions of Theorem 14.6 are therefore satisfied, so every principal congruence on a semisimple double MS-algebra is complemented.

Using the ordered set of subdirectly irreducible double MS-algebras, we see that in order to establish the result it suffices to produce examples of algebras that belong to the subvarieties $\mathbf{SID}_4$, $\mathbf{SID}_8$, $\mathbf{SID}_{10}$ in which there is a principal congruence that is not complemented. Suitable examples are provided by the subdirectly irreducible algebras SID_4, SID_8, SID_{10} that generate the subvarieties in question (see page 198). By Theorem 13.1, in each of these Con L is the three-element chain

$$\omega \prec \Phi_+^\circ \prec \iota.$$

Moreover, Φ_+° a principal congruence; in the first it is $\vartheta(d,g)$, in the second it is $\vartheta(e,h)$, and in the third it is $\vartheta(c,f)$. ◇

It is of course natural to ask when a principal congruence has a complement that is also principal.

Theorem 14.9 *The subvariety* $\mathbf{SID}_2$ *of trivalent Lukasiewicz algebras is the largest subvariety of* **DMS** *in which every principal congruence has a principal complement.*

Proof A trivalent Lukasiewicz algebra is a semisimple double Stone algebra and, as shown at the end of Chapter 13, is characterised by the axioms $(G, 2_d)$. It follows by (G) and Theorem 14.7 that every principal congruence in such an algebra has a principal complement.

To complete the proof it suffices, again by considering the ordered set of subdirectly irreducible double MS-algebras, to provide examples of algebras in $\mathbf{SID}_3$ and $\mathbf{SID}_4$ in which the property fails.

Now $\mathbf{SID}_3$ is the subvariety of Kleene algebras. In this, consider the six-element chain $0 < a < b < c < d < 1$ with

$$0^\circ = 1, \quad a^\circ = d, \quad b^\circ = c^\circ = c, \quad d^\circ = a, \quad 1^\circ = 0.$$

The congruence $\vartheta(a,b)^c$ is described by the partition

$$\{\{0,a\},\{b,c\},\{d,1\}\}$$

and is not principal.

As for $\mathbf{SID}_4$, this is the class of double Stone algebras and here, by Theorem 14.8, a principal congruence need not have a complement. ◇

We now turn our attention to the problem of determining which double MS-algebras have a boolean lattice of congruences. For this purpose, we require several analogues of results in Chapter 8. The first is the following characterisation of the least extension to a double MS-algebra L of a congruence on its skeleton $S(L)$.

Theorem 14.10 *If $L \in$ **DMS** and $\varphi \in \mathrm{Con}\, S(L)$ then the smallest lattice congruence on L that extends φ is given by*

$$\overline{\varphi} = \bigvee_{(x^{\circ\circ},y^{\circ\circ})\in\varphi} \vartheta_{\mathrm{lat}}(x^{\circ\circ},y^{\circ\circ}).$$

Moreover, $\overline{\varphi} \in \mathrm{Con}\, L$.

Proof This is exactly as in Theorem 8.22. ◇

Precisely as in Theorem 8.23, we also have

Theorem 14.11 *The mapping $f : \mathrm{Con}\, S(L) \to \mathrm{Con}\, L$ given by $f(\varphi) = \overline{\varphi}$ is a lattice morphism.* ◇

Proceeding in the opposite direction, we clearly have

Theorem 14.12 *If $\vartheta \in \mathrm{Con}\, L$ is complemented then so is $\vartheta|_{S(L)} \in \mathrm{Con}\, S(L)$.* ◇

Precisely as in Theorem 8.25, we also have

Theorem 14.13 *For every $L \in$ **DMS** the lattices $\mathrm{Con}\, L$ and $\mathrm{Con}\, S(L)$ have isomorphic centres.* ◇

We can now establish the following result.

Theorem 14.14 *If $L \in$ **DMS** then $\mathrm{Con}\, L$ is boolean if and only if L is semi-simple and finite. In this case, $\mathrm{Con}\, L$ is also finite.*

Proof $\Rightarrow$: Suppose that $\mathrm{Con}\, L$ is boolean. If $\vartheta \in \mathrm{Con}\, S(L)$ then its extension $\overline{\vartheta} \in \mathrm{Con}\, L$ is complemented and therefore, by Theorem 14.12, so is

$$\vartheta = \overline{\vartheta}\,|_{S(L)} \in \mathrm{Con}\, S(L).$$

Hence $\mathrm{Con}\, S(L)$ is boolean. It follows by Theorem 14.13 that

$$\mathrm{Con}\, L \simeq \mathrm{Con}\, S(L).$$

Since $S(L)$ must then be finite by Theorem 8.15, it follows that Con L is also finite, and the intersection of all maximal congruences in Con L is ω. But every maximal congruence contains Φ_+°. Hence $\Phi_+^\circ = \omega$ and therefore, by Theorem 11.4, L is semisimple. Suppose now, by way of obtaining a contradiction, that L is infinite. Since the elements of $S(L)$ are precisely the biggest elements in the Φ°-classes, it follows from the fact that $S(L)$ is finite that there must be at least one Φ°-class that is infinite. But since $\Phi_+ \wedge \Phi^\circ = \Phi_+^\circ = \omega$, the relation Φ_+ must separate any two elements in this infinite Φ°-class. There are, consequently, infinitely many Φ_+-classes. Since the elements of $S(L)$ are precisely the smallest elements in the Φ_+-classes, this contradicts the fact that $S(L)$ is finite. We conclude that L must be finite.

$\Leftarrow$: Suppose now that L is semisimple and finite. Then, by Theorems 8.15 and 14.13, the centre of Con L is a finite boolean algebra. By Theorem 14.8, there are therefore only finitely many principal congruences on L. Since every $\varphi \in$ Con L can be written

$$\varphi = \bigvee_{(a_i,b_i)\in\varphi} \vartheta(a_i, b_i),$$

such a supremum involves only finitely many principal congruences each of which is complemented. It follows that φ is also complemented and hence that Con L is boolean. ◇

15 Singles and Doubles

In Chapter 5 we showed that every variety of MS-algebras is characterised by adjoining to the basic axioms of **MS** at most three from a list of ten identities. If a double MS-algebra $(L; {}^\circ, {}^+)$ is such that $(L; {}^\circ)$ satisfies some of these identities then it is natural to ask when $(L^{op}; {}^+)$ satisfies the same identities. More particularly, if $(L; {}^\circ, {}^+) \in \mathbf{DMS}$ and $\mathbf{V}(L; {}^\circ) = \mathbf{R}$, what are the possible subvarieties $\mathbf{V}(L^{op}; {}^+)$?

Theorem 15.1 *The following axioms are self-dual in* **DMS** :

$$(1),\ (\alpha),\ (\gamma),\ (2),\ (2_d),\ (5),\ (9),\ (11_d),\ (12_d).$$

Proof It is clear that (1) is self-dual in **DMS**. That the others are also self-dual in **DMS** is a consequence of the following observations.

(*a*) *If* $(L; {}^\circ)$ *satisfies* (α) *then so does* $(L^{op}; {}^+)$.

If $a = a^{\circ\circ}$ for every $a \in L$ then, by (D2), $a^+ = a^{\circ\circ+} = a^{\circ\circ\circ} = a^\circ$ and hence $a^{++} = a^{\circ+} = a^{\circ\circ} = a$.

(*b*) *If* $(L; {}^\circ)$ *satisfies* (γ) *then so does* $(L^{op}; {}^+)$.

By (γ) we have $a^+ \wedge a^{+\circ} \leqslant b \vee b^\circ$ hence $a^{++} \vee a^{+\circ+} \geqslant b^+ \wedge b^{\circ+}$, and therefore $a^{++} \vee a^+ \geqslant b^+ \wedge b^{\circ\circ}$. Since $a^{++} \leqslant a$ and $b \leqslant b^{\circ\circ}$, we obtain $a \vee a^+ \geqslant b^+ \wedge b$.

(*c*) *If* $(L; {}^\circ)$ *satisfies* (2) *then so does* $(L^{op}; {}^+)$.

This was shown above.

(*d*) *If* $(L; {}^\circ)$ *satisfies* (2_d) *then so does* $(L^{op}; {}^+)$.

In fact, if $a \wedge a^\circ = 0$ for every $a \in L$ then $a^+ \wedge a^{+\circ} = 0$ which gives $a^{++} \vee a^{+\circ+} = 1$ so that, by (D1), $a^{++} \vee a^+ = 1$ and hence $a \vee a^+ = 1$.

(*e*) *If* $(L; {}^\circ)$ *satisfies* (5) *then so does* $(L^{op}; {}^+)$.

By (5) we have $a^+ \wedge a^{+\circ} \leqslant b^{+\circ\circ} \vee b^{+\circ}$, i.e., by (D1), $a^+ \wedge a^{++} \leqslant b^+ \vee b^{++}$. It follows that $a^{++} \vee a^+ \geqslant b^{++} \wedge b^+$ and hence that $a \vee a^+ \geqslant b^{++} \wedge b^+$.

(*f*) *If* $(L; {}^\circ)$ *satisfies* (9) *then so does* $(L^{op}; {}^+)$.

Writing a^+ for a in $a \wedge a^\circ \leqslant a^{\circ\circ} \vee b \vee b^\circ$, we obtain $a^+ \wedge a^{++} \leqslant a^+ \vee b \vee b^\circ$ whence, applying $^+$ to this, we have $a^{++} \wedge b^+ \wedge b^{\circ\circ} \leqslant a^{++} \vee a^+$. Consequently $a^{++} \wedge b^+ \wedge b \leqslant a \vee a^+$, which is $(9^\star)$.

(*g*) *If* $(L; {}^\circ)$ *satisfies* (11_d) *then so does* $(L^{op}; {}^+)$.

Observe first that (11_d), namely $a^\circ \wedge a^{\circ\circ} \leqslant a \vee b^\circ \vee b^{\circ\circ}$, is equivalent to

$$(11'_d) \qquad (a \wedge a^\circ) \vee b^\circ \vee b^{\circ\circ} \in S(L).$$

In fact, if (11_d) holds then

$$\begin{aligned}(a \wedge a^\circ) \vee b^\circ \vee b^{\circ\circ} &= (a \vee b^\circ \vee b^{\circ\circ}) \wedge (a^\circ \vee b^\circ \vee b^{\circ\circ})\\ &= (a \vee a^\circ \vee b^\circ \vee b^{\circ\circ}) \wedge (a^{\circ\circ} \vee b^\circ \vee b^{\circ\circ}) \wedge (a^\circ \vee b^\circ \vee b^{\circ\circ})\\ &= (a^{\circ\circ} \vee b^\circ \vee b^{\circ\circ}) \wedge (a^\circ \vee b^\circ \vee b^{\circ\circ}) \in S(L).\end{aligned}$$

Conversely, if $(a \wedge a^\circ) \vee b^\circ \vee b^{\circ\circ} \in S(L)$ then

$$(a \wedge a^\circ) \vee b^\circ \vee b^{\circ\circ} = (a^{\circ\circ} \wedge a^\circ) \vee b^\circ \vee b^{\circ\circ}.$$

Since the left-hand side is $\leqslant a \vee b^\circ \vee b^{\circ\circ}$, and the right-hand side is $\geqslant a^{\circ\circ} \wedge a^\circ$, we obtain (11_d).

Writing b^{++} for b in $(11'_d)$ we obtain

$$(a \wedge a^\circ) \vee b^+ \vee b^{++} = (a^{\circ\circ} \wedge a^\circ) \vee b^+ \vee b^{++}.$$

Taking the join of each side with b gives

$$(A) \qquad (a \wedge a^\circ) \vee b \vee b^+ = (a^{\circ\circ} \wedge a^\circ) \vee b \vee b^+.$$

Conversely, writing $b^{\circ\circ}$ for b in (A) we obtain $(11'_d)$. Hence (11_d) and (A) are equivalent.

Similarly, we have that $(11^\star_d)$ is equivalent to

$$(11^{\star\prime}_d) \qquad (a \vee a^+) \wedge b^{++} \wedge b^+ \in S(L).$$

Writing $b^{\circ\circ}$ for b in this, we obtain

$$(a \vee a^+) \wedge b^{\circ\circ} \wedge b^\circ \in S(L),$$

which implies that

$$(B) \qquad (a \vee a^+) \wedge b^{\circ\circ} \wedge b^\circ = (a^{\circ\circ} \vee a^+) \wedge b^{\circ\circ} \wedge b^\circ.$$

Conversely, writing b^{++} for b in this and using a similar argument, we obtain $(11^{\star\prime}_d)$. Thus $(11^\star_d)$ and (B) are equivalent.

It suffices, therefore, to show that if (B) fails then so does (A). Now if (B) fails there exist $t \geqslant t^+$ and $s^{\circ\circ} \leqslant s^\circ$ such that

$$v^{\circ\circ} = t^{\circ\circ} \wedge s^{\circ\circ} > t \wedge s^{\circ\circ} = v.$$

We then have $v \vee t = t$, and $v^{\circ\circ} \vee t > t$ (since otherwise $v^{\circ\circ} \vee t = t$ which gives $v^{\circ\circ} \leqslant t$ whence, since $v^{\circ\circ} \leqslant s^{\circ\circ}$, we would have $v^{\circ\circ} \leqslant v$, which contradicts $v < v^{\circ\circ}$). Thus we have

$$v \vee t < v^{\circ\circ} \vee t.$$

Since $v \wedge v^\circ = t \wedge s^{\circ\circ} \wedge (t^\circ \vee s^\circ) = t \wedge s^{\circ\circ} = v$, we conclude that (A) fails.

(*b*) *If* $(L; {}^\circ)$ *satisfies* (12_d) *then so does* $(L^{\mathrm{op}}; {}^+)$.

It is readily seen that (12_d), namely $a^{\circ\circ} \wedge b^{\circ} \wedge b^{\circ\circ} \leqslant a \vee a^{\circ}$, is equivalent to

$$(a \vee a^{\circ}) \wedge b^{\circ} \wedge b^{\circ\circ} \in S(L)$$

and therefore gives

$$[(a \vee a^{\circ}) \wedge b^{\circ} \wedge b^{\circ\circ}]^{\circ} = [(a \vee a^{\circ}) \wedge b^{\circ} \wedge b^{\circ\circ}]^{+}.$$

Writing b^{+} for b in this, we obtain

$$(\dagger) \qquad (a^{\circ} \wedge a^{\circ\circ}) \vee b^{+} \vee b^{++} = (a^{+} \wedge a^{\circ\circ}) \vee b^{+} \vee b^{++}.$$

Now (12_d) implies (11_d) which is equivalent to

$$(a \wedge a^{\circ}) \vee b^{\circ} \vee b^{\circ\circ} \in S(L),$$

from which we obtain, writing b^{+} for b,

$$(\dagger\dagger) \qquad (a^{\circ\circ} \wedge a^{\circ}) \vee b^{+} \vee b^{++} = (a^{++} \wedge a^{\circ}) \vee b^{+} \vee b^{++}.$$

It now follows from $(\dagger)$, $(\dagger\dagger)$ and the inequalities

$$a^{++} \wedge a^{\circ} \leqslant a^{++} \wedge a^{+} \leqslant a^{\circ\circ} \wedge a^{+}$$

that we have the identity

$$(a^{++} \wedge a^{+}) \vee b^{++} \vee b^{+} = (a^{\circ\circ} \wedge a^{+}) \vee b^{++} \vee b^{+},$$

which gives

$$(a \wedge a^{+}) \vee b^{++} \vee b^{+} \in S(L)$$

which means that $(L^{\mathrm{op}}; {}^{+})$ satisfies (12_d). ◇

It is, of course, possible to establish Theorem 15.1 by means of duality. To show, for example, that (11_d) is self-dual in **DMS**, suppose that $(X; g, h)$ is the dual space of $(L; {}^{\circ}, {}^{+})$. We have to show that if $(X; g)$ satisfies

$$g^2 \nparallel g \ \bigvee\ g^2 \geqslant g^0$$

then $(X; h)$ satisfies

$$h^2 \nparallel h \ \bigvee\ h^0 \geqslant h^2.$$

Suppose that this were not so. Since $g^2 \nparallel g$ is equivalent to $h^2 \nparallel h$, there must exist $p \in X$ such that

(*a*) $g^2(p) \parallel g(p)$,
(*b*) $g^2(p) \geqslant p$ (hence $g^2(p) = p$ by axiom (1)),
(*c*) $h^2(p) \parallel h(p)$,
(*d*) $p \not\geqslant h^2(p)$ (hence $p < h^2(p)$ since in **DMS** we have $x \leqslant h^2(x)$).

Consider the element $q = h^2(p)$. We have $g(q) = gh^2(p) = g(p)$ and so $g^2(q) = g^2(p)$. It follows by (a) that $g^2(q) \nparallel g(q)$; and by (b) and (d) that

$$g^2(q) = g^2(p) = p \not\geqslant h^2(p) = q.$$

The fact that q does not satisfy the axiom (11_d) therefore provides the required contradiction.

Theorem 15.2 *Let* $(L;^\circ,^+)$ *be a double MS-algebra. Then*

(*a*) *if* $(L;^\circ)$ *satisfies* (6_d) *then* $(L^{op};^+)$ *satisfies* (3_d);

(*b*) *if* $(L;^\circ)$ *satisfies* (3_d) *then* $(L^{op};^+)$ *satisfies* (6_d).

Proof (a) : It is readily seen that (6_d), namely $a^{\circ\circ} \leqslant a \vee b^\circ \vee b^{\circ\circ}$, is equivalent to

$$a \vee b^\circ \vee b^{\circ\circ} \in S(L).$$

Writing b^{++} for b in this, we obtain

$$a \vee b^+ \vee b^{++} \in S(L)$$

which, on taking $b = a$, gives

$$a \vee a^+ \in S(L)$$

and therefore $a^{++} \vee a^+ \geqslant a$. It follows that $(L^{op};^+)$ satisfies (3_d).

The proof of (b) is similar. ◇

Before solving the problem stated at the beginning of this chapter, we consider the following strong concept.

Definition We shall say that a subvariety **R** of **MS** is *fertile* if every algebra $(L;^\circ) \in \mathbf{R}$ can be made into a double MS-algebra.

Theorem 15.3 *The fertile subvarieties of* **MS** *are* **B**, **K**, *and* **M**.

Proof That **M** is fertile follows from Example 12.2. Since **B** and **K** are subclasses of **M**, they are also fertile. To show that these are the only fertile subvarieties of **MS**, it suffices to show that **S** and $\mathbf{K}_1$ are not fertile. For the first of these, consider the Stone algebra $\mathbf{1} \oplus \mathbf{2}^2$; and for the second, consider the $\mathbf{K}_1$-algebra

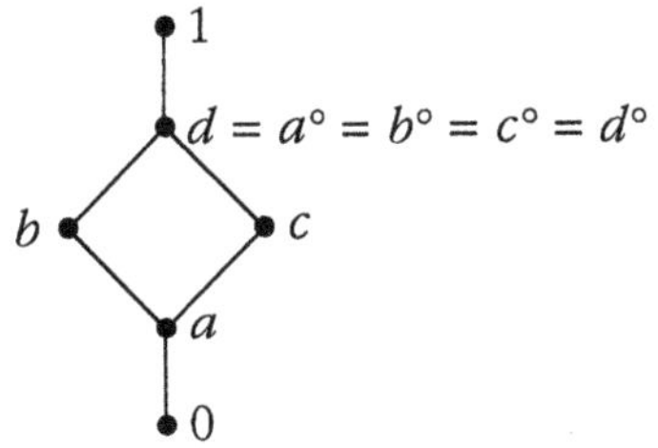

By Corollary 2 of Theorem 12.7, neither of these can be made into a double MS-algebra. ◇

Definition We shall say that a subvariety $\mathbf{R}$ of $\mathbf{MS}$ is *barren* if no algebra L that properly belongs to $\mathbf{R}$ (i.e. $\mathbf{V}(L;{}^\circ) = \mathbf{R}$) can be made into a double MS-algebra.

Theorem 15.4 *The subvarieties* $\mathbf{K}_3$ *and* $\mathbf{M} \vee \mathbf{K}_3$ *are barren.*

Proof Suppose, by way of obtaining a contradiction, that $\mathbf{M} \vee \mathbf{K}_3$ were not barren, i.e. that $\mathbf{V}(L;{}^\circ) = \mathbf{M} \vee \mathbf{K}_3$ and L can be made into a double MS-algebra. Then $(L;{}^\circ)$ satisfies (6_d) and so, by Theorem 15.2, $(L^{\text{op}};{}^+)$ satisfies (3_d), hence (12_d). But, by Theorem 15.1, (12_d) is self-dual. Hence $(L;{}^\circ)$ also satisfies (12_d), and this contradicts the hypothesis that $L \notin \mathbf{S} \vee \mathbf{M} \vee \mathbf{K}_1$.

Similarly, if $\mathbf{V}(L;{}^\circ) = \mathbf{K}_3$ and L can be made into a double MS-algebra then $(L;{}^\circ)$ satisfies (6_d), and so $(L^{\text{op}};{}^+)$ satisfies (3_d) and hence (12_d) which is self-dual. Thus $(L;{}^\circ)$ satisfies $(1,5,6_d,12_d)$, hence also $(1,\delta,6_d,12_d)$. Since, as was shown in Chapter 5, $(1,\delta,12_d)$ is equivalent to $(1,\gamma)$, it follows that $(L;{}^\circ)$ satisfies $(1,6_d,\gamma)$. This contradicts the hypothesis that $L \notin \mathbf{S} \vee \mathbf{K}_1$. ◇

Definition A subvariety $\mathbf{R}$ of $\mathbf{MS}$ will be called *bistable* if, for every double MS-algebra $(L;{}^\circ,{}^+)$, whenever $\mathbf{V}(L;{}^\circ) = \mathbf{R}$ we have $\mathbf{V}(L^{\text{op}};{}^+) = \mathbf{R}$.

Theorem 15.5 *The following subvarieties of* $\mathbf{MS}$ *are bistable*:

$$\mathbf{M} \vee \mathbf{S},\ \mathbf{M},\ \mathbf{S} \vee \mathbf{K},\ \mathbf{K},\ \mathbf{S},\ \mathbf{B}$$

[*i.e. those that belong to the ideal* A *of* $\Lambda(\mathbf{MS})$ *generated by* $\mathbf{M} \vee \mathbf{S}$] *and*

$$\mathbf{K}_1 \vee \mathbf{K}_2,\ \mathbf{M} \vee \mathbf{K}_1 \vee \mathbf{K}_2,\ \mathbf{K}_2 \vee \mathbf{K}_3,\ \mathbf{M} \vee \mathbf{K}_2 \vee \mathbf{K}_3,\ \mathbf{M}_1$$

[*i.e. those that belong to the filter* B *of* $\Lambda(\mathbf{MS})$ *generated by* $\mathbf{K}_1 \vee \mathbf{K}_2$].

Proof All of the subvarieties listed are characterised by axioms that are self-dual or by both the axioms (3_d) and (6_d). By Theorems 15.1 and 15.2, $(L^{\text{op}};{}^+)$ belongs to the same subvariety as $(L;{}^\circ)$.

Now, suppose that $\mathbf{V}(L;{}^\circ) = \mathbf{M} \vee \mathbf{S}$. Then $(L^{\text{op}};{}^+) \in \mathbf{M} \vee \mathbf{S}$. If $\mathbf{V}(L^{\text{op}};{}^+) = \mathbf{M}$ or $\mathbf{S} \vee \mathbf{K}$ then $(L;{}^\circ) \in \mathbf{M}$ or $\mathbf{S} \vee \mathbf{K}$, contradicting the fact that $\mathbf{V}(L;{}^\circ) = \mathbf{M} \vee \mathbf{S}$. The same argument is valid for all subvarieties in A, and for $\mathbf{M}_1$ and $\mathbf{M} \vee \mathbf{K}_2 \vee \mathbf{K}_3$ since $\mathbf{M} \vee \mathbf{K}_3$ is barren.

Suppose that $\mathbf{V}(L;{}^\circ) = \mathbf{K}_2 \vee \mathbf{K}_3$. If $(L^{\text{op}};{}^+) \in \mathbf{K}_1 \vee \mathbf{K}_2$ then $(L;{}^\circ) \in \mathbf{K}_1 \vee \mathbf{K}_2$, a contradiction.

It is not possible to have $\mathbf{V}(L;{}^\circ) = \mathbf{K}_3$ since $\mathbf{K}_3$ is barren.

Finally, suppose that $\mathbf{V}(L;{}^\circ) = \mathbf{M} \vee \mathbf{K}_1 \vee \mathbf{K}_2$. If $(L^{\text{op}};{}^+) \in \mathbf{K}_2 \vee \mathbf{M}$ then $(L;{}^\circ)$ satisfies $(1,6_d)$ whence we have the contradiction $(L;{}^\circ) \in \mathbf{M} \vee \mathbf{K}_3$. If $(L^{\text{op}};{}^+) \in \mathbf{S} \vee \mathbf{M} \vee \mathbf{K}_1$ then $(L;{}^\circ)$ satisfies $(1,3_d)$ whence we have the contradiction $(L;{}^\circ) \in \mathbf{K}_2 \vee \mathbf{M}$. ◇

Theorem 15.6 *The subvarieties* $\mathbf{K}_1, \mathbf{K}_2, \mathbf{S} \vee \mathbf{K}_1, \mathbf{S} \vee \mathbf{M} \vee \mathbf{K}_1$, $\mathbf{K}_1 \vee \mathbf{M}$, $\mathbf{K}_2 \vee \mathbf{M}$ *are neither barren nor bistable. More precisely,*

(1) *if* $\mathbf{V}(L;{}^\circ) = \mathbf{K}_1$ *then* $\mathbf{V}(L^{\mathrm{op}};{}^+) = \mathbf{K}_2$;
(2) *if* $\mathbf{V}(L;{}^\circ) = \mathbf{K}_2$ *then* $\mathbf{V}(L^{\mathrm{op}};{}^+) \in \{\mathbf{K}_1, \mathbf{K}_1 \vee \mathbf{S}\}$;
(3) *if* $\mathbf{V}(L;{}^\circ) = \mathbf{S} \vee \mathbf{K}_1$ *then* $\mathbf{V}(L^{\mathrm{op}};{}^+) = \mathbf{K}_2$;
(4) *if* $\mathbf{V}(L;{}^\circ) = \mathbf{S} \vee \mathbf{M} \vee \mathbf{K}_1$ *then* $\mathbf{V}(L^{\mathrm{op}};{}^+) = \mathbf{K}_2 \vee \mathbf{M}$;
(5) *if* $\mathbf{V}(L;{}^\circ) = \mathbf{K}_1 \vee \mathbf{M}$ *then* $\mathbf{V}(L^{\mathrm{op}};{}^+) = \mathbf{K}_2 \vee \mathbf{M}$;
(6) *if* $\mathbf{V}(L;{}^\circ) = \mathbf{K}_2 \vee \mathbf{M}$ *then* $\mathbf{V}(L^{\mathrm{op}};{}^+) \in \{\mathbf{K}_1 \vee \mathbf{M}, \mathbf{S} \vee \mathbf{M} \vee \mathbf{K}_1\}$.

Proof (1) If $\mathbf{V}(L;{}^\circ) = \mathbf{K}_1$ then $(L;{}^\circ)$ satisfies $(1, 4_d, \gamma)$, hence also $(1, 6_d, \gamma)$. Then $(L^{\mathrm{op}};{}^+)$ satisfies $(1, 3_d, \gamma)$ and so belongs to $\mathbf{K}_2$. Since $\mathbf{K}_2$ covers only the bistable subvariety $\mathbf{S} \vee \mathbf{K}$, we have $\mathbf{V}(L^{\mathrm{op}};{}^+) = \mathbf{K}_2$.

(2) If $\mathbf{V}(L;{}^\circ) = \mathbf{K}_2$ then $(L^{\mathrm{op}};{}^+)$ satisfies $(1, 6_d, \gamma)$, hence belongs to $\mathbf{S} \vee \mathbf{K}_1$ which covers the bistable subvariety $\mathbf{S} \vee \mathbf{K}$. It follows that $\mathbf{V}(L^{\mathrm{op}};{}^+)$ is either $\mathbf{K}_1$ or $\mathbf{K}_1 \vee \mathbf{S}$.

(3) If $\mathbf{V}(L;{}^\circ) = \mathbf{S} \vee \mathbf{K}_1$ then $(L^{\mathrm{op}};{}^+)$ satisfies $(1, 3_d, \gamma)$ and so belongs to $\mathbf{K}_2$. In fact, $\mathbf{V}(L^{\mathrm{op}};{}^+) = \mathbf{K}_2$ since $\mathbf{K}_2$ covers only $\mathbf{S} \vee \mathbf{K}$, which is bistable.

(4) If $\mathbf{V}(L;{}^\circ) = \mathbf{S} \vee \mathbf{M} \vee \mathbf{K}_1$ then $(L^{\mathrm{op}};{}^+)$ satisfies $(1, 3_d)$ and so belongs to $\mathbf{K}_2 \vee \mathbf{M}$. The latter covers $\mathbf{M} \vee \mathbf{S}$, which is bistable, and $\mathbf{K}_2$. If $(L^{\mathrm{op}};{}^+) \in \mathbf{K}_2$ then $(L;{}^\circ) \in \mathbf{K}_1 \vee \mathbf{S}$, contradicting the fact that $\mathbf{V}(L;{}^\circ) = \mathbf{S} \vee \mathbf{M} \vee \mathbf{K}_1$.

(5) If $\mathbf{V}(L;{}^\circ) = \mathbf{M} \vee \mathbf{K}_1$ then $(L;{}^\circ)$ satisfies $(1, 4_d)$, hence $(1, 6_d)$, and $(L^{\mathrm{op}};{}^+)$ satisfies $(1, 3_d)$ and belongs to $\mathbf{K}_2 \vee \mathbf{M}$. The latter covers $\mathbf{M} \vee \mathbf{S}$, which is bistable, and $\mathbf{K}_2$. If $(L^{\mathrm{op}};{}^+) \in \mathbf{K}_2$ then $(L;{}^\circ) \in \{\mathbf{K}_1, \mathbf{K}_1 \vee \mathbf{S}\}$ which is impossible by the minimality of $\mathbf{K}_1 \vee \mathbf{M}$.

(6) If $\mathbf{V}(L;{}^\circ) = \mathbf{M} \vee \mathbf{K}_2$ then $(L^{\mathrm{op}};{}^+)$ satisfies $(1, 6_d)$ and belongs to $\mathbf{M} \vee \mathbf{K}_3$. But the latter is barren. It follows that $(L^{\mathrm{op}};{}^+)$ belongs to $\mathbf{S} \vee \mathbf{M} \vee \mathbf{K}_1$, $\mathbf{M} \vee \mathbf{K}_1$, or $\mathbf{S} \vee \mathbf{K}_1$. The last of these is excluded since $\mathbf{V}(L^{\mathrm{op}};{}^+) = \mathbf{S} \vee \mathbf{K}_1$ implies $\mathbf{V}(L;{}^\circ) = \mathbf{K}_2$.

It remains to show that there exist double MS-algebras that satisfy the preceding conditions. For this purpose, consider first the lattice L_1 with Hasse diagram

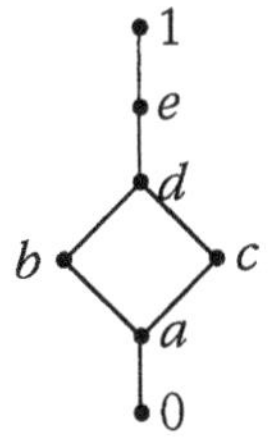

Of the ways in which L_1 can be made into a double MS-algebra, the following four are relevant :

	x	0	a	b	c	d	e	1	
(1)	x°	1	e	e	e	e	e	0	$\mathbf{K}_1$
	x^+	1	1	1	1	1	e	0	$\mathbf{K}_2$
(2)	x°	1	d	c	b	a	0	0	$\mathbf{K}_2$
	x^+	1	d	c	b	a	a	0	$\mathbf{K}_1$
(3)	x°	1	e	c	e	c	c	0	$\mathbf{S} \vee \mathbf{K}_1$
	x^+	1	1	1	e	e	c	0	$\mathbf{K}_2$
(6)	x°	1	d	b	c	a	0	0	$\mathbf{M} \vee \mathbf{K}_2$
	x^+	1	d	b	c	a	a	0	$\mathbf{M} \vee \mathbf{K}_1$

Next, consider the lattice $L_2 \simeq L_1^d$, namely

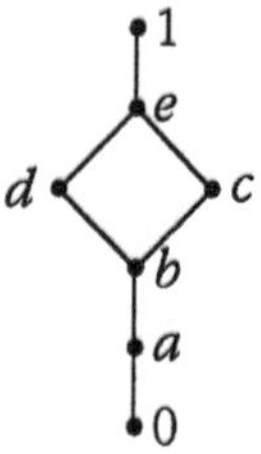

Of the ways in which L_2 can be made into a double MS-algebra, the following two are relevant :

	x	0	a	b	c	d	e	1	
(2′)	x°	1	c	a	a	0	0	0	$\mathbf{K}_2$
	x^+	1	c	c	a	c	a	0	$\mathbf{S} \vee \mathbf{K}_1$
(5)	x°	1	e	e	c	d	b	0	$\mathbf{M} \vee \mathbf{K}_1$
	x^+	1	1	e	c	d	b	0	$\mathbf{M} \vee \mathbf{K}_2$

Finally, let $L_1^{(6)}$ be the double MS-algebra consisting of L_1 with the operations as described in (6) above, and consider

$$L_3 = L_1^{(6)} \times \mathrm{SID}_4.$$

The double MS-algebra $(L_3; {}^\circ, {}^+)$ is such that

$$\mathbf{V}(L_3; {}^\circ) = \mathbf{K}_2 \vee \mathbf{M} \vee \mathbf{S} = \mathbf{K}_2 \vee \mathbf{M}, \qquad \mathbf{V}(L_3; {}^+) = \mathbf{K}_1 \vee \mathbf{M} \vee \mathbf{S}.$$

Similarly, with

$$L_4 = L_2^{(5)} \times \mathrm{SID}_4$$

we have

$$\mathbf{V}(L_4;{}^\circ) = \mathbf{K}_1 \vee \mathbf{M} \vee \mathbf{S}, \qquad \mathbf{V}(L_4;{}^+) = \mathbf{K}_2 \vee \mathbf{M}. \quad \diamond$$

We know by Corollary 1 of Theorem 12.7 that if an MS-algebra $(L;{}^\circ)$ can be made into a double MS-algebra $(L;{}^\circ,{}^+)$ then this can be done in one and only one way. The problem that we shall consider now is that of relating the subvarieties of **MS** to which $(L;{}^\circ)$ belongs to the subvarieties of **DMS** to which $(L;{}^\circ,{}^+)$ belongs.

Given a double MS-algebra $(D;{}^\circ,{}^+)$ we shall denote by $\mathbf{V}(D)$ the smallest subvariety of **MS** to which $(D;{}^\circ)$ belongs. Consider the relation Θ defined on $\Lambda(\mathbf{DMS})$ by

$$(\mathbf{D}_1,\mathbf{D}_2) \in \Theta \iff (\forall D_1 \in \mathbf{D}_1)(\forall D_2 \in \mathbf{D}_2) \quad \mathbf{V}(D_1) = \mathbf{V}(D_2)$$
$$\iff \begin{cases} \text{every algebra in } \mathbf{D}_1 \text{ satisfies the same} \\ {}^\circ\text{-axioms as every algebra in } \mathbf{D}_2. \end{cases}$$

Clearly, Θ is an equivalence relation. It is in fact a lattice congruence on $\Lambda(\mathbf{DMS})$. For, if $(\mathbf{D}_1,\mathbf{D}_2) \in \Theta$ and $\mathbf{X} \in \Lambda(\mathbf{DMS})$ then, on the one hand, the ${}^\circ$-axioms satisfied by $\mathbf{D}_1, \mathbf{X}$ are the same as those of $\mathbf{D}_2, \mathbf{X}$, so that we have $(\mathbf{D}_1 \wedge \mathbf{X},\ \mathbf{D}_2 \wedge \mathbf{X}) \in \Theta$; and, on the other hand, any ${}^\circ$-axiom that holds in $\mathbf{D}_1 \vee \mathbf{X}$ holds in both $\mathbf{D}_1$ and $\mathbf{X}$, hence in $\mathbf{D}_2$ and $\mathbf{X}$, and consequently in $\mathbf{D}_2 \vee \mathbf{X}$, so that we also have $(\mathbf{D}_1 \vee \mathbf{X},\ \mathbf{D}_2 \vee \mathbf{X}) \in \Theta$.

Since $\Lambda(\mathbf{DMS})$ is a finite lattice, the Θ-classes are intervals and $\Lambda(\mathbf{DMS})/\Theta$ can be ordered by writing

$$[\mathbf{D}_1]\Theta \leqslant [\mathbf{D}_2]\Theta \iff (\exists \mathbf{X} \in [\mathbf{D}_1]\Theta)\,(\exists \mathbf{Y} \in [\mathbf{D}_2]\Theta) \quad \mathbf{X} \leqslant \mathbf{Y}.$$

We shall denote by $\Lambda^\star(\mathbf{MS})$ the sublattice of $\Lambda(\mathbf{MS})$ obtained by deleting the barren subvarieties, which by Theorem 15.4 are $\mathbf{K}_3$ and $\mathbf{M} \vee \mathbf{K}_3$. If we ignore as usual the trivial subvariety, this is then a 17-element distributive lattice, and we have a lattice isomorphism

$$\Lambda(\mathbf{DMS})/\Theta \simeq \Lambda^\star(\mathbf{MS}).$$

Our objective now is to determine, for each variety in $\Lambda^\star(\mathbf{MS})$, the corresponding Θ-class, i.e. the corresponding interval of $\Lambda(\mathbf{DMS})$. We shall then determine the cardinality of each of these intervals and use the results to compute the cardinality of $\Lambda(\mathbf{DMS})$.

Theorem 15.7 *The subvarieties in* $\Lambda^*(\mathbf{MS})$ *are associated with the following intervals in* $\Lambda(\mathbf{DMS})$:

$$
\begin{aligned}
\mathbf{B} &\leftrightarrow I_1 = \{\mathbf{SID}_1\}\\
\mathbf{K} &\leftrightarrow I_2 = \{\mathbf{SID}_3\}\\
\mathbf{M} &\leftrightarrow I_3 = \{\mathbf{SID}_7\}\\
\mathbf{S} &\leftrightarrow I_4 = [\mathbf{SID}_2, \mathbf{SID}_4]\\
\mathbf{S}\vee\mathbf{K} &\leftrightarrow I_5 = [\mathbf{SID}_2\vee\mathbf{SID}_3, \mathbf{SID}_4\vee\mathbf{SID}_3]\\
\mathbf{K}_1 &\leftrightarrow I_6 = [\mathbf{SID}_6, \mathbf{SID}_{10}]\\
\mathbf{S}\vee\mathbf{K}_1 &\leftrightarrow I_7 = [\mathbf{SID}_2\vee\mathbf{SID}_6, \mathbf{SID}_4\vee\mathbf{SID}_{10}]\\
\mathbf{K}_2 &\leftrightarrow I_8 = [\mathbf{SID}_5, \mathbf{SID}_4\vee\mathbf{SID}_8]\\
\mathbf{M}\vee\mathbf{K}_1 &\leftrightarrow I_9 = [\mathbf{SID}_6\vee\mathbf{SID}_7, \mathbf{SID}_{10}\vee\mathbf{SID}_7]\\
\mathbf{S}\vee\mathbf{M} &\leftrightarrow I_{10} = [\mathbf{SID}_2\vee\mathbf{SID}_7, \mathbf{SID}_4\vee\mathbf{SID}_7]\\
\mathbf{K}_1\vee\mathbf{K}_2 &\leftrightarrow I_{11} = [\mathbf{SID}_5\vee\mathbf{SID}_6, \mathbf{SID}_4\vee\mathbf{SID}_{11}\vee\mathbf{SID}_{12}]\\
\mathbf{M}\vee\mathbf{K}_2 &\leftrightarrow I_{12} = [\mathbf{SID}_5\vee\mathbf{SID}_7, \mathbf{SID}_4\vee\mathbf{SID}_7\vee\mathbf{SID}_8]\\
\mathbf{S}\vee\mathbf{M}\vee\mathbf{K}_1 &\leftrightarrow I_{13} = [\mathbf{SID}_2\vee\mathbf{SID}_6\vee\mathbf{SID}_7, \mathbf{SID}_4\vee\mathbf{SID}_{10}\vee\mathbf{SID}_7]\\
\mathbf{K}_2\vee\mathbf{K}_3 &\leftrightarrow I_{14} = [\mathbf{SID}_{13}, \mathbf{SID}_{17}\vee\mathbf{SID}_{18}]\\
\mathbf{M}\vee\mathbf{K}_1\vee\mathbf{K}_2 &\leftrightarrow I_{15} = [\mathbf{SID}_5\vee\mathbf{SID}_6\vee\mathbf{SID}_7, \mathbf{SID}_4\vee\mathbf{SID}_{11}\vee\mathbf{SID}_{12}\vee\mathbf{SID}_7]\\
\mathbf{M}\vee\mathbf{K}_2\vee\mathbf{K}_3 &\leftrightarrow I_{16} = [\mathbf{SID}_7\vee\mathbf{SID}_{13}, \mathbf{SID}_7\vee\mathbf{SID}_{17}\vee\mathbf{SID}_{18}]\\
\mathbf{M}_1 &\leftrightarrow I_{17} = [\mathbf{SID}_{14}, \mathbf{SID}_{21}]
\end{aligned}
$$

Proof In order to obtain the biggest element in each of the intervals, the method is routine so we shall illustrate it by an example. Consider the subvariety $\mathbf{M}\vee\mathbf{K}_2\vee\mathbf{K}_3$ which has the equational basis $(1, 11_d)$. Of the subdirectly irreducible double MS-algebras we observe that SID_{14} does not satisfy these °-axioms (since its 'MS-part' is M_1), whereas SID_7, SID_{17}, and SID_{18} do satisfy these axioms. It follows that the biggest element in the Θ-class of $\mathbf{M}\vee\mathbf{K}_2\vee\mathbf{K}_3$ is $\mathbf{SID}_7\vee\mathbf{SID}_{17}\vee\mathbf{SID}_{18}$.

To obtain the smallest elements in the intervals, observe that if $\mathbf{X}\leftrightarrow[\mathbf{P},\mathbf{Q}]$ then we have

$$[\mathbf{P},\mathbf{Q}] = \mathbf{Q}^{\downarrow} \setminus \bigcup_{k\in K} I_k$$

where $(I_k)_{k\in K}$ denotes the family of ideals generated by the intervals which are associated with the subvarieties of **MS** that are covered by **X**. Note that for every **X** there are at most three such subvarieties. Consequently, the determination of the smallest elements in the intervals can be done by traversing $\Lambda(\mathbf{MS})$ upwards.

If $\mathbf{Q}$ is the supremum in $\Lambda(\mathbf{DMS})$ of subvarieties that are generated by linearly ordered ideals, the task is particularly simple. For example, consider the subvariety $\mathbf{X} = \mathbf{M} \vee \mathbf{K}_2$. We know that in this case $\mathbf{Q} = \mathbf{SID}_4 \vee \mathbf{SID}_7 \vee \mathbf{SID}_8$. Since $\mathrm{SID}_4^{\downarrow}$, $\mathrm{SID}_7^{\downarrow}$, and $\mathrm{SID}_8^{\downarrow}$ are chains we can depict $\mathbf{Q}^{\downarrow}$ as follows (in which $\mathbf{n}$ denotes $\mathbf{SID}_n$) :

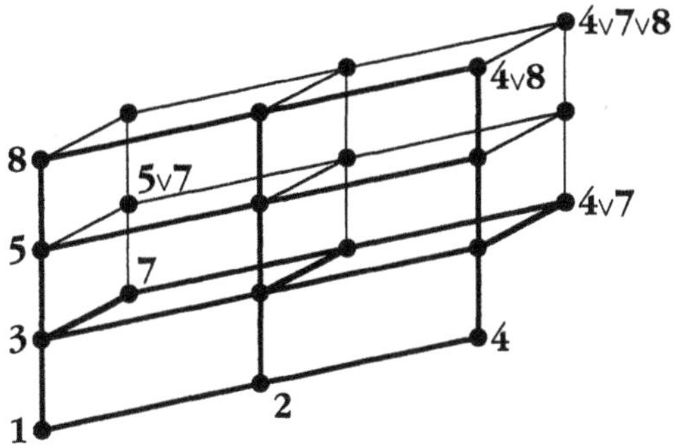

Now the subvariety $\mathbf{M} \vee \mathbf{K}_2$ covers $\mathbf{K}_2$ and $\mathbf{M} \vee \mathbf{S}$. Hence $\mathbf{P}$ is the smallest element of

$$\mathbf{Q}^{\downarrow} \setminus ((\mathbf{SID}_4 \vee \mathbf{SID}_8)^{\downarrow} \cup (\mathbf{SID}_4 \vee \mathbf{SID}_7)^{\downarrow}).$$

From the diagram, this is $\mathbf{SID}_5 \vee \mathbf{SID}_7$.

If $\mathbf{Q}$ fails to have the above property then the determination of $\mathbf{P}$ is somewhat more intricate, but can be done by considering only the ordered set of subdirectly irreducible double MS-algebras. For example, consider $\mathbf{X} = \mathbf{K}_1 \vee \mathbf{K}_2$. Here we know that $\mathbf{Q} = \mathbf{SID}_4 \vee \mathbf{SID}_{11} \vee \mathbf{SID}_{12}$. Now the subvariety $\mathbf{K}_1 \vee \mathbf{K}_2$ covers $\mathbf{K}_2$ and $\mathbf{S} \vee \mathbf{K}_1$. Hence $\mathbf{P}$ is the smallest element of

$$\mathbf{Q}^{\downarrow} \setminus ((\mathbf{SID}_4 \vee \mathbf{SID}_{10})^{\downarrow} \cup (\mathbf{SID}_4 \vee \mathbf{SID}_8)^{\downarrow}).$$

By considering the diagram

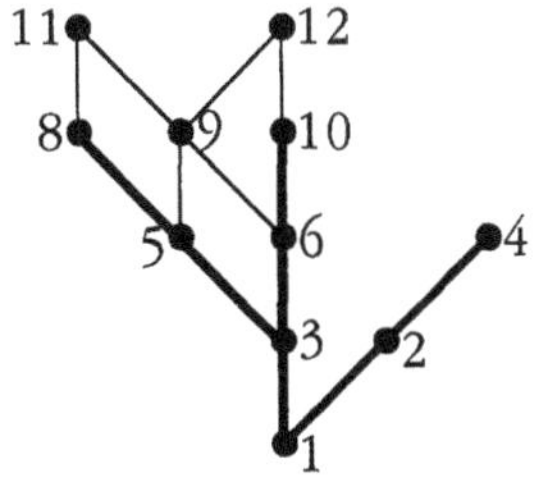

we see that $\mathbf{P} = \mathbf{SID}_5 \vee \mathbf{SID}_6$. ♢

The partitioning of $\Lambda(\mathbf{DMS})$ into 17 intervals makes possible the determination of its cardinality. This is as follows (again with the trivial subvariety ignored).

Theorem 15.8 $|\Lambda(\mathbf{DMS})| = 380$.

Proof In a distributive lattice we have $[x \vee z, y \vee z] \simeq [x, y]$. It follows by Theorem 15.7 therefore that we have the following isomorphisms :

$$I_4 \simeq I_5 \simeq I_{10}, \quad I_6 \simeq I_9, \quad I_8 \simeq I_{12}, \quad I_7 \simeq I_{13}, \quad I_{11} \simeq I_{15}, \quad I_{14} \simeq I_{16}.$$

Now there are three singletons, namely I_1, I_2, I_3. To determine the cardinality of the other intervals is routine, and we illustrate the method by considering

$$I_{11} = [\mathbf{SID}_5 \vee \mathbf{SID}_6, \mathbf{SID}_4 \vee \mathbf{SID}_{11} \vee \mathbf{SID}_{12}].$$

We require to compute the number of down-sets of the ordered set

$$(4^{\downarrow} \cup 11^{\downarrow} \cup 12^{\downarrow}) \setminus (5^{\downarrow} \cup 6^{\downarrow}),$$

i.e. the ordered set

Using Theorem 5.5, we can see that the number of down-sets is 39. Thus we have $|I_{11}| = 39 = |I_{15}|$.

In summary, we have

interval	*cardinality*
I_1, I_2, I_3	1
$I_4, I_5, I_6, I_9, I_{10}$	2
I_7, I_{13}	4
I_8, I_{12}	6
I_{11}, I_{15}	39
I_{14}, I_{16}	41
I_{17}	187

and conclude from this that $|\Lambda(\mathbf{DMS})| = 380$. ◇

Given a double MS-algebra $(L; ^{\circ}, ^{+})$ let us now consider the operation $x \mapsto x^{\sim}$ defined on L by setting

$$(\forall x \in L) \qquad x^{\sim} = (x \wedge x^{+}) \vee x^{\circ} = (x \vee x^{\circ}) \wedge x^{+}.$$

Our objective now is to determine when this operation makes L into an Ockham algebra. First we observe that

$$(\forall x \in L) \qquad x \vee x^{\sim} = x \vee x^{\circ}, \quad x \wedge x^{\sim} = x \wedge x^{+}.$$

Hence we see that $x^{\sim}$ is the complement of x in the interval $[x \wedge x^{+}, x \vee x^{\circ}]$. Clearly, $0^{\sim} = 1$ and $1^{\sim} = 0$. We also have

$$\begin{cases} x^{\circ\sim} = (x^{\circ} \wedge x^{\circ +}) \vee x^{\circ\circ} = x^{\circ\circ}; \\ x^{+\sim} = (x^{+} \vee x^{+\circ}) \wedge x^{++} = x^{++}, \end{cases}$$

whereas

$$\begin{cases} x^{\sim\circ} = (x^{\circ} \vee x^{++}) \wedge x^{\circ\circ} = (x^{\circ} \wedge x^{\circ\circ}) \vee x^{++}; \\ x^{\sim +} = (x^{+} \vee x^{++}) \wedge x^{\circ\circ} = (x^{+} \wedge x^{\circ\circ}) \vee x^{++}. \end{cases}$$

Theorem 15.9 *If $L \in$ **DMS** then, for every $x \in L$,*

(*a*) $x \geqslant x^{\sim\sim} \iff x \wedge x^{\circ} \in S(L)$;

(*b*) $x \leqslant x^{\sim\sim} \iff x \vee x^{+} \in S(L)$.

Proof Using the above, we have

$$\begin{cases} x^{\sim} \vee x^{\sim\circ} = (x \wedge x^{+}) \vee x^{\circ} \vee x^{++} = (x \vee x^{\circ}) \wedge (x^{+} \vee x^{++}); \\ x^{\sim} \wedge x^{\sim +} = (x \vee x^{\circ}) \wedge x^{+} \wedge x^{\circ\circ} = (x \wedge x^{+}) \vee (x^{\circ} \wedge x^{\circ\circ}), \end{cases}$$

and consequently

$$x^{\sim\sim} = \begin{cases} (x^{\sim} \wedge x^{\sim +}) \vee x^{\sim\circ} = (x \wedge x^{+}) \vee (x^{\circ} \wedge x^{\circ\circ}) \vee x^{++}; \\ (x^{\sim} \vee x^{\sim\circ}) \wedge x^{\sim +} = (x \vee x^{\circ}) \wedge (x^{+} \vee x^{++}) \wedge x^{\circ\circ}. \end{cases}$$

It follows that

$$\begin{cases} x \geqslant x^{\sim\sim} \iff x \geqslant x^{\circ} \wedge x^{\circ\circ} \iff x \wedge x^{\circ} \in S(L) \\ x \leqslant x^{\sim\sim} \iff x \leqslant x^{+} \vee x^{++} \iff x \vee x^{+} \in S(L). \end{cases} \qquad \diamond$$

Corollary $x = x^{\sim\sim} \iff x \wedge x^{\circ}, x \vee x^{+} \in S(L)$. $\diamond$

A remarkable fact concerning the operation $x \mapsto x^{\sim}$ is the following.

Theorem 15.10 *If $L \in$ **DMS** then*

$$(\forall x \in L) \qquad x^{\sim} = x^{\sim\sim\sim}.$$

Proof Using the above observations, we have

$$\begin{aligned} x^{\sim} \wedge x^{\sim\circ} &= (x \vee x^{\circ}) \wedge x^{+} \wedge (x^{\circ} \vee x^{++}) \wedge x^{\circ\circ} \\ &= x^{+} \wedge (x^{\circ} \vee x^{++}) \wedge x^{\circ\circ} \\ &= (x^{\circ\circ} \vee x^{\circ}) \wedge x^{+} \wedge (x^{\circ} \vee x^{++}) \wedge x^{\circ\circ} \\ &= x^{\sim\circ\circ} \wedge x^{\sim\circ}. \end{aligned}$$

Similarly we can show that

$$x^{\sim} \vee x^{\sim +} = x^{\sim ++} \vee x^{\sim +}.$$

It now follows by the Corollary to Theorem 15.9 that $x^{\sim} = x^{\sim\sim\sim}$. ◇

In view of Theorem 15.10 it is natural to consider the question of precisely when $(L; {}^{\sim})$ is an Ockham algebra. In order to answer this, we require the following result.

Theorem 15.11 *If* $L \in \mathbf{DMS}$ *then the following conditions are equivalent*:

(1) $(\forall x, y \in L) \quad (x \vee y)^{\sim} = x^{\sim} \wedge y^{\sim}$;

(2) $(\forall x, y \in L) \quad (x \wedge y)^{\sim} = x^{\sim} \vee y^{\sim}$;

(3) $(\forall x, y \in L) \quad x \wedge x^{+} \wedge y^{+} \leqslant y \vee y^{\circ}$;

(4) $(\forall x, y \in L) \quad y \wedge y^{+} \leqslant x \vee x^{\circ} \vee y^{\circ}$.

Proof (1) $\Leftrightarrow$ (3) : The equality $(x \vee y)^{\sim} = x^{\sim} \wedge y^{\sim}$ holds if and only if

$$[x \vee y \vee (x^{\circ} \wedge y^{\circ})] \wedge x^{+} \wedge y^{+} = (x \vee x^{\circ}) \wedge x^{+} \wedge (y \vee y^{\circ}) \wedge y^{+},$$

which is equivalent to the inequality

$$[x \vee y \vee (x^{\circ} \wedge y^{\circ})] \wedge x^{+} \wedge y^{+} \leqslant (x \vee x^{\circ}) \wedge x^{+} \wedge (y \vee y^{\circ}) \wedge y^{+},$$

which is equivalent to

$$(x \wedge x^{+} \wedge y^{+}) \vee (y \wedge x^{+} \wedge y^{+}) \vee (x^{\circ} \wedge y^{\circ}) \leqslant (x \vee x^{\circ}) \wedge x^{+} \wedge (y \vee y^{\circ}) \wedge y^{+}.$$

Clearly, this holds if and only if (3) holds.

Dually, we can show that (2) $\Leftrightarrow$ (4).

We now show that (3) $\Leftrightarrow$ (4), i.e. that (3) is self-dual. For this purpose, it suffices to prove only that (3) $\Rightarrow$ (4), a dual proof providing the converse implication.

Suppose then that (3) holds. Applying $^{\circ}$, we obtain

$$y^{\circ} \wedge y^{\circ\circ} \leqslant x^{\circ} \vee x^{++} \vee y^{++}$$

which gives

$$(y^{\circ} \wedge y^{\circ\circ}) \vee x \vee x^{\circ} \leqslant (y^{\circ} \wedge y^{++}) \vee x \vee x^{\circ}$$

whence $(y^{\circ} \wedge y) \vee x \vee x^{\circ} = (y^{\circ} \wedge y^{\circ\circ}) \vee x \vee x^{\circ}$, which is axiom $(17) = (9, 12_d)$ of Chapter 5. Now by Theorem 15.1 the axioms (9) and (12_d) are self-dual in **DMS**. Hence so is (17), and therefore we have $(y \vee y^{+}) \wedge x \wedge x^{+} = (y^{++} \vee y^{+}) \wedge x \wedge x^{+}$ which gives $y^{+} \vee y^{++} \geqslant y \wedge x \wedge x^{+}$, whence

$$y^{+} \vee y^{++} \geqslant y^{\circ\circ} \wedge x^{\circ\circ} \wedge x^{+}.$$

Applying $^{+}$ to this, we obtain

$$(\star) \qquad y^{++} \wedge y^{+} \leqslant y^{\circ} \vee x^{\circ} \vee x^{++}.$$

Applying $^+$ also to (3), we obtain

$$(\star\star) \qquad y^+ \wedge y^{\circ\circ} \leqslant x^+ \vee x^{++} \vee y^{++}.$$

It follows that

$$\begin{aligned} y \wedge y^+ &\leqslant y^{\circ\circ} \wedge y^+ \\ &\leqslant x^+ \vee x^{++} \vee (y^{++} \wedge y^+) \quad \text{by } (\star\star) \\ &\leqslant x^+ \vee x^{++} \vee y^\circ \quad \text{by } (\star). \end{aligned}$$

Consequently,

$$\begin{aligned} y \wedge y^+ &= y \wedge y^+ \wedge (x^+ \vee x^{++} \vee y^\circ) \\ &= (y \wedge y^+ \wedge x^+) \vee (y \wedge y^+ \wedge x^{++}) \vee (y \wedge y^+ \wedge y^\circ) \\ &\leqslant x \vee x^\circ \vee x^{++} \vee y^\circ \quad \text{by (3)} \\ &= x \vee x^\circ \vee y^\circ, \end{aligned}$$

and we have (4). ◇

Theorem 15.12 *If* $L \in \mathbf{DMS}$ *then* $(L; {}^\sim)$ *is an Ockham algebra if and only if*

$$L \in \mathbf{SID}_2 \vee \mathbf{SID}_7 \vee \mathbf{SID}_{11} \vee \mathbf{SID}_{12}.$$

In this case we have necessarily $(L; {}^\sim) \in \mathbf{M} \vee \widetilde{\mathbf{K}}_1$.

Proof In view of Theorem 15.10, we have that $(L; {}^\sim) \subset \mathbf{O}$ if and only if the mapping $x \mapsto x^\sim$ is a dual lattice endomorphism; and by Theorem 15.11 this is the case if and only if L satisfies the axiom

$$x \wedge x^+ \wedge y^+ \leqslant y \vee y^\circ.$$

Now a routine verification shows that this is satisfied by the subdirectly irreducible double MS-algebras $\mathbf{SID}_2$, $\mathbf{SID}_7$, $\mathbf{SID}_{11}$, and $\mathbf{SID}_{12}$ but not by $\mathbf{SID}_4$, $\mathbf{SID}_{13}$, or $\mathbf{SID}_{14}$. The first statement now follows.

To show that when $(L; {}^\sim)$ is an Ockham algebra it necessarily belongs to the subvariety $\mathbf{M} \vee \widetilde{\mathbf{K}}_1$ of $\mathbf{P}_{3,1}$, we observe that the mapping $x \mapsto x^\sim$ satisfies in general the inequalities

$$(4)\ x \wedge x^\sim \leqslant x^{\sim\sim}, \qquad (4_d)\ x \vee x^\sim \geqslant x^{\sim\sim}.$$

These follow from the expressions for $x^{\sim\sim}$ obtained in the proof of Theorem 15.9 and the equalities $x \wedge x^+ = x \wedge x^\sim$ and $x \vee x^\circ = x \vee x^\sim$. Thus $(L; {}^\sim)$ satisfies $(\widetilde{4})$ which is an equational basis for $\mathbf{M} \vee \widetilde{\mathbf{K}}_1$. ◇

The lattice of (non-trivial) subvarieties of $\mathbf{M} \vee \widetilde{\mathbf{K}}_1$ is

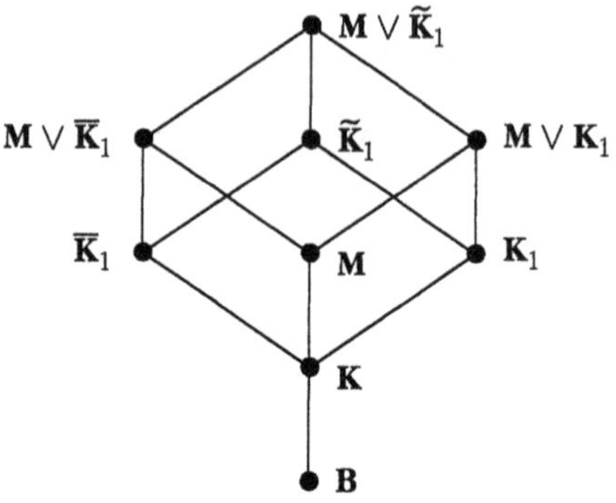

Theorem 15.13 *Let $L \in$ **DMS** be such that $(L;^{\sim}) \in$ **O**. Then the various subvarieties are related as follows :*

$(L;^{\sim})$ *belongs to* **MS**-*subvariety*	$(L;^{\circ},^{+})$ *belongs to* **DMS**-*subvariety*	$(L;^{\circ})$ *belongs to* **MS**-*subvariety*
$\mathbf{M} \vee \widetilde{\mathbf{K}}_1$	$\mathbf{SID}_2 \vee \mathbf{SID}_7 \vee \mathbf{SID}_{11} \vee \mathbf{SID}_{12}$	$\mathbf{M} \vee \mathbf{K}_1 \vee \mathbf{K}_2$
$\mathbf{M} \vee \mathbf{K}_1$	$\mathbf{SID}_2 \vee \mathbf{SID}_7 \vee \mathbf{SID}_{10}$	$\mathbf{S} \vee \mathbf{M} \vee \mathbf{K}_1$
$\mathbf{M} \vee \overline{\mathbf{K}}_1$	$\mathbf{SID}_2 \vee \mathbf{SID}_7 \vee \mathbf{SID}_8$	$\mathbf{M} \vee \mathbf{K}_2$
$\widetilde{\mathbf{K}}_1$	$\mathbf{SID}_2 \vee \mathbf{SID}_{11} \vee \mathbf{SID}_{12}$	$\mathbf{K}_1 \vee \mathbf{K}_2$
$\mathbf{K}_1$	$\mathbf{SID}_2 \vee \mathbf{SID}_{10}$	$\mathbf{S} \vee \mathbf{K}_1$
$\overline{\mathbf{K}}_1$	$\mathbf{SID}_2 \vee \mathbf{SID}_8$	$\mathbf{K}_2$
$\mathbf{M}$	$\mathbf{SID}_2 \vee \mathbf{SID}_7$	$\mathbf{S} \vee \mathbf{M}$
$\mathbf{K}$	$\mathbf{SID}_2 \vee \mathbf{SID}_3$	$\mathbf{S} \vee \mathbf{K}$
$\mathbf{B}$	$\mathbf{SID}_1$	$\mathbf{B}$

Proof For example, consider $(L;^{\sim}) \in \widetilde{\mathbf{K}}_1$. An equational basis of

$$\widetilde{\mathbf{K}}_1 = \widetilde{\mathbf{L}} \wedge (\widetilde{\mathbf{K}}_1 \vee \mathbf{M})$$

is $(\gamma, \widetilde{4})$. Since, as observed above, (4) and (4_d) hold for $(L;^{\sim})$ we deduce that

$$(L;^{\sim}) \in \widetilde{\mathbf{K}}_1 \iff x \wedge x^{\sim} \leqslant y \vee y^{\sim} \iff x \wedge x^{+} \leqslant y \vee y^{+}.$$

It is readily seen that the biggest subvariety of $\mathbf{SID}_2 \vee \mathbf{SID}_7 \vee \mathbf{SID}_{11} \vee \mathbf{SID}_{12}$ in which this holds is $\mathbf{SID}_2 \vee \mathbf{SID}_{11} \vee \mathbf{SID}_{12}$. By Theorem 15.7, the corresponding subvariety of **MS** to which $(L;^{\circ})$ belongs is then $\mathbf{K}_1 \vee \mathbf{K}_2$. ◇

Bibliography

(*a*) Books

[1] R. Balbes and Ph. Dwinger : *Distributive Lattices*, University of Missouri Press, Missouri, 1974.

[2] G. Birkhoff : *Lattice Theory*, 3rd edition, Coll. Pub., XXV, American Mathematical Society, Providence, R.I., 1967.

[3] T. S. Blyth and M. F. Janowitz : *Residuation Theory*, Pergamon Press, 1972.

[4] G. Boole : *An Investigation into the Laws of Thought*, London, 1854. Reprinted by Open Court Publishing Co., Chicago, 1940.

[5] S. Burris and H. P. Sankappanavar : *A Course in Universal Algebra*, Graduate Texts in Mathematics, **78**, Springer Verlag, 1981.

[6] D. A. Cohen : *Basic techniques of combinatorial theory*, Wiley, 1978.

[7] W. H. Cornish : *Antimorphic Action. Categories of Algebraic Structures with Involutions or Anti-endomorphisms*, Research and Exposition in Mathematics, **12**, Heldermann Verlag, Berlin, 1986.

[8] H. B. Curry : *Foundations of Mathematical Logic*, McGraw-Hill series in higher mathematics, New York, 1963.

[9] B. A. Davey and H. A. Priestley : *Introduction to Lattices and Order*, Cambridge University Press, 1990.

[10] P. J. Davis : *Circulant Matrices*, Wiley, 1979.

[11] G. Grätzer : *Lattice Theory; First concepts and Distributive Lattices*, Freeman, 1971.

[12] G. Grätzer : *General Lattice Theory*, Birkhäuser, 1978.

[13] G. Grätzer : *Universal Algebra*, 2nd edition, Springer Verlag, 1979.

[14] F. Harary : *Graph Theory*, Addison-Wesley, 1969.

[15] B. Jónsson : *Topics in Universal Algebra*, Springer Verlag, 1972.

[16] R. N. McKenzie, G. F. McNulty and W. F. Taylor : *Algebras, Lattices and Varieties*, vol. I, Wadworth and Brooks, 1987.

(*b*) Research Papers

[17] M. E. Adams : Principal congruences in de Morgan algebras, *Proc. Edinburgh Math. Soc.*, **30**, 1987, 415–421.

[18] M. E. Adams and R. Beazer : Congruence relations on de Morgan algebras, *Algebra Universalis*, **26**, 1989, 103–125.

[19] M. E. Adams and R. Beazer : The intersection of principal congruences on de Morgan algebras, *Houston J. Math.*, **16**, 1990, 59–70.

[20] M. E. Adams and H. A. Priestley : Kleene algebras are almost universal, *Bull. Austral. Math. Soc.*, **34**, 1986, 343–373.

[21] M. E. Adams and H. A. Priestley : De Morgan algebras are universal, *Discrete Math.*, **66**, 1987, 1–13.

[22] M. E. Adams and H. A. Priestley : Equational bases for varieties of Ockham algebras (preprint).

[23] C. J. Ash : Pseudovarieties, generalised varieties and similarly described classes, *J. Algebra*, **92**, 1985, 104–115.

[24] R. Beazer : On some small varieties of distributive Ockham algebras, *Glasgow Math. J.*, **25**, 1984, 175–181.

[25] R. Beazer : Injectives in some small varieties of Ockham algebras, *Glasgow Math. J.*, **25**, 1984, 183–191.

[26] R. Beazer : Congruence pairs for algebras abstracting Kleene and Stone algebras, *Czechoslovak Math. J.*, **35**, 1985, 260–268.

[27] N. D. Belnap and J. H. Spencer : Intensionally complemented distributive lattices, *Portugal. Math.*, **25**, 1966, 99–104.

[28] J. Berman : Distributive lattices with an additional unary operation, *Aequationes Math.*, **16**, 1977, 165–171.

[29] J. Berman and P. Köhler : Cardinalities of finite distributive lattices, *Mitt. Math. Sem. Giessen*, **121**, 1976, 103–124.

[30] G. Birkhoff : On the combination of subalgebras, *Proc. Camb. Phil. Soc.*, **29**, 1933, 441–464.

[31] T. S. Blyth : Some examples of distributive Ockham algebras with de Morgan skeletons, in *Lattices, Semigroups and Universal Algebra*, Lisbon 1988, J. Almeida, G. Bordalo and Ph. Dwinger (eds), Plenum Press, 1990, 21–28.

[32] T. S. Blyth and J. C. Varlet : Sur la construction de certaines MS-algèbres, *Portugal. Math.*, **39**, 1980, 489–496. [published 1985]

[33] T. S. Blyth and J. C. Varlet : On a common abstraction of de Morgan algebras and Stone algebras, *Proc. Roy. Soc. Edinburgh*, **94A**, 1983, 301–308.

[34] T. S. Blyth and J. C. Varlet : Subvarieties of the class of MS-algebras, *Proc. Roy. Soc. Edinburgh*, **95A**, 1983, 157–169.

[35] T. S. Blyth and J. C. Varlet : Fixed points in MS-algebras, *Bull. Soc. Roy. Sci. Liège*, **53**, 1984, 3–8.

[36] T. S. Blyth and J. C. Varlet : Congruences on MS-algebras, *Bull. Soc. Roy. Sci. Liège*, **53**, 1984, 341–362.

[37] T. S. Blyth and J. C. Varlet : Double MS-algebras, *Proc. Roy. Soc. Edinburgh*, **98A**, 1984, 37–47.

[38] T. S. Blyth and J. C. Varlet : Subdirectly irreducible double MS-algebras, *Proc. Roy. Soc. Edinburgh*, **98A**, 1984, 241–247.

[39] T. S. Blyth and J. C. Varlet : Corrigendum – Sur la construction de certaines MS-algèbres, *Portugal. Math.*, **42**, 1984, 469–471. [published 1986]

[40] T. S. Blyth and J. C. Varlet : Amalgamation properties in the class of MS-algebras, *Proc. Roy. Soc. Edinburgh*, **100A**, 1985, 139–149.

[41] T. S. Blyth and J. C. Varlet : MS-algebras definable on a distributive lattice, *Bull. Soc. Roy. Sci. Liège*, **54**, 1985, 167–182.

[42] T. S. Blyth and J. C. Varlet : The ideal lattice of an MS-algebra, *Glasgow Math. J.*, **30**, 1988, 137–143.

[43] T. S. Blyth and J. C. Varlet : Polarised Ockham algebras with locally complemented skeletons, *Bull. Soc. Roy. Sci. Liège*, **57**, 1988, 151–156.

[44] T. S. Blyth and J. C. Varlet : On the dual space of an MS-algebra, *Math. Pannonica*, **1**, 1990, 95–109.

[45] T. S. Blyth and J. C. Varlet : Principal congruences on some lattice-ordered algebras, *Discrete Math.*, **81**, 1990, 323–329.

[46] T. S. Blyth and J. C. Varlet : On the MS-algebras of the unit cube, *Portugal. Math.*, **48**, 1991, 405–421.

[47] T. S. Blyth, G. Hansoul and J. C. Varlet : The dual space of a double MS-algebra, *Algebra Universalis*, **28**, 1991, 26–35.

[48] T. S. Blyth, A. S. A. Noor and J. C. Varlet : Subvarieties of double MS-algebras into which an MS-algebra can be extended, *Bull. Soc. Roy. Sci. Liège*, **56**, 1987, 47–53.

[49] T. S. Blyth, A. S. A. Noor and J. C. Varlet : Congruences on double MS-algebras, *Bull. Soc. Roy. Sci. Liège*, **56**, 1987, 143–152.

[50] T. S. Blyth, P. Goossens and J. C. Varlet : MS-algebras arising from fences and crowns, *Contributions to General Algebra*, **6**, Verlag Hölder-Pichler-Tempsky, Wien 1988, pp. 31–48.

[51] T. S. Blyth, A. S. A. Noor and J. C. Varlet : Ockham algebras associated with double MS-algebras, *Bull. Soc. Roy. Sci. Liège*, **57**, 1988, 591–597.

[52] T. S. Blyth, A. S. A. Noor and J. C. Varlet : Ockham algebras with de Morgan skeletons, *J. Algebra*, **117**, 1988, 165–178.

[53] T. S. Blyth, Jie Fang and J. C. Varlet : MS-algebras arising from hat racks, double fences and double crowns, *Bull. Soc. Roy. Sci. Liège*, **58**, 1989, 63–84.

[54] T. S. Blyth, A. S. A. Noor and J. C. Varlet : Equational bases for subvarieties of double MS-algebras, *Glasgow Math. J.*, **31**, 1989, 1–16.

[55] T. S. Blyth, Jie Fang and J. C. Varlet : Subdirectly irreducible Ockham algebras, *Contributions to General Algebra*, **7**, Verlag Hölder-Pichler-Tempsky, Wien 1991, pp. 37–48.

[56] T. S. Blyth, Jie Fang and J. C. Varlet : Fixed point separating congruences on Ockham algebras, *Bull. Soc. Roy. Sci. Liège*, **63**, 1993, 147–162.

[57] G. Bordalo : A duality between unary algebras and their subuniverse lattices, *Portugal. Math.*, **46**, 1989, 431–439.

[58] G. Bordalo and H. A. Priestley : Relative Ockham algebras; their order-theoretic and algebraic characterisation, *Glasgow Math. J.*, **32**, 1990, 47–66.

[59] G. Bordalo and H. A. Priestley : Negation operations definable on finite distributive lattices, *Portugal. Math.*, **49**, 1992, 37–49.

[60] W. H. Cornish and P. R. Fowler : Coproducts of de Morgan algebras, *Bull. Austral. Math. Soc.*, **16**, 1977, 1–13.

[61] W. H. Cornish and P. R. Fowler : Coproducts of Kleene algebras, *J. Austral. Math. Soc.*, **27A**, 1979, 209–220.

[62] B. A. Davey : On the lattice of subvarieties, *Houston J. Math.*, **5**, 1979, 183–192.

[63] B. A. Davey and D. Duffus : Exponentiation and duality, in *Ordered Sets*, I. Rival (ed.), NATO Advanced Study Institute Series, Series C, vol. 83, D. Reidel, 1982, pp. 43–95.

[64] B. A. Davey and H. A. Priestley : Lattices of homomorphisms, *J. Austral. Math. Soc.*, **40A**, 1986, 364–406.

[65] B. A. Davey and H. A. Priestley : Generalised piggyback dualities and applications to Ockham algebras, *Houston J. Math.*, **13**, 1987, 151–197.

[66] A. Day : A note on the congruence extension property, *Algebra Universalis*, **1**, 1971, 234–235.

[67] Jie Fang : Contributions to the theory of Ockham algebras, Ph.D. thesis, University of St Andrews, 1991.

[68] M. S. Goldberg : Distributive p-algebras and Ockham algebras; a topological approach, Ph.D. thesis, LaTrobe University, 1979.

[69] M. S. Goldberg : Distributive Ockham algebras; free algebras and injectivity, *Bull. Austral. Math. Soc.*, **24**, 1981, 161–203.

[70] M. S. Goldberg : Topological duality for distributive Ockham algebras, *Studia Logica*, **42**, 1983, 23–31.

[71] M. Habib and R. H. Möhring : On some complexity properties of N-free posets with bounded decomposition diameter, *Discrete Math.*, **63**, 1987, 157–182.

[72] J. A. Kalman : Lattices with involution, *Trans. Amer. Math. Soc.*, **87**, 1958, 485–491.

[73] T. Katrinak and K. Mikula : On a construction of MS-algebras, *Portugal. Math.*, **45**, 1988, 157–163.

[74] H. Lakser : Principal congruences of pseudocomplemented distributive lattices, *Proc. Amer. Math. Soc.*, **37**, 1973, 32–36.

[75] G. C. Moisil : Recherches sur l'algèbre de la logique, *Ann. Sci. Univ. Jassy*, **22**, 1935, 1–117.

[76] A. Monteiro: Matrices de Morgan caractéristiques pour le calcul propositionnel classique, *An. Acad. Brasil Ci.*, **52**, 1960, 1–7.

[77] A. S. A. Noor : Bistable subvarieties of MS-algebras, *Proc. Roy. Soc. Edinburgh*, **105A**, 1987, 127–128.

[78] H. A. Priestley : Ordered topological spaces and the representation of distributive lattices, *Proc. London Math. Soc.*, **2**, 1972, 507–530.

[79] H. A. Priestley : Ordered sets and duality for distributive lattices, *Ann. Discrete Math.*, **23**, 1984, 39–60.

[80] H. A. Priestley : The determination of subvarieties of certain congruence-distributive varieties, *Algebra Universalis* (to appear).

[81] H. A. Priestley : Natural dualities for varieties of distributive-lattice-ordered algebras (preprint).

[82] M. Ramalho and M. Sequeira : On generalised MS-algebras, *Portugal. Math.*, **44**, 1987, 315–328.

[83] B. Sands : Counting antichains in finite partially ordered sets, *Discrete Math.*, **35**, 1981, 213–228.

[84] H. P. Sankappanavar : A characterisation of principal congruences of de Morgan algebras and its applications, *Math. Logic in Latin America*, North Holland, 1980.

[85] H. P. Sankappanavar : Distributive lattices with a dual endomorphism, *Z. Math. Logik Grundlag. Math.*, **31**, 1985, 385–392.

[86] D. Schweigert and M. Szymanska : On distributive correlation lattices, *Coll. Math. Soc. Janos Bolyai*, **33**, 1980, 697–721.

[87] D. Schweigert and M. Szymanska : Polynomial functions of correlation lattices, *Algebra Universalis*, **16**, 1983, 355–359.

[88] D. Schweigert and M. Szymanska : A completeness theorem for correlation lattices, *Zeitschr. f. math. Logik und Grundlagen d. Math.*, **29**, 1983, 427–434.

[89] M. Sequeira : Varieties of Ockham algebras satisfying $x \leqslant f^4(x)$, *Portugal. Math.*, **45**, 1988, 369–391.

[90] M. Sequeira : Algebras-MS generalizadas e algebras-$\mathbf{K}_{n,1}$ duplas, doctoral thesis, University of Lisbon, 1989.

[91] M. Sequeira : Double MS_n-algebras and double $\mathbf{K}_{n,m}$-algebras, *Glasgow Math. J.*, **35**, 1993, 189–201.

[92] M. H. Stone : The theory of representations for boolean algebras, *Trans. Amer. Math. Soc.*, **40**, 1936, 37–111.

[93] W. T. Trotter Jr : Dimension of the crown S_n^k, *Discrete Math.*, **8**, 1974, 85–103.

[94] A. Urquhart : Lattices with a dual homomorphic operation, *Studia Logica*, **38**, 1979, 201–209.

[95] A. Urquhart : Lattices with a dual homomorphic operation II, *Studia Logica*, **40**, 1981, 391–404.

[96] J. C. Varlet : Algèbres de Lukasiewicz trivalentes, *Bull. Soc. Roy. Sci. Liège*, **37**, 1968, 399–408.

[97] J. C. Varlet : Fermetures multiplicatives, *Bull. Soc. Roy. Sci. Liège*, **38**, 1969, 101–115.

[98] J. C. Varlet : Considérations sur les algèbres de Lukasiewicz trivalentes, *Bull. Soc. Roy. Sci. Liège*, **39**, 1969, 462–469.

[99] J. C. Varlet : On the greatest boolean and stonean decomposition of a p-algebra, *Colloquia Math. Soc. Janos Bolyai*, **29**, 1977, 781–791.

[100] J. C. Varlet : A strengthening of the notion of essential extension, *Bull. Soc. Roy. Sci. Liège*, **48**, 1979, 432–437.

[101] J. C. Varlet : Congruences on de Morgan algebras, *Bull. Soc. Roy. Sci. Liège*, **50**, 1981, 332–343.

[102] J. C. Varlet : Relative de Morgan algebras, *Discrete Math.*, **46**, 1983, 207–209.

[103] J. C. Varlet : Fixed points in finite de Morgan algebras, *Discrete Math.*, **53**, 1985, 265–280.

[104] J. C. Varlet : MS-algebras; a survey, in *Lattices, Semigroups and Universal Algebra*, Lisbon 1988, J. Almeida, G. Bordalo and Ph. Dwinger (eds), Plenum Press, 1990, 299–313.

[105] J. Vaz de Carvalho : The subvariety $\mathbf{K}_{2,0}$ of Ockham algebras, *Bull. Soc. Roy. Sci. Liège*, **53**, 1984, 393–400.

[106] J. Vaz de Carvalho : Congruences on algebras of $\mathbf{K}_{n,0}$, *Bull. Soc. Roy. Sci. Liège*, **54**, 1985, 301–303.

[107] J. Vaz de Carvalho : Sobre as variedades $\mathbf{K}_{n,0}$ e algebras de Lukasiewicz generalizadas, doctoral thesis, University of Lisbon, 1986.

Notation index

$P(E)$	1	$T_i(L)$	38
$\mid$	1	$T(L)$	39
$\mathbb{Z}_+$	1	$K(L)$	39
$\leqslant,\ \geqslant$	1	$\mathbf{K}_\omega$	42
S^{op}	1	$(X; \tau, \leqslant)$	52
$\prec$	1	$\mathcal{O}(X)$	52
$P \mathbin{\dot\cup} Q$	1	$I_p(L)$	52
$P \oplus Q$	1	$(I_p(L); \tau, \leqslant)$	52
$P \mathbin{\overline{\oplus}} Q$	2	$\mathbf{D}_{01}$	53
$\wedge,\ \vee$	2	$\mathbf{P}$	53
$\mathbb{N}_0 = \{1, 2, 3, \ldots\}$	2	$\mathbf{Q}$	53
a'	4	$(X; \tau, g) = (X; g)$	53
$Z(L)$	4	$g^\omega(Q),\ g^\omega\{x\}$	54
$a^\star$	5	$G(X)$	54
$\omega,\ \iota$	6	ϑ_Q	54
$\sim a$	6	$\mathbf{P}_{m,n}$	54
$(L; \wedge, \vee, f, 0, 1) = (L; f)$	8	$\mathbf{B}$	55
$\mathbf{O}$	8	$\mathbb{N}_\infty$	55
$\mathbf{K}_{p,q}$	8	$B,\ M,\ K,\ S,\ S_1,\ K_1,\ K_2$	70
$\mathbf{M}$	8	$K_3,\ M_1,\ L,\ N,\ B_1$	71
$\overline{a}$	10	$\Lambda(\mathbf{V}),\ \Lambda(\mathbf{O}),\ \Lambda_f(\mathbf{O})$	75
$S(L)$	11	$\mathbf{K},\ \mathbf{S},\ \overline{\mathbf{S}},\ \mathbf{M}_1 = \mathbf{MS},\ \overline{\mathbf{M}}_1 = \overline{\mathbf{MS}}$	75
$M(L)$	11	(n_d)	83
$X^\downarrow,\ x^\downarrow,\ X^\uparrow,\ x^\uparrow$	13	$(\tilde{n}),\ \tilde{\mathbf{A}}$	90
$\mathbf{n}$	14	$\mathbf{S}_1,\ \mathbf{K}_1,\ \overline{\mathbf{K}}_1,\ \mathbf{K}_2,\ \overline{\mathbf{K}}_2,\ \mathbf{K}_3,\ \overline{\mathbf{K}}_3,\ \mathbf{L},\ \overline{\mathbf{L}},\ \mathbf{N},\ \overline{\mathbf{N}}$	91
$\mathbf{MS}$	17	$\mathbf{V}(L)$	102
$\mathrm{MS}(L)$	17	$\mathrm{Fix}\, L$	105
$\vartheta(a, b),\ \vartheta_{\mathrm{lat}}(a, b)$	20	$\mathcal{P}(L)$	108
ϑ_a	20	$\lfloor x \rfloor$	108
$\Phi_n,\ \Phi_\omega$	23	$L^\wedge,\ L^\vee$	109
$\mathbf{K}^{\leqslant}_{p,0}$	25	$\mathcal{F}(L)$	115
$\mathbf{V}_B(L)$	26	$\alpha^\star,\ \alpha_\star$	117
$I_{2n},\ I^{2n+1},\ I_\infty,\ I^\circ$	27	Ψ	118
C_a	30		

Index

The manufacturer's authorised representative in the EU for product safety is Oxford University Press España S.A. of el Parque Empresarial San Fernando de Henares, Avenida de Castilla, 2 – 28830 Madrid (www.oup.es/en or product.safety@oup.com). OUP España S.A. also acts as importer into Spain of products made by the manufacturer.

www.ingramcontent.com/pod-product-compliance
Lightning Source LLC
LaVergne TN
LVHW041205250326
834689LV00001BA/20

9780198599388